1979

CELLULAR RADIOBIOLOGY

CELLULAR
RADIOBIOLOGY

TIKVAH ALPER

Formerly Director of Medical Research Council's Experimental
Radiopathology Unit at Hammersmith Hospital, London; Senior
Radiobiology Adviser, Cancer Research Campaign's Gray
Laboratory, Mount Vernon Hospital, Northwood

CAMBRIDGE UNIVERSITY PRESS

CAMBRIDGE

LONDON · NEW YORK · MELBOURNE

Published by the Syndics of the Cambridge University Press
The Pitt Building, Trumpington Street, Cambridge CB2 1RP
Bentley House, 200 Euston Road, London NW1 2DB
32 East 57th Street, New York, NY 10022, USA
296 Beaconsfield Parade, Middle Park, Melbourne 3206, Australia

First published 1979

Photoset and printed in Malta
by Interprint Limited

Library of Congress Cataloguing in Publication Data

Alper, T
Cellular radiobiology.

Includes bibliographical references and indexes.
1. Radiobiology. I. Title.
QH652.A63   574.1′915   78–68331

ISBN 0 521 22411 X hard covers
ISBN 0 521 29479 7 paperback

To the memory of my teacher, Lise Meitner, who, in her warm personality, combined deep concern for people and for scientific accuracy, but none for prestige and fame

CONTENTS

PREFACE

The title of this book, chosen for brevity, may convey the false impression that all aspects of cellular radiobiology are dealt with. This has not been possible; I am well aware of several topics that have not been mentioned at all, and of others that have received too cursory treatment. Even so, I found the undertaking to be almost too much and decided at one point that my coverage of the field would be so inadequate that I could not complete the book. I was persuaded not to give up by Dr Jack Fowler, director of the Cancer Research Campaign's Gray Laboratory and by my husband, Dr Max Sterne. They have both given me not only great moral support but also help in many very practical ways: for example Jack Fowler has read and criticized some of the chapters, Max Sterne took on the onerous task of preparing the bibliography for press.

I am grateful to many friends and colleagues who have read and criticized chapters, provided as yet unpublished material, or helped in other ways. I should particularly like to thank Drs David Bewley, Don Chapman, Bill Cramp, Jim Fischer, Neil Gillies, Shirley Hornsey, Barry Michael and Hugh Thomlinson.

My thanks are due also to Jean Shorthose for her patient and meticulous work in typing the manuscript.

1

The significance of cell death

Almost as soon as ionizing radiations were discovered, towards the end of the nineteenth century, some of their biological effects were recognized. Indeed, much of the practical justification for the support of radiobiological research is the same now as then. By understanding how radiation exerts its damaging effects on organisms, it may be learned how to utilize it most effectively in the treatment of disease, principally cancer; and how to provide effective protection against its hazards. Thus in the first half of the twentieth century radiobiological research was fostered mainly by forward-looking radiotherapists on the one hand and, on the other, by physicists who painfully became aware of the biological damage that could be wrought by the new radiations against which the sensory equipment of mammals provides no protective warning signals. But the resources of early radiobiologists were slender, compared with those available today. So the materials of research were cheap to work with, like ascaris or arbacia eggs, bean roots, bacteria and fungi and, occasionally, small colonies of animals used for implanting tumour fragments. The pressing need to learn what happens when the whole organism is exposed to radiation arose when release of nuclear energy became an important factor not only for war, but for peace-time use. Thus it came about that, in the late 1940s, facilities were made available in several countries for extensive studies of effects of radiation on many species of animals – not only on the magnitude of lethal doses, but also on every measurable biochemical and physiological effect.

Anyone who has recently become interested in radiobiology might well suppose that, from the outset, studies on the killing of animals by radiation would have been directed towards the death or survival of 'target cells' in 'target tissues', as the origin of the syndromes being investigated, since this is now so basic a part of the framework of radiobiological thinking. But this represents a dramatic conceptual change since the middle of the century. Whereas it was recognized very early that the sterilization of cancerous tumours depended on the cytocidal

1

action of radiation, the death of an animal was believed to ensue as a consequence of various complex biochemical changes. Whether effects on whole animals or individual animal tissues were studied, it seemed an impossibly complex set of effects to unravel.

However, a rather simple experiment, in which only 100 mice were used, suggested to Quastler in 1945 that there might be a coherent pattern in the relationship between animal survival time and dose (fig. 1.1). That pattern is now well established. Although in that and other papers Quastler himself described many pathological changes and might have been bemused by their apparent complexity, the relationship between survival time and dose induced him to conclude with these perspicacious remarks: 'Our observations have not disclosed any specific information about the mechanism of roentgen death. They have strongly suggested that more than one mechanism is involved, different reactions dominating in different dosage ranges. It seems, however, that their number is limited, probably to three. Thus it can be hoped that the problem, while certainly not simple, is not exceedingly complex.'

Fig. 1.1. Survival time of mice after whole body irradiation. Abscissa: dose in roentgens; ordinate: survival time in hours and days; circles – individual observations. (After Quastler, 1945.) Note: 100 roentgens corresponds to about 0.93 grays.

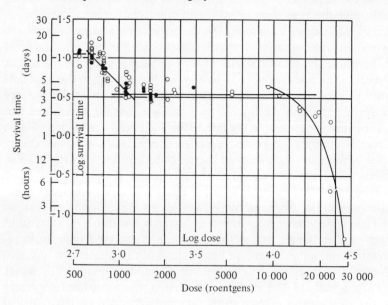

Cell death as the basis of lethal effects on animals: 'haemopoietic death'

Quastler's prediction about the killing of animals by radiation was subsequently well substantiated. Detailed information on the relationship of different modes of death to 'disturbance(s) in cellular kinetics' was given in a book by Bond, Fliedner & Archambeau (1965). However, before the cellular bases of death by irradiation were established, Quastler's comparatively simple pattern of survival time as a function of dose was found to be modifiable in seemingly bizarre ways. For example, animals subjected to the 30-day mean lethal dose for their species would survive for much longer if, say, the head were shielded, or a bit of one limb. On the conceptual basis that animals died because of the complexity of effects on their biochemistry, these observations were mystifying, until it was found that the grafting of haemopoietic tissue from isogenic non-irradiated animals into lethally irradiated ones would enable the recipients to survive longer, so the shielding from direct irradiation of a small piece of bone marrow amounted to an autograft of the crucial tissue. However, it was initially a matter of hot debate whether the unirradiated tissue carried some vital humoral factor, or whether cells still capable of dividing actually repopulated the whole system.

There were several researches from which it was deduced that the latter explanation was the correct one, but the matter was finally clinched by Ford *et al.* (1956) who showed that when cells containing a marker chromosome were grafted into an irradiated animal, the marked cells themselves multiplied. So it came to be accepted that animals died within two to four weeks after being irradiated by doses of a few hundred grays because too many of the proliferative cells in the blood-forming tissues had lost their ability to give rise to the functioning cells on which life depends: so that a graft of good blood-forming tissue would provide stem cells enabling the animals to survive the critical period.

Intestinal death

While the cellular basis for haemopoietic death was being established, work proceeded on the second mode, by which mice died within four to five days after doses in the range 10 to 100 grays. This, too, was found to have a cellular basis. The proposal came, also from Quastler (1956), that the 'intestinal syndrome' resulted from destruction of proliferative capacity of those stem cells which give rise to the epithelial cells lining the intestine. The complex physiological consequences of radiation damage to the intestinal tract did, perhaps, for some time impede acceptance of the close causal relationship between the killing of those cells and animal deaths, by that syndrome, and no demonstration has been provided which is quite as direct as that of Ford and his colleagues in respect of rescue from haemopoietic death. However, if a causal relationship between two different biological end-points is postulated, one way of testing the hypothesis is to observe whether dose-effect relationships are modified to the same extent by the same changes in irradiation techniques. Several different conditions of irradiation were used by Hornsey (1973a) to examine the relationship between intestinal death in mice and the killing of the stem cells in the crypts of Lieberkühn. Within experimental error she established that in every condition a reduction of the stem cell population to about 1% of normal was

Fig. 1.2. Upper diagram: fractional survival of stem cells in the crypts of Lieberkühn, mouse jejunum. Lower diagram: animals surviving more than five days after whole body irradiation. Symbols in the two sets of curves refer to the same conditions for irradiation. Vertical dotted lines show that survival of the same fraction of cells (about 6×10^{-3} of those normally present) enabled about 50% of the animals to survive, whatever conditions were used. Curves on the extreme left are for irradiation by fast neutrons; on the extreme right, for irradiation by high-energy electrons while the mice were anoxic. (After Hornsey, 1973a; redrawn by Dr Hornsey.)

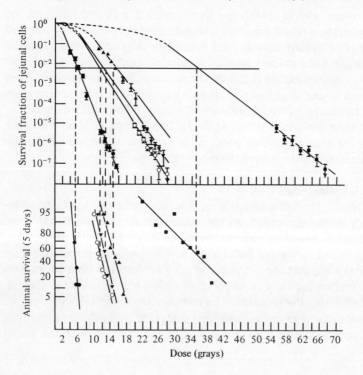

correlated with the death of 20–80% of the animals at four to five days after irradiation (fig. 1.2).

Late effects

Thus there is convincing evidence of a cellular basis for the apparently complex physiological effects leading to two modes of radiation-induced death, but it is now widely accepted that many so-called 'late' effects on individual tissues in irradiated animals can likewise be traced to the killing of cells whose normal function is to proliferate. For example, late damage to the vascular system is thought to be attributable to loss of that function by endothelial cells lining the blood vessels. In radiotherapy, normal tissues are unavoidably irradiated, along with the tumour that has to be sterilized, so the magnitude of the dose that can be delivered is, in many cases, ultimately limited by the requirement that the proliferative capacity of cells in the relevant tissues must not be too severely damaged.

Work on protection against the hazards of ionizing radiation is concerned partly with the possibility of accidents in which persons may be exposed to doses large enough to be comparable with those given in radiotherapy; so the same sort of knowledge about cell killing is basic. With much lower doses, like those received in the course of some occupations, the main risk is that cancers, or somatic gene mutations, may be induced: there is no general agreement at present as to whether cancers are induced because somatic mutations arise. However, the expression of a mutation (or the development of a cancer) requires that the affected cell retains its capacity to proliferate; and the probability that an induced mutation will be expressed may therefore be strongly dependent on the precise mode by which radiation dose is related to loss of viability. Furthermore, the initial events following on the absorption in the cell of radiation energy must be the same, whatever the final biological outcome, so a better understanding of the mechanisms resulting in cell death must throw light on the comparatively much rarer result demonstrated as a mutation. Both for its practical importance and its scientific interest, then, the loss by cells of proliferative capacity will be the main effect of ionizing radiation dealt with in this book.

2

Relationships between dose and effect

Little can be understood about the biological effects of a perturbing agent unless they are related to the 'dose' of the agent delivered to the test object, and this may present great problems. With chemical substances, for example, there are often uncertainties about diffusion, penetration and possible changes in the nature of the substance under investigation before the active molecules reach the point at which they act. Some drugs may bind preferentially in unknown ways to macromolecules or to specific structures within the cell; others may require to undergo conversion by a metabolic pathway. Such problems do not arise when ionizing radiation is applied from an outside source, since the duration of its action is precisely known; and ionizing radiation energy is deposited at random in biological material, with no specific absorption in one or other macromolecule. Nevertheless, accuracy in the measurement of radiation dose is more difficult to achieve than is sometimes assumed.

A thorough understanding of the principles of radiation dosimetry requires knowledge of the physics of ionizing radiation and of its interaction with matter. Very useful accounts of aspects relevant to radiobiology were given by Hutchinson & Pollard (1961a) and by Johns & Cunningham (1974). A standard work on radiation dosimetry is that edited by F. H. Attix and W. C. Roesch (1968–1972).

The term 'dose' used in radiobiology is more strictly defined as 'absorbed dose', i.e. energy imparted (or absorbed) per unit mass. The Système International (SI) unit is one joule per kilogram and this is named one gray (abbreviated Gy). Previously the unit of dose was named the rad, defined as energy absorption of 100 erg/g. Since 1 joule $= 10^7$ ergs, the gray is equal to 100 rads.

In practice, the radiation dose absorbed by cells and tissues has to be calculated from measurements made by observing physical or chemical changes in one or other appropriate device placed in a suitable position within the beam. The commonest method of 'measuring' dose (or dose rate) is by reference to the amount of ionization caused in air or other

6

gases, for which standardized ionization chambers are used. Such measurements have to be referred back to the absolute measurement of absorbed energy, which is converted into heat in the absorber, and can therefore be estimated from observations on the rise in temperature of a suitable substance in carefully controlled conditions. Dose (or dose rate) measurements made by the use of ionization chambers or other devices have to be converted to the required parameter, energy absorbed per unit mass of the irradiated biological material, by using factors that depend on the atomic composition of the material being irradiated and the mode by which it will 'stop' the radiation of the particular kind under consideration.

There are special difficulties in the precise dosimetry of some forms of radiation, particularly those which exert their effects by giving rise to secondary ionizing particles of differing qualities, like beams of neutrons or negative π-mesons. There are also special difficulties in estimating the dose received by biological materials in some circumstances: for example, when cells are exposed as a single layer and there is back-scatter from the material to which they are attached. For these reasons, the precision of estimates of dose in biological materials is often less than

Fig. 2.1. To show that the description of one population as 'more radiosensitive' than another may be inadequate, without more specification.

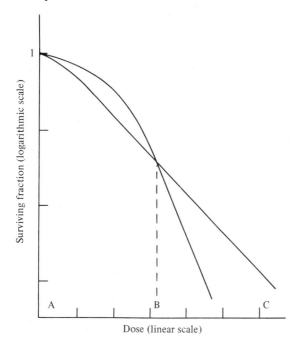

Surviving fraction (logarithmic scale)

Dose (linear scale)

can be achieved in the physical measurements on which they are based. Fortunately a great many radiobiological quantities of importance can be measured with greater accuracy than that which relates to the estimate of absolute absorbed dose. This is the case when the effects of some modifying agent are examined, provided the agent itself has no more than a negligible influence on absorbed dose. The greatest caution is demanded in those experiments in which measurement is made of the relative effectiveness of radiations of different quality, i.e. measurements of 'Relative Biological Effectiveness', abbreviated RBE. Such measurements do require a knowledge of the accuracy of estimation of absolute dose delivered to the test system, and errors calculated in measuring RBE values are often under-estimated for that reason.

Fig. 2.2. In each case the doses required to achieve the same effects are twice as great for curves B1 B2 B3 as A1 A2 A3; but with the same dose D, A2/B2 has the values 2, 1.3, 4 and 1 respectively in a, b, c and d.

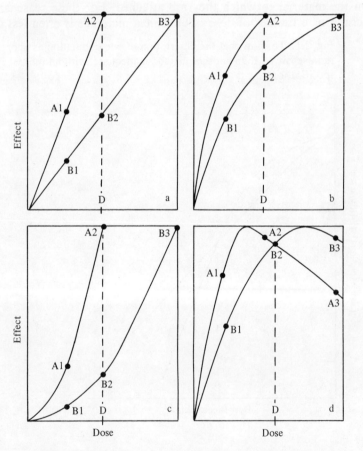

Studying how radiation response is modified by one means or another is an essential tool for relating cause to effect: for example, in establishing the cellular basis (if any) for a given effect on an animal tissue, or in testing a hypothesis about the mechanism of an effect observed at the cellular level.

The importance of relating effect to dose is illustrated by figs. 2.1 and 2.2. Figure 2.1 shows the hypothetical response of two organisms, or of the same organism in two different conditions. Examples of this kind of relationship can readily be found. It is clear that 'radiosensitivity' as delineated by one curve could be judged to be greater or less than, or equal to, that delineated by the other, according to whether the single dose were in the range AB, BC or of magnitude B. This pair of curves illustrates also the difficulty in associating any meaning with the word 'radiosensitivity' unless the circumstances are precisely defined.

The hypothetical dose–effect curves depicted in fig. 2.2 are likewise representative of actual experimental observations. The pairs of curves illustrate effects of radiation with and without a modifying agent present: the sensitizing action of oxygen might be considered as a specific example. In each case, the presence of the agent has been assumed to reduce the dose necessary to achieve a given effect by the factor two.

Fig. 2.3. Curves B and C represent modifications of A after repair procedures operating by different modes. Neither can be described as 'more effective' than the other.

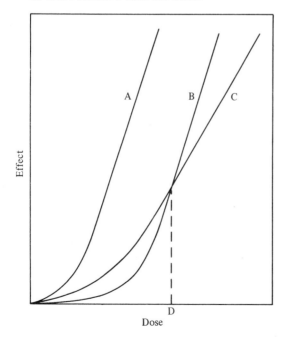

Although the 'dose-reducing' action of the agent is the same for each pair of curves, the differences in effect at a given dose are strongly dependent on the nature of the response. At the particular dose chosen to illustrate this point, we may calculate the ratios of effects at dose D for the pairs of curves of fig. 2.2a, b, c and d. These ratios would be respectively 2, 1.3, 4 and 1, whereas in fact the agent would be acting uniformly in all cases.

Consider now a comparison of the effectiveness of two modifying agents or procedures, both of which may be described as increasing the cells' capacity for repair. In fig. 2.3, the hypothetical unmodified dose-effect curve is A, and the two agents or procedures result in the dose-effect curves labelled respectively B and C. Their relative effectiveness would clearly be judged differently, depending on whether the single dose delivered were less than, equal to, or greater than D.

The loss of viability of members of an irradiated cell population is usually assessed by the scoring of surviving (colony-forming) cells; counting those that have been killed would usually not be practicable. It is therefore conventional to describe dose–effect relationships in terms of survival as a function of dose. For the visual presentation of survival

Fig. 2.4. Survival of *Brucella abortus*, irradiated as a freeze-dried preparation (Sterne, Alper & Trim, unpublished).

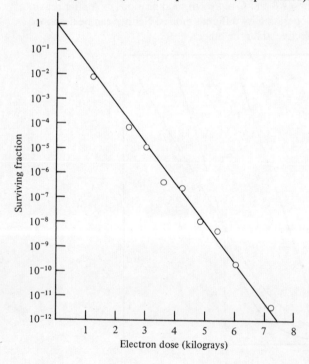

Surviving fraction

Electron dose (kilograys)

data, it is usual to plot surviving fraction on a logarithmic scale. A linear one would be impracticable for visual examination of results for surviving fractions less than 1% and, as shown by fig. 2.4, it is possible, with appropriate material, to measure surviving fractions lower than that by many powers of 10. Furthermore, considerable theoretical importance attaches to the question whether a survival curve, or a substantial region thereof, follows an exponential form. However, that question can often not be definitively answered, particularly from the results of experiments on mammalian cells *in vitro* (Chapter 3, Chapter 14).

3

Techniques for observing survival of mammalian cells

It is convenient to divide the cells in higher organisms into two broad classes: those that will never be called upon to proliferate, because they are fully differentiated and have a specific function; and those whose main function is to proliferate, and so provide functioning cells as end product. Effects of radiation on fully differentiated cells are difficult to detect, mainly because the doses required measurably to disturb their functions are usually large, compared with those which recognizably affect the more primitive cells. With the latter, the most easily observable consequences of irradiation are: (1) delay in division, probably associated with temporary delay in bio-synthetic procedures; (2) the loss of ability to proliferate for as many generations as would be normal; (3) the induction of chromosomal aberrations or 'lethal mutations'.

Interphase and mitotic death of cells

Some animal organs contain functioning cells that hardly ever proliferate, unless they are given the appropriate stimulus: examples are the liver and the thyroid gland. Many years ago the liver was considered by radiotherapists to be a 'radioresistant' organ, since quite large doses of radiation did not detectably affect its functions. However, if portions of the liver are removed surgically, or by disease, the remaining cells are stimulated to proliferate until the amount of lost tissue is made up. It was observed by Weinbren (see Doniach & Logothetopoulos, 1955) that, in patients who suffered hepatectomy after previous irradiation, there was a great deal of morphological damage in the dividing cells, even if considerable time had elapsed between irradiation and the stimulus to cell division. Analogous observations have been made when the circulating lymphocytes of previously irradiated patients have been stimulated *in vitro* to go into division. That non-cycling cells can 'remember' damage, sometimes over years, may be unique to the mode by which ionizing radiation exerts its effects. Lymphocytes are a class of cells in mammalian organisms that may be regarded from the radiobiologist's point of view as being between primitive and fully differentiated cells in respect of their life history, but more akin to the former in their response to radiation. These cells, constituents of the haemopoietic system, play a very important part in immunological responses. They do not proliferate regularly, but may do so in response to mitogenic stimuli (usually immunological, *in vivo*) that cause the initiation of biochemical syntheses leading to mitosis and cell division. If lymphocytes are exposed to radiation after stimulation, the effects in general terms are similar to those on normally proliferating cells. However, if there is no such stimulus, i.e. the observations are made on non-proliferating cells, the doses required to effect detectable damage are in the same range as those within which proliferative death is normally observed. With lymphocytes this damage is detectable by virtue of changed morphology, the nuclei of the cells becoming pyknotic. In an irradiated population of lymphocytes, the proportion of cells that show pyknotic nuclei is related not only to the dose, but also to the time after irradiation at which the observations are made, and to various environmental factors. Cells which are judged by microscopic observation to be 'dead' may be regarded as having suffered from loss of proliferative capacity, like primitive cells in which this loss would be identified with failure to divide. It is usual to refer to the form of death that occurs before mitosis, and is recognized by a pyknotic nucleus, as 'interphase death', whereas the death that occurs during or after the first mitosis after irradiation is called 'mitotic death'.

Counting cells that survive mitotic death

In general, cells that are fated to suffer 'mitotic death' cannot be morphologically distinguished, soon after irradiation, from those that have retained their capacity to proliferate, so a functional test must be used. The most direct method for perceiving the ability of a single cell to proliferate is to wait until it gives rise to a colony. This is a very old-established technique with micro-organisms, going back nearly one hundred years to methods developed by Robert Koch for studying micro-organisms, and is known as 'viable counting'. The earliest experiments relating the mode of loss of proliferative capacity to radiation dose were therefore done with micro-organisms. Methods of culturing higher cells *in vitro* developed over many years, until Puck & Marcus, in 1955, successfully grew discrete colonies of mammalian cells from single ones. From then on, many experiments have been done with mammalian cells *in vitro* that are quite analogous with experiments on micro-organisms. It is natural that the terminology used by microbiologists should have been taken over for radiobiological experiments with cultured mammalian cells. With micro-organisms, viability and colony-forming ability are synonymous terms; this definition has been extensively used, whatever bacteriocidal or fungicidal processes might have been under investigation. It is therefore conventional, as well as convenient, to refer to non-colony-forming higher cells as 'dead', and, in that context, to the lethal effect of radiation; and to describe treated cells that have retained their ability to give rise to a colony as 'survivors'. Both ultraviolet and ionizing radiation are notable for destroying proliferative capacity while leaving other 'living' processes undisturbed. For example, both lower and higher cells that are no longer able to originate colonies often continue to respire at normal rates; they usually continue to synthesize DNA and other macromolecules; and viruses will grow in irradiated cells after very large doses that would leave no detectable viable cells. Thus the terms defined in this section should be regarded as referring only to the function of proliferation.

Other definitions of cell death

The sense in which the phrase 'cell death' is used by radiobiologists sometimes gives rise to confusion, because in other fields it has a different connotation. To some, a cell is still 'alive' if certain dyes fail to enter; to immunologists, a specific test of the 'killing' of target cells by immunologically committed thymocytes is that the target cells release a radioactive label with which their proteins have been tagged, which means that the plasma membrane must have been torn open. Pathologists describe cells as dead when they are seen to be in the process of lysis or disintegration. The justification for confining the words 'death' and 'survival' to the proliferative function, in the present context, is firstly the historical background, as described above; secondly the advantage of a short word or phrase over a long one; thirdly, the importance of loss or retention of proliferative capacity in the contexts of radiotherapy and radiation hazards.

The definition of 'cell death' as 'loss of reproductive capacity' automatically includes other definitions, though the converse is not true. Cells which are seen to be disintegrating will not reproduce, nor will those which allow dye to cross the plasma membrane, nor will a cell be able to reproduce if it has been torn open and its protein released into surrounding medium. Indeed a radiation-killed cell may fail to show evidence of morphological change for long periods, given a suitable environment. Even with this restricted use of the phrase 'cell death', the difference between viability and non-viability may require further definition. In tissue culture, irradiated cells may go through a few divisions, but still be defined as 'killed' because the number of progeny is limited. In some experiments it may therefore be necessary to lay down an arbitrary standard by which to define 'surviving' cells. A 'survivor' *in vitro* is usually defined as a cell that has given rise to a colony of at least a minimum number of cells – say, fifty or sixty; that is, there must have been at least five or six cell divisions. However, relationships between dose and effect are on the whole not much changed by the choice of larger or smaller numbers of cells per colony as indicative of survival.

Cell survival *in vivo*

Ways have also been devised of assessing the viability of populations of irradiated mammalian cells *in vivo*. If colonies are to be counted directly, it is necessary that the progeny of a single cell remain in a group discrete from other groups. Till & McCulloch (1961) reported that discrete nodules, each containing a homogeneous population, appeared on the spleens of mice that had been exposed to radiation doses that were just not sufficient to kill them. These nodules could be seen and counted only because so few of the relevant cells in the haemopoietic tissue had retained their ability to proliferate. Later Kember (1965) used microscopic observation to count clones arising from surviving cells in irradiated cartilage, and Withers (1967a) developed techniques that enabled nodules arising from viable cells in the irradiated epidermis of mice to be counted. Withers & Elkind (1969, 1970), used this approach also to study effects of radiation on the intestinal mucosa.

The viability of malignant cells, which may in some cases be irradiated *in vitro*, may be judged by their ability to originate solid tumours *in vivo*. This technique was first used by Silini & Hornsey (1961), working with a line of mouse ascites tumour cells. The radiobiological responses of several tumour lines have been measured by counting small tumours that appear on the surfaces of the lungs of rats, after suspensions of malignant cells have been injected into the tail veins (Williams & Till, 1966).

The techniques described so far have all depended on the observation of visible colonies of cells arising from one cell, or a few, that have survived irradiation. Again as in microbiological experiments, the fraction of viable cells in a population may be estimated by less direct methods. An 'end point dilution' technique was devised by Hewitt (1958) and used to study the effect of radiation on the malignant cells of leukaemic mice. By measuring the numbers of cells required to give 50% of tumour takes in recipient mice, Hewitt & Wilson (1959) deduced the fraction of cells that remained viable in donor mice after given doses of radiation. Radiation responses of several cell lines have been investigated by analogous methods.

Some constraints in delineating survival curves for mammalian cells

Whatever technique is used, it is customary for cell survival curves to be described in terms of the constants of one or other fairly simple equation that is judged, or assumed, to fit the data points. Those in common use are set out in Chapter 5. However, the reliability of curve-fitting is subject to limitations imposed by the techniques, particularly those used with mammalian cells: observations with naturally free-living cells, like bacteria, algae and some fungi are much less subject to those constraints.

Some of the sources of error inherent in techniques *in vitro* with mammalian cells were discussed by Boag (1975). Within those limits of error, data may not infrequently be equally well fitted by two or more formal survival curve equations, although these may be based theoretically on quite different assumptions (e.g. Hall, 1975). Reliability in testing the fitting of an equation depends to a great degree on the range of survival over which observations are made. In particular, a decision as to whether a survival curve has a final exponential region can, in some instances, be made only if data are available for reduction in survival to less than 10^{-3} (see Uretz, 1955). An example is afforded by results with the green alga *Chlamydomonas reinhardii*. Bryant (1973) specifically tested for, and confirmed, exponential survival in the range 10^{-2} to about 10^{-6}, but a complete range of data points above a survival level of 10^{-3} were best fitted by an equation that could never approximate to exponential survival (Bryant & Lansley, 1975). With mammalian cells *in vitro* it is rare for experimenters to get data for surviving fractions substantially less than 10^{-3}. For one thing, the preferred methods of growing cultures and preparing cells for irradiation do not lend themselves to providing populations large enough to give adequate numbers of survivors when these have been reduced to only 10^{-4} or 10^{-5} of the original population. But in any case it is difficult, with such cells, to avoid artefacts caused by the inescapable fact that the observation of a surviving fraction of $1/n$ cells requires the plating of a total of n cells for every colony-former. Most of the proliferatively dead cells are still able to attach to a surface and to metabolize, so when a great number of them are plated there may be too little space and nutrients for adequate growth of the daughters of any surviving cell. It has been shown experimentally that such conditions can result in the failure of some potential survivors to give rise to countable colonies (Bender & Gooch, 1962; Berry, 1964; Bedford & Hall, 1966). The range of observations can be extended by increasing the number of plates used per dose-point, at the lower survival levels, but there are practical limits to the number of plates that can be handled in a single experiment.

Limitations in the opposite sense are inherent in the methods used for delineating survival curves for cells *in situ* in organized tissues. Clones arising from single surviving cells in spleens, cartilage or skin cannot be identified until the majority of the clonogenic population have been killed; and in any case the surviving fraction cannot be calculated if the number of initially viable cells in the irradiated area is not known: this applies to survival curves from the counting of clones in volumes of cartilage or areas of skin that have been irradiated. Thus the initial region of the survival curves cannot be delineated. The most complete data on survival of proliferating cells in normal tissue are those for the stem cells in the crypts of Lieberkühn: these give rise to the population of cells that normally cover the villi in the intestinal tract. After sufficiently high doses, of the order of 16 Gy, the survival of single cells is identified with nodules of mucosal epithelium arising in irradiated sections of the alimentary canal. It is possible to get data relating to survival at lower doses by counting the number of regenerating crypts per section of gut, and these can be referred back to the average number of crypts per unirradiated section. Where counts have been made by both techniques, the curves can be fitted together satisfactorily at intermediate doses (e.g. Withers & Elkind, 1970). But the low dose region of the curve for regenerating crypts does not provide information on the survival curve for single cells, since the complete destruction of a crypt depends on the killing of all the stem cells therein. In summary, the observation of cell survival *in situ*, in animal tissues, does not lend itself to the direct measurement of certain survival curve parameters for which the initial regions of the curves are required; whereas mammalian cell survival curves taken *in vitro* do not in general provide a basis for firm conclusions about models that can be tested only by observations at low levels of survival. Despite these constraints, hypotheses on the mechanisms for cell killing are frequently tested against experimental data with mammalian cells. Such hypotheses as have been put forward commonly lead to fairly simple equations for surviving fraction as a function of dose. Since the assumptions made almost always embody the concept that radiation kills cells by destroying the function of specific targets, it is useful to consider first both the target concept and the more formal 'target theory'.

4

The target concept and target theory

Two classical works on basic mechanisms in radiobiology appeared almost simultaneously about thirty years ago: Lea's *Actions of Radiations on Living Cells* (1946), and Timoféeff-Ressovsky & Zimmer's *Das Trefferprinzip in der Biologie* (1947). Zimmer (1961) elaborated on the distinctions between 'hit theory', as dealt with in great detail in the earlier book of which he was co-author, and 'target theory', which was, in effect, the main theme of Lea's book. In either case, the basic concepts apply to those effects which are attributable to a single 'primary ionization' or 'energy deposition event' in a 'target', although, in theory at least, some biological consequences of radiation may require the occurrence of two or more events in the same target (i.e. more than one 'hit').

It has been stated that 'Target theory is out of date', or even that it has been 'disproved'. Such views stem from a lack of understanding of the restrictions on the rigorous applicability of the theory; or, perhaps, from confusion between the theory, which is a tool for calculation, and the concept, which is implicitly adopted by most radiobiologists who have interested themselves in mechanisms of radiation effects on cells.

The target concept in cellular radiobiology implies the existence of a vital structure, damage to which will impair the biological function under test. The role of 'target' is conferred by the requirement for the integrity of the structure if the function is to be carried out, not by its 'sensitivity'. If our concern is with loss of the reproductive capacity of cells, it is not meaningful to apply target theory directly, for several reasons. One is that almost all cells are biochemically equipped to bypass or repair potentially critical lesions. Another reason for inapplicability of specific target theory, in our present state of knowledge, is the intimate cohesion between structures that might well be individual targets for radiation damage, like the bacterial membrane and genophore, or the nuclear membrane and the condensed chromatin in eukaryotic cells, during most of their cell cycle. Figure 4.1 is a schematic diagram, from Davies & Haynes (1975), showing how two layers of nuclear membrane in an animal cell can approximate, trapping between them the adjacent chromatin (nucleoprotein). Plate 1 (see p. 20), from the same paper, illustrates how a sheet, such as those shown diagrammatically in fig. 4.1, can form a pocket. The electron micrograph at the higher power (plate 2, p. 20) shows the regularity of the structure, the units of which are seen end-on, between the two layers of membrane. It has been identified as condensed chromatin by Davies and his co-workers, who have observed these 'envelope-limited sheets' in electron micrographs of animal cells of many different origins. The trapping of chromatin between layers of the nuclear membrane could hardly be such a frequent occurrence unless the two structures were normally tightly adherent. Such close cohesion could well afford the possibility of energy transfer between macromolecules that are chemically different, as well as of mutual influence for repair or fixation of damage.

In his book, Lea assumed that target theory was directly applicable not only to the inactivation of enzymes and viruses, but also to the induction by radiation of chromosomal aberrations, regarding these as

Fig. 4.1. To show the ultra-structure of a thin section through a nucleus. om, outer, and im, inner membranes of the nucleus. C, cytoplasm; n, nucleus. C** n**C represents a section from cytoplasm through nucleus to cytoplasm, etc. *B* and *C* show how thread-like nucleoprotein units in chromatin are packed side-by-side into one or more layers at the surface of the nucleus, so that a series of bands is seen in thin sections. Dense regions, sections through the threads, have diameters shown by d, d'; a' is the separation between them. (After Davies & Haynes, 1975.)

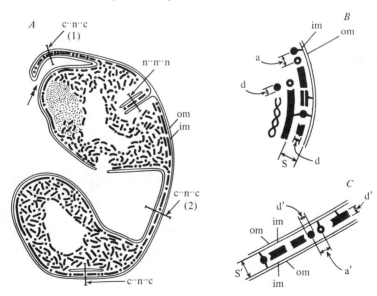

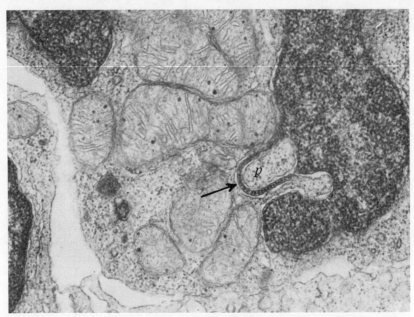

Plate 1. Electron micrograph of a lymphocyte from the intestine of a fish, p is a pocket formed as shown in fig. 4.1*A* (×29 000) (from Davies & Haynes, 1975).

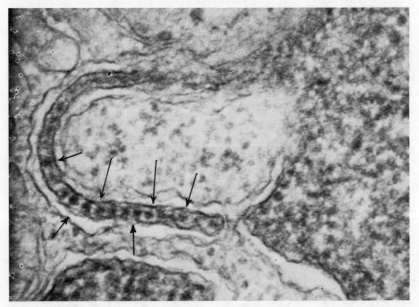

Plate 2. Electron micrograph of the sheet arrowed in plate 1, showing end-on views of structural units (× 120 000).

being due solely to 'hits' within the chromatid or chromosome. Most analyses of data on chromosome aberrations are still made in accordance with that concept, but fig. 4.1 and plates 1 and 2 suggest strongly that energy absorption events within the nuclear membrane might play a part in the production of breaks, and that their restitution might depend on the contact between membrane and DNA. Even if radiation-induced mutations are engendered primarily by energy absorption strictly within the genome (i.e. the nuclear DNA) there is undoubtedly a sequence of biochemical events, including 'repair' of some pre-mutational lesions, before a mutation is expressed (e.g. Wolstenholme & O'Connor, *Mutation as a Cellular Process*, 1969). For such reasons, it is at present unprofitable to apply formal target theory to observations on cells. In contrast, the theory can be and has been used with success when applied to investigations into the loss of biochemical or biological function by macromolecules irradiated extracellularly: provided that the correct conditions are used.

Exponential inactivation

When 'single-hit' action is involved, the 'target volume' is defined as that volume within which a single energy-absorption event of adequate magnitude will cause 'inactivation' or loss of function. When this mode of action is involved, the dose–response curve is governed by the consideration that the energy-deposition events, or hits, occur at random, in conformity with the Poisson distribution. Thus, when a whole population of macromolecules has been exposed to a dose D_0 that would give an average of one lethal event per member of the population, a fraction e^{-1} would each have experienced no events, and the remainder would each have experienced one or more. Thus e^{-1} members of the population would be survivors and $1 - e^{-1}$ would have been inactivated. The next increment D_0 would leave a fraction e^{-1} as the surviving fraction of the e^{-1} survivors of the first dose D_0, giving surviving fraction e^{-2}, and so on. After dose D the surviving fraction f will be given by the equation

$$\left. \begin{array}{l} f = e^{-D/D_0} \text{ or } e^{-\lambda D} \\ \ln\ f = -D/D_0 \text{ or } -\lambda D \end{array} \right\} \tag{4.1}$$

where $\lambda = 1/D_0$

If the fraction of macromolecules remaining active, after a given dose, is plotted on a logarithmic scale as a function of dose, the curve will be a straight line passing through the point zero dose, 100% activity. D_0 and λ are customarily known respectively as the inactivation dose and the inactivation constant. A dose–response curve of this form is described as exponential.

It has been pointed out that an inactivation (or survival) curve that is exponential, within experimental error, need not necessarily be evidence of one-hit inactivation, since it is possible to contrive dose–response curves that appear as straight lines, on semi-logarithmic plots, from postulated inhomogeneous populations that suffer inactivation or killing by a variety of modes (Dittrich, 1960). None-the-less, inactivation curves for proteins and viruses are found to be exponential, almost without exception; and the few exceptions can be accounted for (one example is for viruses irradiated in suspension, in which toxic products, particularly hydrogen peroxide, may accumulate). It would be implausible to ascribe all these numerous examples of exponential inactivation to chance distributions of a variety of modes of inactivation, rather than to a basic mechanism of action, namely one-hit inactivation; and it is justifiable, therefore, to use the parameter D_0 of equation 4.1 to estimate the size of the target. This is the essence of target theory, which has proved successful in that respect. With appropriate arrangements, the theory can also be used to give indications of target shape, but the correct conditions are difficult to achieve experimentally.

Conditions for applicability of target theory

Before we consider the methods for calculating target size, given an exponential inactivation curve, and the confirmatory evidence for the validity of the theory, it is appropriate to discuss some irradiation conditions in which target theory may fail to give the right answers. It may be that some scepticism about the value of the theory has stemmed from inadequate appreciation of the required precautions.

There is good evidence that radiobiologically important consequences of the deposition of energy within the target may be the loss of an electron, or the detachment of a hydrogen atom, so that the macromolecule exists, for a short space of time, as a radical, i.e. a molecule with an unpaired electron. This is chemically very reactive, and its first reaction may be of such a nature as to render the macromolecule incapable of functioning. The radiation target volume will correspond with the true volume (or molecular weight) of the macromolecule only if a hit anywhere within it will inactivate it. There are conditions in which this correspondence would not be observed:

(1) If some inactivation were due to energy absorption outside the macromolecule: for example, if it were suspended in a liquid medium, and could be affected by the radiolysis products of energy absorption in the medium (so-called 'indirect effects'). The inactivation dose would then be too small, and the apparent radiation target volume too large.

(2) If the function under test did not depend on the integrity of the complete macromolecule: for example, the combining site of a γ-globulin molecule remains active, even when it is enzymically detached from the

complete macromolecule. Thus a radical created in another part of the macromolecule might be harmless from the functional point of view. (3) If the radical-molecule recaptured its lost electron before it was chemically changed ('charge recombination').

If it is desired to test the theory, or use radiation as a tool for estimating molecular weight, condition (1) is easily avoided by irradiating the test material in the dry state (Hutchinson & Pollard, 1961b). It has often been assumed that equivalence to the dry state may be attained by the addition to an aqueous suspension of large amounts of impurities, usually proteins, which will react with, or scavenge, the radiolysis products of water, and so prevent them from inactivating the test macromolecules. But this technique cannot be relied upon to eliminate all consequences of events outside the test molecule, and there have been few comparisons made of inactivation doses for material irradiated dry as well as in an aqueous suspension thought to contain sufficient scavenging material to eliminate indirect action. For bacteriophages, 4% of nutrient broth is usually accepted as being sufficient radical-scavenging material to eliminate indirect action; but Hotz (1968) irradiated the phages T1 (double-stranded DNA) and ΦX174 (single-stranded DNA) after drying, as well as in 4% nutrient broth, and found inactivation doses that were respectively twice and four times as great for the dried material. From similar experiments on the DNA extracted from these phages differences in inactivation doses for dried and suspended materials were even greater: by factors of about thirteen and four respectively for DNA of T1 and ΦX174. It might be argued that 4% added material (mostly polypeptides) is not in fact adequate for complete radical scavenging. In our own experiments on the radiation target size of a γ-globulin, *Clostridium welchii* Type A α-antitoxin (Alper, Sterne & Double, unpublished), we irradiated horse antiserum dried, and in suspension containing 20% protein, which is much more than the amount usually assumed to give full protection against indirect effects. The inactivation dose was five times as great with dried material (fig. 4.2). If we had assumed that the γ-globulin in suspension had been fully protected against indirect action, the target size would have been calculated as nearly five times too large.

Another hazard of deducing radiation target sizes from experiments with the test macromolecules in suspension, with scavenger material present, is that some protection may be afforded by the scavenger beyond what is due to the mopping up of the active radicals formed in the water. It has been shown that compounds with free –SH groups may react with the target radicals in such a way as to leave their function unimpaired, as will be discussed below. When mixed proteins are present as scavengers, such reactions may occur even when the material is irradiated at temperatures well below freezing point.

Condition (2) cannot be avoided: but if it is possible to determine the size of the test object by independent means, and the radiation target size is apparently much less, this may well suggest that the function under test depends on only part of the object under investigation. Examples are afforded by some radiation inactivation studies on proteins comprising subunits which can function independently when the molecules are dissociated (Blum & Alper, 1971).

Early studies on the radiation target sizes of viruses demonstrated good correspondence with the independently determined virus size, for small viruses, but an increasing difference as the virus size increased. Since the capacity of a virus to reproduce depends on the integrity of the genetic material, we would now predict that the radiation target size should correspond with that of the nucleic acid core, if target theory holds. The protein coat can evidently stand quite a lot of damage without effect on the ability of the virus (or its nucleic acid) to enter the host cell.

Condition (3), in particular, has not achieved sufficiently wide recognition as a reason for some failures in correspondence between radiation target size and independently determined molecular weight. There are some circumstances in which the probability p of inactivation by a single

Fig. 4.2. Inactivation of *C. welchii* Type A α-antiserum irradiated in liquid form and dry in the presence and absence of oxygen (Alper, Sterne & Double, unpublished).

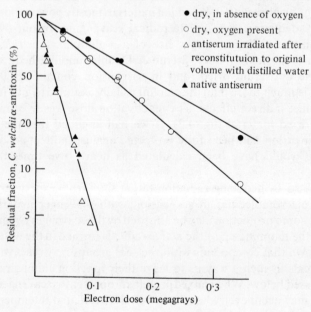

● dry, in absence of oxygen
○ dry, oxygen present
△ antiserum irradiated after reconstitutuion to original volume with distilled water
▲ native antiserum

event might be considerably reduced, as Lea (1946) explained. Only when p is near to one will the calculated target size and molecular weight agree. Since inactivation is the consequence of the formation of a radical, the nature of its first chemical reaction (the metionic reaction: Alper, 1956) will determine whether or not the macromolecule will suffer loss of function. If the metionic reaction results in recapture of an electron, or addition of a hydrogen atom, the target may be restored to normal functioning, and if such restitutory reaction occurs once in every n reactions, p will equal $(n - 1)/n$. Oxygen combines avidly with active radicals, so if it is present during irradiation in sufficiently high concentration to compete successfully with any restitutory metionic reaction, p is likely to be near to one, whereas it may be significantly less than one if no oxygen is present. A protective action of anoxia has been demonstrated with proteins as test objects in several investigations, for example with enzymes (e.g. Alexander, 1957; Shalek & Gillespie, 1960) and with dried γ-globulin (fig. 4.2). Several strains of virus, irradiated dry, were found by Alper & Haig (1968) to be protected by hypoxia: for the same degree of inactivation, doses to the anoxic viruses had to be larger by factors ranging from 1.6 to 2.5.

With enzymes irradiated dry, and *in vacuo*, inactivation doses were higher when compounds containing $-SH$ groups were added to suspensions before drying (Norman & Ginoza, 1958; Braams, 1960). Braams observed protection also of enzymes dried with yeast extract present. Ginoza & Norman (1957) reported considerable protection of tobacco mosaic virus ribonucleic acid when the material was frozen or dried from a suspension containing 2% glutathione, but it is not clear whether or not oxygen was present during irradiation.

Hydrogen donation to radicals was suggested as a restitutory or 'protective' mechanism by Alexander & Charlesby (1954) and evidence in support comes from investigations by microwave spectroscopy, used for observing electron spin resonance (abbreviated ESR) or paramagnetic resonance. The observations depend on the precession or spin of unpaired electrons in a magnetic field, the frequency of precession varying with the strength of field. If an oscillatory field is now applied at right angles by means of a beam of electromagnetic radiation and the frequency equals that of the precession, the electrons will resonate and absorb energy which in practice is supplied by a microwave beam. With suitable arrangements, signals from the ESR give information about the nature of the radical to which the unpaired electrons belong. Differences in ESR spectra for macromolecules irradiated with and without –SH compounds present were noted, for example, by Gordy & Miyagawa (1960) and Ormerod & Alexander (1963). It was deduced that the changes occurred by interaction between the radicals formed in the compound under test and the –SH group, Ormerod & Alexander attributing the change to hydrogen donation, and Gordy & Miyagawa to the 'migration of the electron hole'. In both cases the materials were irradiated in the absence of oxygen, and an important observation of the latter authors was that the presence of the – SH compound altered the nature of the secondary radical produced by the later admission of oxygen, which suggests competition between oxygen, as damage-fixing agent, and restitutory metionic reactants.

It is possible that restitutory metionic reaction with –SH or other groups accounts for the results of Alper & Haig (1968), which differ from those of Hotz (1968) for bacteriophages T_1 and ΦX174. He failed to find any protective effect of anoxia for the phages dried from nutrient broth suspensions. The context in which Alper & Haig carried out their experiments required that the viruses be dried down in a suspension of cellular material, so the protection they observed in anoxia might be attributable to a restitutory action of –SH or other compounds when no oxygen was present.

It is important to note that, in general, anoxia fails to protect either proteins or nucleic acids against radiation-induced loss of biological function when they are irradiated in aqueous suspension (table 4.1). This has an important bearing on the interpretation of the sensitizing action of oxygen on the radiation-induced killing of wet cells (Chapters 6, 15). When a sulphydryl compound (cysteamine) was added in high concentration to a suspension of bacteriophage particles, they were protected against inactivation to a greater extent when irradiated in the absence than in the presence of oxygen; the maximum ratio of doses to give the same effect was 1.8 (Howard-Flanders 1960). This result has been widely interpreted as showing how oxygen can sensitize intracellular DNA, in contrast with its failure to sensitize extracellularly irradiated DNA and RNA targets in suspension (table 4.1). It is assumed that the sensitization is attributable to the competition of oxygen with intracellular –SH which would otherwise donate a hydrogen atom to the potentially lethal radical formed in the DNA. Competition between hydrogen donation and oxygen addition to radicals formed in dry materials does not seem to have been quantified; but in wet systems hydrogen donation has been observed to be a poor competitor with oxygen. Ratios of reaction rates of oxygen and RSH with model radicals in aqueous solutions were quoted by Adams (1972) as ranging from ten to several hundreds.

The calculation of target size

We now discuss briefly the use of D_0, the inactivation dose, of equation 4.1 in calculating the size, or molecular weight, of a target macromolecule irradiated extracellularly, provided the conditions allow of a probability near to one that a single energy deposition event of appropriate size will inactivate the whole molecule.

If the events were deposited spatially at random, and the targets were spherical in shape, the target volume could be calculated directly as D_0^{-1}, if D_0 were expressed in units of average event size per unit volume. However, this would be an approximation even for spherical targets, since the events are left in the tracks of the ionizing particle: secondary or tertiary electrons, if we are concerned with γ-rays, X-rays, or a beam of energetic electrons. The approximation would not be very great if the target macromolecules were so small that their diameter was less than the average spacing between events, even in a particle track; but the size of larger targets would be underestimated, because some of them would have

Table 4.1. *Ratios of radiation doses to yield the same biological or biochemical effects in anoxic and well-oxygenated dilute aqueous solutions or suspensions of sub-cellular test systems, from experiments in which pH values were near 7*

Test system	Test of damage	Dose ratio, anoxic: oxygenated	Reference
Bacteriophage S13	Loss of infective activity	0.50	Alper (1955)
DNA of ΦX174	Loss of infective activity	0.56	Van der Schans, quoted by Blok & Loman (1973)
Polyuracil	Loss of phenylalanine synthesizing ability	0.74	Ekert & Grunberg-Manago (1966)
Transfer-RNA, *E. coli*	Loss of transfer activity, Lysyl-t-RNA	0.47	Ekert & Latarjet (1971)
Transforming DNA (Pneuomo-coccus) (a)	Loss of transform-ing activity	0.8–1.1	DeFilippes & Guild (1959)
Lysozyme	Loss of enzyme activity	0.80	Brustad (1967)
Ribonuclease (b)	Loss of enzyme activity	0.74	Adams *et al.* (1971b)
γ-Chymotrypsin (b)	Loss of esterase activity	0.43	Adams *et al.* (1973)
Lima bean pro-tease inhibitor	Loss of inhibitory activity towards β-trypsin	0.73	Lynn & Raoult (1976)

(a) The concentration of DNA used for these measurements was high enough for the contribution from indirect action to be reduced.
(b) Anoxia achieved by flushing with argon; in other experiments with oxygen-free nitrogen.

experienced more events, during the passage of the particle, than would be predicted on the basis of random distribution of the events deposited by a given dose; and others would have experienced correspondingly fewer events. The converse would be true if target size were estimated rather in terms of D_0 measured as the total number of particle tracks engendered by the radiation: each track passing through a target would then score a hit. This would provide a reasonably good estimate for targets which were large, compared with the average spacing between the events left in the track, but the size of small targets would be underestimated.

Lea's 'associated volume' method (1946) was designed to give more accurate estimates of target size (assuming the targets to be spherical) for all cases in which one-hit inactivation occurs, including the range intermediate between those sizes for which the two methods described above would be adequate. In essence, his method was to associate a spherical volume with each 'primary ionization' and to take into account the overlapping of those imaginary spheres when the ionizations were closer together than the radii of the associated volumes. Thus he calculated an 'overlapping factor', F, from the 'mean separation of consecutive primary ionizations'. The parameters he used were based on observations on the effects of ionizing radiation on gases. His assumption of the energy deposited in a 'primary ionization', about 100 electron-volts (eV) (or 1.6×10^{-17} J) was likewise based on such observations. According to Hutchinson & Pollard (1961a), theoretical considerations lead to the calculation that the energy deposited per primary ionization in water varies from 71 eV (11×10^{-18} J) per event for electrons of energy 1 keV (1.6×10^{-16} J) to 105 eV (1.7×10^{-17} J) per event for electrons of energy 10 MeV (1.6×10^{-12} J).

Measurements of the energy loss in tissue-equivalent plastic films were made by Rauth & Simpson (1964). Their method was to shoot electrons of energies of 5, 10 or 20 keV through the films, then capture and count those which had lost less than a predetermined fraction of their initial energy. On the basis of certain assumptions, they concluded that the most probable energy loss per inelastic collision per electron in the plastic films (i.e. per energy-loss event) was 22 eV, while the average energy loss was about 60 eV. Energy losses of up to several hundred electron-volts per event were detected.

Provided the theory relating target size to inactivation dose is valid, and target shapes approximate sufficiently to the spherical, measurement of inactivation doses for targets of a wide range of known molecular weights should make it possible to estimate the average energy loss per event in biological material. For this purpose it is desirable to have a range of measurements from experiments in which consistent conditions of irradiation were used. Such a range is available from work done by my colleagues and me over many years, using the Medical Research Council's 8-MeV linear accelerator at Hammersmith Hospital as the source of the electron beam (Alper, Sterne & Double, unpublished; Alper & Haig, 1968; Blum & Alper, 1971). Plots of target molecular weights, calculated from the Lea theory, against independently determined molecular weights for the same macromolecules, are shown in fig. 4.3. Each point has been plotted with the target molecular weights calculated both directly from the Lea theory (based on an average of 100 eV per event) and on the same theory

Fig. 4.3. Correlation between target molecular weight and molecular weight independently determined: i.e. of sub-units, for oligomeric proteins, and nucleic acid cores, for viruses. Upper and lower blocks represent calculations from Lea theory based respectively on 100 eV and 60 eV per energy-deposition. Horizontal dimensions of blocks represent 95% confidence intervals, or ranges, for the estimates of molecular weight. Vertical dimensions give 95% confidence intervals for estimates based on inactivation doses. A, γ-globulin, Fab' fragment. B, urease, subunit. C, glutamate dehydrogenase, subunit. D, μ bacteriophage (RNA). E, yellow fever virus (RNA). F, T3 bacteriophage (DNA). G, *Herpes* virus (DNA). (Data from Alper & Haig, 1968; Blum & Alper, 1971; Alper, Sterne & Double, unpublished.)

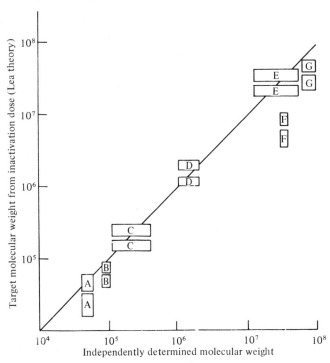

adapted to an average of 60 eV per event, in accordance with the measurements of Rauth and Simpson on electrons of much lower energies. The range, or confidence limits, on the independently determined molecular weights, as well as on our own determinations of the inactivation doses, are rather too wide to allow of a firm conclusion on the size of the average energy loss event, but the calculations based on 100 eV are a better fit to the theoretical line of equality. As quoted above, it is to be expected that the energy loss per event would be significantly greater for the electrons of the energy we used (about 6 MeV, in the irradiated material) than for those used in the experiments of Rauth and Simpson.

Evidence was adduced many years ago of the general validity of target theory, provided appropriate conditions for irradiation were used (see Hutchinson & Pollard, 1961b). Previous apparent discrepancies between 'target molecular weights' and 'molecular weights' of some proteins are now accounted for by their oligomeric nature: for example, when we tested radiation damage to γ-globulin by loss of ability to combine with the appropriate antigen, we found that the target molecular weight was only about one-third of that of the whole molecule. We could subsequently identify the target molecular weight with that of the Fab' fragment, the subunit which carries the combining site, and which can function even when it is dissociated from the whole molecule by enzymic digestion. Similarly oligomeric enzymes are readily dissociated into functioning subunits, and the target molecular weights have been found equal to those of the subunits, not of the whole associated molecules (Blum & Alper, 1971).

In reviewing the results of irradiation experiments on viruses, Ginoza (1967) put double-stranded DNA viruses in a special category, since, for these, the target molecular weights had been calculated by various authors as being considerably less than those of the nucleic acid cores as measured by conventional methods. In our own experiments, the target molecular weight of the double-stranded DNA virus *Herpes simplex* was not significantly too small; that of the bacteriophage T1 was too small, but not to the great degree found in the work quoted by Ginoza. Some of the discrepancies he reported might be attributable to failure to observe appropriate conditions: for example, irradiation *in vacuo* might reduce the probability of inactivation by a single event, as described above. Furthermore, detailed investigation of the nucleic acid cores of some viruses has disclosed that redundant lengths of polynucleotides may be present, and these would be included in estimates of molecular weight of the nucleic acid, but would not form part of the target.

All the results taken together justify the general conclusion that target theory is supported by the results of irradiating macromolecules ranging in molecular weight from about 10^4 to 10^8 daltons, when loss of biological function is the test of damage. In all cases inactivation curves have been of exponential form, in accordance with the expectations for single-hit action.

Relationship of inactivation dose to radiation quality

The discussion so far has related only to effects of sparsely ionizing radiation, that is, radiation of 'low LET'. The abbreviation LET stands for Linear Energy Transfer, that is, the amount of energy transferred by a charged particle to the medium through which it passes, per unit of track length. For any particle of given energy, the higher the LET, the closer together will be the events in which it gives up energy: for example, an α-particle (or helium ion) of energy 5 meV will transfer about 95 keV to each μm of tissue through which it passes; an electron of the same energy will transfer less than 0.2 keV per μm. Thus, if the average sizes of energy-deposition events are nearly the same for the two particles, the events will be 500 times as close together in the α-particle track, which has much the higher LET.

The effectiveness of radiation is commonly measured in terms of the dose required to yield a given effect (say, inactivation to e^{-1}, for single-hit action), while the dose is measured as the total energy absorbed per unit mass of the irradiated material, therefore as the total number of events per unit mass (or per molecule). Where inactivation of a molecule can occur as the result of a single hit or energy-deposit the occurrence of several events per molecule, in each passage of a particle, will clearly be 'wasteful', i.e. the radiation will be less effective than if the events were spaced, on average, further apart than the diameter of the molecule. Thus a criterion of single-hit action is that radiation should become less effective (i.e. the inactivation dose should increase) as the LET increases.

Natural α-particle emitters, and small sources of neutrons (mixtures of radium and beryllium) were used for some years to collect information on the comparative biological effectiveness of radiations of different quality. Evidence of the decrease in effectiveness, with increase in LET, was given by Lea (1946) in his discussion of those biological entities which were inactivated by single-hit action. The subsequent development of high-energy machines has made available, in a few centres, beams of particles of different atomic masses, which can be accelerated to the desired energies. No exception has been reported to the rule that biological molecules, irradiated extracellularly, are the less effectively inactivated, the higher the LET of the radiation. This statement applies when loss of function is the test of damage. Examples are shown in fig. 4.4, relating to proteins and nucleic acid targets. It would be useful if information were to hand also about membranes, which might act as an intracellular target for the killing of cells. The kind of study illustrated by fig. 4.4

has not as yet been made with any model membrane system. For these, we have only the report of Watkins & Deacon (1973), who found that neutrons were less effective than electrons in damaging lysosomal membranes.

The rule, reduced effectiveness of radiation as the LET increases, applies to 'inactivation' strictly in the sense of loss of biological or biochemical function: damage measured by physico-chemical tests does not necessarily conform.

The two phenomena, exponential inactivation, and reduction in effectiveness with increase in LET, are together diagnostic for single-hit action. Another criterion, quoted by various authors, is independence of the rate at which the radiation is delivered, but that criterion requires qualification (Chapter 13).

Many physical and chemical detectors of radiation are 'one-hit' in their response (Katz, 1973), and this appears to be true also of biological detectors irradiated extracellularly. On this basis, we might expect intracellular targets likewise to behave as one-hit detectors. Two general observations seem to conflict with that expectation: survival curves for eukaryotic cells are seldom of exponential form; and the effectiveness of radiation in killing cells almost always increases with increasing LET. As will be shown in Chapter 15, it may nevertheless be possible to reconcile those facts with the hypothesis that the basic response of intracellular targets is that of one-hit detectors.

Fig. 4.4. Decrease in effectiveness of radiation with increasing density of ionization, for inactivation of bacteriophages and proteins irradiated dry.

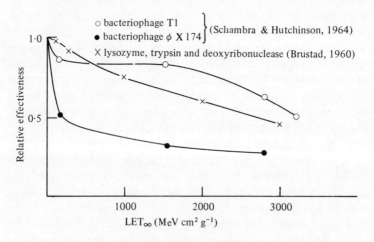

o bacteriophage T1
● bacteriophage φ X 174 } (Schambra & Hutchinson, 1964)
X lysozyme, trypsin and deoxyribonuclease (Brustad, 1960)

Relative effectiveness

1·0

0·5

1000   2000   3000

LET_{∞} (MeV cm^2 g^{-1})

5

Cell survival curve shapes: algebraic expressions

The special form of dose–effect curve known as a survival curve is based on viable counting, that is, the counting of colonies originating from single cells (see Chapter 3). The 'surviving fraction', i.e. the fraction of irradiated cells retaining their colony-forming capacity after exposure to radiation is plotted (usually on a logarithmic scale) as a function of radiation dose (usually on a linear scale).

Methods for the viable counting of free-living cells like bacteria and fungi were already well established before the existence of ionizing radiation was discovered, so numerous survival curves for micro-organisms had been constructed long before Puck & Marcus (1955) achieved the first viable counting of mammalian cells *in vitro*. So many survival curves for bacteria had been found to be exponential, or nearly so, that the view was sometimes expressed that survival curves of other forms were attributable to experimental artefact. The inclination today is rather to regard a survival curve of strictly exponential form as exceptional. Figure 2.4 is an example of a survival curve for bacteria that seems to be reasonably well described as of exponential form, right down to a surviving fraction of 10^{-11}. This pertains to freeze-dried bacteria irradiated in bottles, each of which contained about 10^{12} organisms.

Survival curves of exponential form

It is customary to characterize survival curves of exponential form by the parameter D_0, which is the dose required to reduce the surviving fraction by the factor e^{-1}, or its reciprocal, often known as the inactivation constant. The theoretical importance of that parameter is given by the simple rationale assumed to underly the effects of radiation, as set out in Chapter 4. But when we are concerned with the killing of cells, even those for which radiation response is strictly of exponential form, it does not necessarily follow that the implied 'single-hit' action involves a single target per cell, the size of which would determine the value of D_0. It may be that there are several vital structures in the cell, the inactivation of

33

any of which would be lethal. For convenience, let us designate these as a, b, c etc . . . . It will be a matter of chance whether a cell is killed by a first event in any one of these; once that event has occurred, any others will be immaterial. Thus it may well be that an exponential survival curve, usually described by equation 5.1:

$$f = e^{-D/D_0} \text{ or } f = e^{-\lambda D} \tag{5.1}$$

where λ is the inactivation constant, would be more accurately expressed as

$$f = e^{-(\lambda_a + \lambda_b + \lambda_c \ldots)D} \text{ or } f = e^{-\Sigma \lambda D} \tag{5.2}$$

It is a further complication that incipiently lethal lesions in cells are subject to biochemical repair which, acting consistently on any one of the damaged targets, would correspondingly reduce the effective value of the relevant inactivation constant. There is clearly insufficient information at present about the targets for cell death, or about the processes which fix or repair potential damage, to enable any useful deductions to be made about cellular target sizes from the magnitude of inactivation doses.

Shouldered survival curves: concept of accumulation of damage

For higher cells irradiated at low LET, survival curves commonly have an initial (low dose) region within which a smaller proportion of cells is killed, per unit of dose, than after higher doses. Such survival curves, with survival plotted logarithmically, have acquired the generic description 'shouldered' (fig. 5.1a, b). This is such a common finding that the occurrence of exponential survival curves with higher cells – particularly diploid mammalian cells – has failed to gain due consideration, when one or other mechanism of cell killing has been proposed basically to account for the shouldered curve. Some examples of exponential survival curves for mammalian cells are therefore listed in table 5.1.

Dose–effect curves in general could be expected to demonstrate a kind of 'threshold' (like the shoulder to survival curves) if one or more of the first energy deposits within the irradiated entity were not in themselves sufficient to cause the effect under observation; that is, if damage had to accumulate before the effect ensued. This concept has for many years been dominant in the interpretation of shouldered survival curves. Lesions leading to the end-point under consideration would have to accumulate if a target required more than one hit for inactivation, or if the irradiated entity contained more than one target, such that any one retaining its integrity were sufficient for success in carrying out the function under test: the ability of the cell to proliferate, when cell killing is under investigation. The theory of the most general case, encompassing the requirement for several hits in each of several targets, which might or might not be alike, was comprehensively treated by Timoféeff-Ressovsky & Zimmer (1947) and more recently by Elkind & Whitmore (1967).

If a population of cells is biologically heterogeneous (and many are), it is unlikely that any simple algebraic expression can be adequate to describe a dose–effect relationship. However, it is convenient to describe cell survival curves by reasonably simple expressions even if these are regarded only as approximations. This facilitates comparisons of curves for cells of different lines, or for cells of the same line irradiated in different conditions or by radiations of different quality, provided the expression used is adequate for defining all the survival curves to be compared. The functions that have most commonly been used as descriptions of

Table 5.1. *Examples of exponential survival curves for asynchronous populations of mammalian cells: X– or γ-rays*

Origin	Cell line or tissue	D_0 Grays	Lowest survival measured	Reference
Human, freshly cultured	skin spleen ovary	~1.0 (a)	10^{-4}	Puck et al. (1957)
Human, freshly cultured	skin lung	1.04 (b) 1.08 (b)	10^{-1} 10^{-1}	Norris & Hood (1962)
Human, freshly cultured	skin fibroblasts	1.32	10^{-3}	Weichselbaum et al.
Human, freshly cultured	skin fibroblasts	1.26	10^{-4}	Cox, Thacker & Goodhead (1977a)
Human	Burkitt's lymphoma	0.62	10^{-3}	C. Sato, quoted by Okada (1975)
Mouse	lymphoma (established in vitro)	0.62	5×10^{-3}	Caldwell, Lamerton & Bewley (1965)
Mouse	leukaemia, L5178Y/S (established in vitro)	0.40	10^{-4}	Ehmann et al. (1974)
Mouse	normal haemopoietic (irradiation in vitro)	1.04	4×10^{-3}	Silini & Maruyama (1965)

Notes
(a) Estimate from later revision of dose measured.
(b) Values of D_0 decreased with age of culture.

shouldered survival curves have embodied the concept of cumulative damage. These functions have usually been associated by their proponents with assumptions as to the mechanism by which radiation kills cells. The more commonly used expressions will be considered here.

Up to the time of writing, the most frequent description of shouldered survival curves has embodied the multi-target, single-hit concept. Suppose that a cell contains n targets, all alike, and that the cell is able to proliferate provided any one of them maintains its integrity. Suppose also that the inactivation constant for each target is $-\lambda$, so that, after dose D, the probability of survival of any target is $e^{-\lambda D}$ and the probability of inactivation is $1 - e^{-\lambda D}$. The probability that all n targets will have been inactivated is $(1 - e^{-\lambda D})^n$, which is also the probability that any cell will have been killed. The probability of cell survival, i.e. the surviving fraction, is then given by

$$f = 1 - (1 - e^{-\lambda D})^n \tag{5.3}$$

Fig. 5.1. Two types of shouldered survival curves.

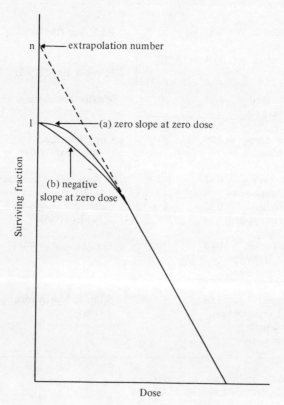

When D becomes large, equation 5.3 approximates to

$$f = ne^{-\lambda D}$$

or $\quad \log_{10} f = \log_{10} n - \lambda D \log_{10} e$

or $\quad \ln f = \ln n - \lambda D$ $\qquad\qquad$ (5.4)

At high dose, therefore, the curve has a terminal exponential region, that is, a straight line region if surviving fraction is plotted on a logarithmic scale. On that plot, the straight terminal region extrapolates back to n on the zero dose axis and n is now commonly known as the extrapolation number, as suggested by Alper, Gillies & Elkind (1960). From the derivation of expression 5.3, it should be equated to the number of cellular targets as defined above.

The rather wide adoption of equation 5.3, or the modified form, equation 5.8, may be attributable to its use by Puck & Marcus (1956) who constructed the first survival curve for mammalian cells, using tissue culture techniques. Their survival curve, for the line of human tumour cells known as HeLa, was described by the expression

$$f = 1 - (1 - e^{-\lambda D})^2$$

However, Puck and Marcus regarded the index 2 as a 'hit number', not a target number. They made use of another general expression, based on the propositions that there are μ sites, damage to *any one* of which is lethal to the cell; and that each site is divided in such a way that it acts as an r-target entity, each of the r targets being inactivated by a single hit. Thus the probability of the survival of any site is given by

$$1 - (1 - e^{-\lambda D})^r$$

and the probability that *all* sites survive, i.e. that any cell survive, is given by

$$f = \left[1 - (1 - e^{-\lambda D})^r \right]^\mu$$ $\qquad$ (5.5)

Puck and Marcus fitted their survival curve with $\mu = 1$, $r = 2$, and speculated on the possibility that the single locus of action could be chromosomal: cell death being attributable to aberrations resulting from two or more chromosome breaks.

The high-dose approximation to equation 5.5 is

$$f = (re^{-\lambda D})^\mu$$

or $\qquad\qquad\qquad\qquad\qquad\qquad\qquad\qquad\qquad$ (5.6)

$$\ln f = \mu \ln r - \mu \lambda D$$

so that the extrapolation number is r^μ.

It is held by some authors that the killing of higher cells is directly correlated with the infliction of 'two-hit' aberrations in certain critical chromosomal sites, that is, there must be a hit in each chromatid at a relevant site. On that view, the r of equations 5.5 or 5.6 is equal to 2, and the μ is equal to the number of critical chromosomal sites. Some cell survival curves have extrapolation numbers of the order of hundreds or even thousands – for example, the green alga *Chlamydomonas reinhardii* yielded a survival curve with a terminal region extrapolating to more than 1500, when the cells were irradiated while resting on a surface (Bryant, 1973). If the survival curve is described by equation 5.3, or its modified form, equation 5.8, it would be necessary to interpret the result as showing that there are more than 1500 critical targets, *all* of which have to be inactivated if a cell is to be killed. However, the interpretation expressed by equation 5.5 leads to the conclusion that there are ten to eleven critical chromosomal sites, since 2^{10} is about 1000; the assumption being that the cell dies if there is a two-hit aberration at any one of these sites. On the face of it, this seems more plausible than the assumption of 1500 or more targets, all but one being redundant for cell division. However, the identification of the critical sites as specifically chromosal becomes less attractive when we take into account the wide variations often seen in extrapolation numbers, which appear to depend not only on the physiological condition of cells and on their stage in the cell cycle, but even on the conditions in which they are exposed to radiation. For example, the very high extrapolation number quoted above for *Chlamydomonas* was observed when they were irradiated on membrane filters resting on a surface. When they were irradiated in suspension, the extrapolation number was about 29, rather less than 2^5. It is difficult to envisage how the change of condition of exposure could effectively halve the number of critical chromosomal sites. The quite common variability in extrapolation numbers to cell survival curves is, indeed, a phenomenon that imposes difficulty in associating those parameters with any specific morphological structures constituting the targets, in any version of a multi-target model.

For the most part, equations 5.3 or 5.5 have been used as descriptions only of the terminal regions of survival curves, that is to say, the high-dose approximations have been used. These are defined by the two para-meters D_0 (or λ, $= 1/D_0$) and the extrapolation number. Alternatively, either one of these may be replaced by the 'quasi-threshold dose', D_Q, which is defined as the intercept of the back-extrapolated terminal straight line on the 100% survival axis (Alper *et al.*, 1962). The three parameters are related by the expression

$$D_Q/D_0 = \ln n$$

so equation 5.4 could be written

$$\ln f = -(D - D_{\mathrm{Q}})/D_0 \tag{5.7}$$

Many survival curves have terminal regions for which a straight line appears to be an adequate fit (when survival is plotted logarithmically), though that region is best defined, of course, when it has been possible to observe low levels of survival, to 10^{-5} or 10^{-6} (e.g. fig. 5.2). But considerable importance attaches to the mode of killing of cells at much lower doses than those at which, in most cases, the logarithm of survival may be judged to approximate to a straight line. The practical need for insight into the effects of low doses comes from the requirements of those who are engaged in research on the hazards of rather low doses; and the 'low dose' region of survival curves is important also to radiotherapists, who customarily treat cancerous tumours by delivering the total radiation dose in many fractions, usually separated by 24 hours or more.

According to equations 5.3 or 5.5, the slope of the survival curve at zero dose should be zero, as in fig. 5.1a. That is to say, the first small element of dose should fail to kill any cells: the higher the value of n, the larger will be the dose range within which there is no observable killing.

Few cell survival curves have that property, most of them being rather of the form illustrated in fig. 5.1b. (see Alper ed. *Cell Survival After Low Doses of Radiation*, 1975.) This, in itself, has not been thought sufficient

Fig. 5.2. Survival curve for well-oxygenated CBA leukaemia cells (from Hewitt & Wilson, 1961).

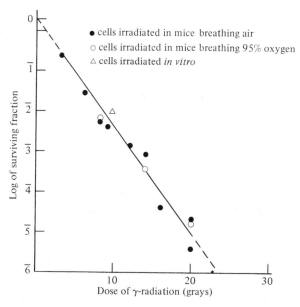

reason for discarding the general principles embodied in the derivation of equations 5.3 or 5.5. An initial non-zero slope would be observed if the irradiated population were inhomogeneous with respect to target number, the number per cell ranging from 1 to some value higher than n. The n of equation 5.3 would then represent a weighted average of all the target numbers, and the total surviving fraction would be given by a summation of all the survivors in the sub-populations, each of which would survive according to equation 5.3, with the appropriate value of n inserted for each sub-population. With mammalian cells cultured *in vitro*, for example, it has been found that the extrapolation number to the survival curve depends strongly on the cycle stage at which the cells are irradiated: survival of cells in mitosis is often exponential (Chapter 12). Nevertheless, survival curves, even for well synchronized populations of cells, do not often display an initial non-zero slope. Another suggestion frequently invoked is that there may be a variety of processes that kill the cells, including a 'one-hit' process. As suggested by Bender & Gooch (1962) the principles embodied in equation 5.5 might then be expressed by the more general form:

$$f = (e^{-\lambda_1 D}) \left[1 - (1 - e^{-\lambda_2 D})^{n_1} \right] \left[1 - (1 - e^{-\lambda_3 D})^{n_2} \right] \text{ etc.}$$

which could be adequately expressed by the approximation

$$f = e^{-\lambda_1 D} \left[1 - (1 - e^{-\lambda_2 D})^n \right] \tag{5.8}$$

Similarly, equation 5.5 could be modified to include a 'single-hit component' giving

$$f = e^{-\lambda_1 D} \left[1 - (1 - e^{\lambda_2 D})^r \right]^\mu \tag{5.9}$$

It is sometimes postulated that the 'one-hit component' is attributable to the action of radiation of high LET, which is part of the spectrum of LET values that characterize the tracks of charged particles engendered by the passage through matter of γ-rays, energetic X-rays or fast electrons. That attribution is associated with the general observation that survival curves tend to approximate more closely to an exponential form as the LET increases (Chapter 9). In other words, it is assumed that the passage of a densely ionizing particle may act as a single 'hit' that would simultaneously inactivate several targets, or simultaneously provide several hits on a target if more than one were required for its inactivation. The postulated 'high LET component of radiation of low LET' comprises the densely ionizing ends of the tracks of secondary or tertiary electrons of low energy. It is unclear, however, whether the spectrum of LET values from high-energy electromagnetic radiation can include a component of sufficiently high LET to account for an initial non-zero slope in all cases.

At very low dose, equation 5.8 approximates to

$$\ln f = -\lambda_1 D$$

and at high dose, to

$$\ln f = -(\lambda_1 + \lambda_2)D + \ln n.$$

If we wish to use the symbol D_0, the inverse of the final slope of the survival curve, we may write

$$\lambda_1 + \lambda_2 = 1/D_0 \text{ and } \lambda_1 = 1/{}_1 D_0$$

so equation 5.8 would be written

$$f = e^{-D/{}_1 D_0} \{1 - [1 - e^{(-D_0^{-1} - {}_1 D_0^{-1})D}]^n\} \tag{5.10}$$

As mentioned above, equation 5.3 is a special case of the more general form for survival of a population such that m hits are required in each of n targets. Equation 5.3 represents the case for one hit in each of n targets; conversely, there may be only one target per member of the population, with m hits required for inactivation. The surviving fraction would then be given by:

$$f = e^{-\lambda D} \sum_{r=0}^{m=1} (\lambda D)^r / r! \tag{5.11}$$

where r represents the inverse of the average size of dose required for an effective hit. Like equations 5.3 and 5.5, equation 5.11 would yield zero slope at zero dose; but the survival curve would not have a terminal exponential region. Nevertheless, with increasing dose, it would become increasingly difficult to distinguish between a straight-line terminal region, on a logarithmic plot, and the form of curve defined by equations 5.3 or 5.5 (Fowler, 1964).

The concept that shoulders to survival curves demonstrate the requirement for more than one hit per target, with radiation of low LET, has found expression in yet another proposal for the description of cell survival curves. Kellerer & Rossi (1972) proposed that this should be

$$f = e^{-(\alpha D + \beta D^2)}$$

or

$$\ln f = -(\alpha D + \beta D^2) \qquad \text{('linear-quadratic equation', 5.12)}$$

They based their formulation on considerations evolved in their 'Theory of Dual Radiation Action', which is discussed briefly in Chapter 11. They considered the theory to be applicable only to eukaryotic cells, and results with mammalian cells *in vitro* were excluded because they considered them to be subject to experimental artefact. These exclusions are rather unsatisfactory for any general theory of radiation action, since most radiobiological phenomena are observed alike with micro-organisms, higher cells in culture, and cells *in vivo*.

Equation 5.12 was proposed also by Chadwick & Leenhouts (1973) on the basis of assumptions other than those made by Kellerer & Rossi (1972): (1) that cell death is attributable solely to the induction of a double-strand break in the nuclear DNA, (2) that, with radiation of low LET, a double-strand break will result from the occurrence of separate energy-deposits at sites near each other in the two strands of the DNA double helix. Those authors did not explicitly confine their analysis to eukaryotic cells.

'Repair' models for shouldered survival curves

All the cell survival curve models cited so far were developed on the assumption that a relatively insensitive initial region to a dose–effect curve signifies the requirement for an accumulation of damaging events, if the given end-point is to be achieved. At the time of writing, no evidence has been adduced, other than the shapes of survival curves for single cells, for the existence of a multiplicity of redundant targets, or of the requirement for two or more 'hits' for the inactivation of an intracellular target. Nevertheless, comparatively little attention has been paid to a group of models based on an alternative set of assumptions: namely, that cells surviving by that mode contain a mechanism for repair that becomes progressively less effective as the dose or the number of incipiently lethal lesions increases, until it ceases to operate. This suggestion was made by Powers (1962), but not formalized.

The simplest form of repair model relates to a basically exponential mode of cell killing. It is postulated that some repair of potentially lethal lesions occurs concurrently with their induction, or in the period before the cells go into division, but the capacity for repair is limited, either because the rate at which lesions are induced is greater than that at which they are repaired, or because the repair mechanism becomes 'saturated', or is itself depleted by radiation. If the initial slope to the survival curve is not zero, and a final slope can be fitted to an experimentally derived survival curve, a reasonably good description of this model can be obtained by a dose function containing three parameters: the basic requirement is that the function equated to the logarithm of surviving fraction approximates to linear functions of dose both at low and at high doses; and that the high dose approximation is the same as equation 5.4, which gives the extrapolation number n. The basic idea of repair models is illustrated in simple form by fig. 5.3.

Suppose that the cell survival curve would be an exponential of slope $-\lambda$ if there were no repair of the type envisaged. Suppose also that the cells were biochemically equipped to repair all but a fraction λ_1/λ of the potentially lethal lesions. The initial slope to the survival curve would be given by

$$\ln f = -\frac{\lambda_1}{\lambda}(\lambda D) = -\lambda_1 D$$

or

$$\ln f = -\lambda D + (\lambda - \lambda_1)D.$$

As the dose increases, the fraction of unrepaired lesions approaches 1, so the survival curve approximates to an exponential of slope $-\lambda$; and survival is given by

$$\ln f = -\lambda D + \ln n, \text{ where } n \text{ is the extrapolation number.}$$

Thus a general equation to express a repair model in simple form could be written

$$\ln f = -\lambda D + (\text{a function of } D) \tag{5.13}$$

where the function of D has the property of approximating to $(\lambda - \lambda_1)D$ at very low dose and to $\ln n$ at high dose. Although several forms of

Fig. 5.3. To illustrate the basic principle of repair models.

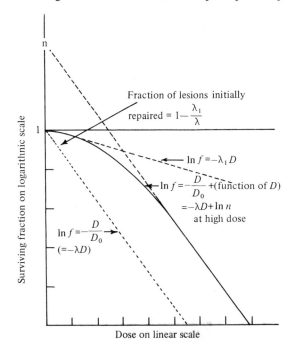

Fraction of lesions initially repaired $= 1 - \dfrac{\lambda_1}{\lambda}$

$\ln f = -\lambda_1 D$

$\ln f = -\dfrac{D}{D_0} + (\text{function of } D)$

$= -\lambda D + \ln n$

at high dose

$\ln f = -\dfrac{D}{D_0}$

$(= -\lambda D)$

Surviving fraction on logarithmic scale

Dose on linear scale

such an equation have been developed, they have hardly been used at all by radiobiologists to describe cell survival curves: in contrast with the extensive use that has been made of equations 5.3, 5.5, 5.8 and 5.12.

It will suffice, therefore, to give an example of the type of function that satisfies the requirements of the repair model as regards the high and the low dose approximations. Haynes (1966) proposed a description of the survival of the repair-proficient bacterial strain, *Escherichia coli* B/r after ultraviolet irradiation, as follows,

$$\ln f = -KD + \alpha(1 - e^{-\beta D}) \tag{5.14}$$

He based his choice of the repair term on the assumption of an exponential approach to 'saturation' of the repair mechanism. The coefficient α represented the natural logarithm of the maximum number of 'hits' that could be repaired. To conform with the symbols used previously, equation 5.14 may be rewritten

$$\ln f = -\lambda D + \ln n \left\{ 1 - \exp\left[-\left(\frac{\lambda - \lambda_1}{\ln n} \right) D \right] \right\}$$

or (see equation 5.10)

$$\ln f = -\frac{D}{D_0} + \frac{D_Q}{D_0} \left\{ 1 - \exp\left[-\left(\frac{{}_1 D_0 - D_0}{{}_1 D_0} \right) \cdot \frac{D}{D_Q} \right] \right\}$$

since $\ln n = \dfrac{D_Q}{D_0}$

As before, $-D_0^{-1}$ and $-{}_1 D_0^{-1}$ represent respectively the final and initial slopes of the survival curve. From the low dose approximation

$$\ln f = -\lambda D + (\lambda - \lambda_1)D$$

or

$$\ln f = -\frac{D}{D_0} + \left\{ \frac{1}{D_0} - \frac{1}{{}_1 D_0} \right\} D$$

it may be seen that $(\lambda - \lambda_1)/\lambda$ represents the fraction of potentially lethal lesions initially repaired. This fraction is given by $\alpha\beta/K$ in the symbols of equation 5.14, so that Haynes' formulation involves an interdependence between the maximum number n ($=$ antilog α) of the events per cell that are repaired, and the fraction initially repaired.

Heterogeneous populations

Each of the equations discussed in previous sections was developed on the assumption of a population homogeneous in its response. When this is not the case the survival curve can be very complex, depending on the type of curve proper to each sub-population. With only two sub-populations, both subject to exponential killing, the resultant curve will be as shown in fig. 5.4 (see also Powers & Tolmach, 1963). The terminal region, or 'resistant tail' as it is often called, will yield the value of D_0 for the more resistant sub-population, and the fraction of the initial total that it represents may be deduced by extrapolating back to the zero-dose axis. The value of D_0 for the more sensitive sub-population may then be calculated by subtracting surviving fractions pertaining to the back-extrapolated line from the composite curve. It is sometimes assumed that the value of D_0 for the most sensitive sub-population is given by the initial slope of a curve like A in fig. 5.4. This will be true only if the resistant sub-population represents such a small fraction that the survival due to its presence, at low doses, is outside the experimental error of the observations, as in curve B.

Fig. 5.4. Hypothetical curves for heterogeneous populations. The resistant fractions can be estimated by extrapolating the 'tails' back to zero dose. In the case represented by curve A the slope for the sensitive fraction would not be given by the apparent 'initial slope' of the curve; it would have to be determined by subtracting surviving fractions pertaining to the resistant 30%.

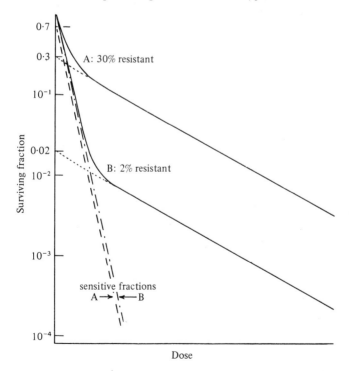

If survival curves proper to all sub-populations are shouldered, and all have terminal exponential regions, of equal slope, the extrapolation number of the composite curve will be a weighted average of the extrapolation numbers proper to each sub-population. If one of the sub-populations survives exponentially, the initial slope of the composite curve cannot be zero.

No generalization can be made about survival curves for any mixture of sub-populations for which individual survival curves differ in respect of more than one parameter, since the shape of the composite curve will depend on the initial sizes of the sub-populations. Nevertheless, a 'resistant tail' to the composite survival curve will betray the presence of a minor sub-population that requires a substantially greater average dose per lethal event than the majority.

Modification of radiation response: changes in survival curve parameters

Whichever equation is used to describe a cell survival curve, the constants will in general be valid only for one set of conditions of irradiation and one set of conditions of the cells themselves. Deposition of energy in (or near) a critical structure results in a sequence of physico-chemical and then biochemical reactions before the ultimate effect is observable, and there are many possibilities of variation in these sequential steps.

Very early in the chain of events, the collision of ionizing particles with target molecules may cause them to lose electrons, that is, to become active radicals; or reactive species may be engendered in the radiolysis of small molecules, like water. Since about 70–80% of cellular material is water, its radiolysis products might be considered to play a major role as agents of radiobiological effects; indeed, that view was widely held in the 1940s and 1950s, and is still propounded, although radiobiological effects have tended latterly to be attributed to a much greater extent primarily to the direct formation of radicals within target structures. Perhaps both pathways are involved.

The first chemical reaction in which target radicals are engaged, the 'metionic reaction' (Alper, 1956), may result in a product that will rob the target of its normal function, or it may be of such a nature that normal function will be unimpaired. The former will be described as metionic fixation of damage. In general, metionic reactions occur within time-spans that are short, compared with such intracellular biochemical processes as may operate to repair or bypass metionically fixed lesions. The chemical environment of the radicals at the time of their formation will to a large extent determine the nature of the metionic reactions, so differing environments will result in shifts towards or away from damage fixation. Many agents that can be introduced into cells are therefore commonly described as radiosensitizers or radioprotectors; conversely, procedures may be adopted that will reduce the concentration of some normal intracellular components. It must be borne in mind that, when the term 'sensitizer' or 'protective agent' is used, there is always an implicit comparison involved: thus the presence of a given agent may be regarded as 'sensitizing'; or its absence may be regarded as 'protective' when the presence of the agent within the cell is regarded as the normal state.

Suppose a modifying agent exerts its effect by reducing the proportion of incipient lesions that are metionically fixed. It follows that there must then be a corresponding increase in the total number of incipient lesions required to bring about a given effect, i.e. the dose must be increased by the same factor as that by which metionic fixation has been reduced. Such an agent is often then said to have a 'dose-reducing' effect. Conversely, an agent that increases the extent of metionic fixation has a 'dose-multiplying' effect. In equations describing effect as a function of dose, the action of such an agent would be expressed by multiplying every constant governing the first power of the dose by the appropriate factor. Thus if a survival curve were expressed by equation 5.8, and all conditions were kept the same, except for the introduction of a metionic reactant for which the uniformly dose-multiplying factor was m, the new survival curve would be given by

$$f = e^{-m\lambda_1 D}\{1 - e^{-m\lambda_2 D})^n\}$$

If equation 5.12 were the appropriate description, the new survival curve would be given by

$$\ln f = -(m\alpha D + m^2\beta D^2)$$

since the second term of the function is in D^2.

Hypothetical examples of exact dose-multiplication were given in fig. 2.2. If curves are plotted of effect (on any kind of scale) against dose on a linear scale, then a change by the appropriate factor in the dose scale should cause the curves to be superposable, when the agent of change has been dose-multiplying.

While restitution of an incipiently lethal lesion may occur as a consequence of metionic reaction, metionically fixed lesions are subject also to repair by biochemical processes, usually assumed to be enzymic in nature. Suppose that a normal enzymic repair process can repair or eliminate $(m - 1)/m$ of all metionically fixed lethal lesions, and we then impose conditions that will inhibit the enzymic action. We should again expect precise dose-multiplication, as is to be expected with metionic fixation. A good example of precise dose-multiplication (by a factor less than one) by a known enzymic process occurs with the phenomenon previously called photoreactivation of cells after exposure to germicidal ultraviolet light. The photoreactivating enzyme is known to be one which, when activated by light, monomerizes the pyrimidine dimers that are the main lethal lesions induced by the UV. Cells containing the enzyme will show increased survival if, after UV irradiation, they are exposed to light of longer wavelength. Provided that exposure to the reactivating light is sufficient to give the maximum effect, the survival curve for the photoreactivated cells will, in general, be superposable upon that for cells held in the dark, if an appropriate adjustment is made to the dose scale. This is well illustrated in the case of the UV-sensitive strain of *Escherichia coli*, B_{s-1}, for which the survival curve after UV has a complicated shape (fig. 5.5).

However, examples of precise dose-multiplication (or dose-reduction) are not common, when modification of radiation effects is attributable to biochemical processes. Several modes of modification of survival curves will be considered in succeeding chapters. A critical examination of the various models for survival curves can be made only after taking these into account (Chapter 14).

Fig. 5.5. An example of dose-multiplication. With dose scales in the ratio 4:1, the survival curve for *Escherichia coli* B_{s-1}, exposed to germicidal UV, and then to photoreactivating light, is superposed on that for organisms kept in the dark after irradiation.

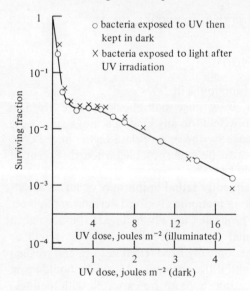

Summary

The advantages of describing cell survival curves by simple algebraic expressions are two-fold: it is helpful to be able to make comparisons between the descriptive parameters applying to different cell lines, or to the same cell line irradiated in different conditions; and a given hypothesis relating to radiobiological mechanisms can, to some extent, be judged in the light of its success as a basis for setting up a model. It must be emphasized, however, that the exigencies of experimental techniques required for constructing cell survival curves impose severe restrictions on the possibilities of regarding one or other model as the most valid, on the basis of curve fitting alone (Chapters 3, 14). Further complications arise when cell populations are radiobiologically heterogeneous, which is often the case.

There are various ways of grouping the models that have been discussed.

(1) Models that fit survival curves with zero initial slope (equations 5.3, 5.5, 5.11) and those that predict non-zero intial slope (equations 5.8, 5.9, 5.12, 5.13).

(2) Models that predict a terminal exponential region (equations 5.2, 5.3, 5.5, 5.8, 5.9, 5.13) and those that predict a continuing increase in slope with increasing dose (equations 5.11, 5.12).

(3) Models that account for shoulders in terms of the requirement for accumulating damage (equations 5.3 to 5.12) and 'repair' models, generalized by equation 5.13.

The constants of the various equations cited will in general be valid only for the relevant experimental conditions. Of the numerous agents that can modify radiation damage, there is a special group that act by the dose-multiplying mode. If such an agent operates, the appropriate descriptive equation for the new survival curve will be like the one taken without the agent, except that all dose terms will be multiplied by the same factor.

6

Radiosensitization by oxygen

Early observations by radiotherapists as well as radiobiologists suggested that anaerobic cells and tissues were less radiosensitive than when normally oxygenated: but the effect was attributed to the biochemical state of anaerobiosis. Yeast cells are 'facultative anaerobes', able to respire by anaerobic glycolysis as well as aerobically, and Anderson & Turkowitz (1941) used this property to test whether it was the biochemical state of the cells during anaerobic metabolism, or the absence of oxygen during irradiation, that resulted in radioprotection; and showed that the latter was correct. Techniques have latterly been developed that enable changes in oxygen environment to be made within very short times, down to a few microseconds, after short pulses of radiation, so the word 'during' requires some modification, as will be discussed below.

It is generally true that the presence of oxygen is sensitizing (or its absence is protective) for the lethal effect of radiation on cells. Indeed, it is sensitizing for most biological manifestations of damage, like the induction of division delay, gene mutations or chromosomal aberrations. Dose–effect curves taken in the presence and absence of oxygen can very often be superposed if appropriate dose scales are chosen (fig. 6.1), and in such cases the oxygen is a dose-multiplying agent, its effect being commonly quantified by the 'oxygen enhancement ratio', abbreviated o.e.r. As shown in Chapter 5, this is what would be expected if molecular oxygen acted by metionic fixation of damage.

The o.e.r. is defined as the ratio of doses to give the same effect for irradiations in the absence and presence of oxygen, and is independent of the level of effect in cases typified by fig. 6.1. Where oxygen does not act as a precise dose-multiplying agent there is no such quantity as the o.e.r. according to the definition given; nevertheless the expression is sometimes used in the manner of the term Relative Biological Effectiveness (RBE) to specify a ratio of doses at a specified level of effect; or, on occasion, to specify ratios of slopes of survival curves, when these can be described at high dose by equation 5.4 (e.g. Littbrand & Révész, 1969).

Oxygen enhancement ratios for the killing of cells of all classes ordinarily fall within the range of about two to four, but when it is measured with precision, the o.e.r. for any one strain may vary with the method of growth and handling after as well as before irradiation. Some of these variations are dealt with below and in subsequent chapters. A special kind of variability is associated with genetically controlled capacity for repair or bypass of damage to one or other cell component.

Oxygen enhancement ratios for repair-deficient mutants

The only evidence on repair of radiation damage that can be firmly correlated with cell survival comes from work on the killing of bacteria by 'germicidal' ultraviolet light, of wavelength in the region 250–270 nm. UV at these wavelengths is specifically absorbed by nucleic

Fig. 6.1. Survival curves for bacteria and mammalian cells irradiated in the presence and absence of oxygen. (Data for mammalian cells provided by Dr B. M. Cullen.)

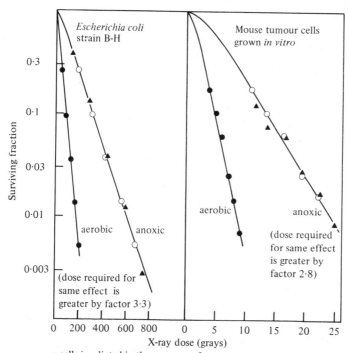

● cells irradiated in the presence of oxygen

○ cells irradiated in the presence of oxygen, with doses multiplied by the appropriate factors: 3·3 for the bacteria, 2·8 for the mammalian cells

▲ cells irradiated with no oxygen present

acid, and cells are killed mainly as a result of lesions in the DNA. Many mutants have been selected on the basis of their greater sensitivity to UV than their wild-type parents, and have been shown, or presumed, to lack the ability to repair lesions specific to DNA. With all classes of cell in which radiation-sensitive mutants have been selected the o.e.r. is less for these than for the wild-type parents. Where several sensitive mutants have been selected from one wild-type strain of bacteria, there is a rough

Table 6.1. *Low values of o.e.r. for some radiosensitive mutant strains*

Strain	o.e.r	Reference
Bacteria		
B. subtilis, BS 15	3.1	
B. subtilis SMBL 4 (a)	1.8	
P. aeruginosa 1 c	3.1 (b)	Alper (1967)
P. aeruginosa 1 HCR⁻ (a)	1.7 (b)	
E. coli B/r WP2	3.0 (b)	
E. coli B/r WP2 HCR⁻ (a)	1.7 (b)	
E. coli B_{S-12} (c)	1.0	Forage & Alper (1970)
E. coli K12 AB1157	3.0	
E. coli K12 AB1886 (c)	2.1	Sapora, Fielden &
E. coli K12 JG 112 (d)	4.3	Loverock (1977).
E. coli K12 JO 307	1.0	
Diploid yeast		
S. cerevisiae, wild type	2.7	
S. 2093, $r^s_{3-1} r^s_{3-1}$	2.1	Averbeck & Ebert (1973)
S. 2094, $r^s_{3-2} r^s_{3-2}$	1.5	
Chlamydomonas reinhardii		
wild type	2.8	Davies (1967)
UVSI (c)	1.0(e)	
Dictyostelium discoideum (slime mould)		
γS-18 (f)	2.6	Deering *et al.* (1970)
γS-13 (f)	1.0	

Notes
(a) Selected for inability to reactivate UV-irradiated bacteriophage.
(b) At dose rate 250 Gy min⁻¹. Higher values of o.e.r. for both strains at dose rate 13 Gy min⁻¹.
(c) Selected for sensitivity to germicidal UV.
(d) Selected for deficiency in DNA polymerase.
(e) o.e.r. = 1 in terms of terminal slopes. UVS1 was protected by oxygen, because the extrapolation number was greater by a factor 10 for anoxically irradiated cells.
(f) Selected for sensitivity to γ-rays; γ_{S-13} was about eight times as sensitive as γ_{S-18}. The wild type was too resistant for accurate determination of the o.e.r.

correlation between o.e.r. and sensitivity to UV or to irradiation under anoxia. An exception to this rule is afforded by a class of mutants of *Escherichia coli* associated with a defect in its ability to synthesize one of the DNA polymerases (table 6.1).

Radiosensitization by oxygen as a function of concentration

Oxygen is normally not just 'present' or 'absent' in cells and tissues, it is present at specific partial pressures. In early work on the oxygen effect it had been noted that radiosensitivity increased with increasing oxygen content until, apparently, the sensitivity became constant. A quantitative relationship was first established semi-empirically by Alper & Howard-Flanders (1956) who irradiated bacteria at low concentration in suspensions containing oxygen at known partial pressures varying from zero to 100% in the gas bubbled through the suspensions. The results were described by the relationship

$$\frac{S_P}{S_N} = \frac{mP + K}{P + K} = r \tag{6.1}$$

where S_N and S_P represent the radiosensitivity for partial pressures zero and P respectively and m and K are constants. The radiosensitivity at P relative to that at zero oxygen content is designated by the symbol r. When P becomes large $r = m$ and this is what is normally understood by the phrase 'oxygen enhancement ratio'. The constant K is a measure of the rate of increase in r with increasing partial pressure of oxygen.

Equation 6.1 is represented graphically in fig. 6.2 with r as a function of P. The function is a hyperbola, which approaches the asymptote $r = m$ at very high partial pressures of oxygen. Theoretically, there is no plateau for r. The slope of the curve at zero point on the abscissa is inversely proportional to the constant K. This is seen by differentiating r

Fig. 6.2. The curve describing equation 6.1. Oxygen at partial pressure P', not accounted for, will cause m to appear smaller and K to appear larger than the true values.

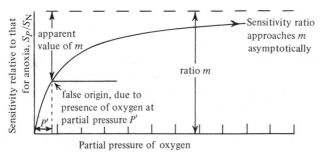

Partial pressure of oxygen

with respect to P:

$$\frac{dr}{dP} = \frac{K(m-1)}{(P+K)^2}$$

$$= \frac{m-1}{K} \text{ when } P = 0 \qquad (6.2)$$

Experimental determination of the constants m and K

In practice the oxygen content of cells is usually varied by passing carrier gases containing pre-determined fractions of oxygen through or over the preparations to be irradiated. The carrier gas is commonly nitrogen, sometimes argon, and it is now well recognized that their oxygen content must be very low, less than 10 parts per million, if 'radiobiological anoxia' is to be attained. The assumption is usually then made that the partial pressure of oxygen (or P_{O_2}) within the cells corresponds with the known content of oxygen in the gas phase. That assumption can rather easily fail to be justified. Unsuspected contamination by oxygen may be caused by leakage of atmospheric air into the system; though this should be detectable if the effluent gas is monitored. More difficult to detect is the presence of contaminating oxygen held in materials used in the experimental equipment. This will be released when the P_{O_2} in the adjacent space has been sufficiently reduced. Chapman *et al.* (1970) reported that the majority of the kinds of plastic used in making petri dishes acted as oxygen reservoirs, from which oxygen was gradually released during the course of an experiment which called for reduction in the P_{O_2}. That phenomenon is particularly relevant to investigations on oxygen effects with mammalian cells *in vitro*, since they are often cultured in plastic petri dishes to which the cells attach. The oxygen released from the plastic therefore comes into immediate contact with the cells, even when vigorous flushing is used. Chapman *et al.* showed that oxygen contamination from that source was not detectable by monitoring the effluent gas.

Oxygen pressure in the immediate vicinity of cells can also be significantly lower than is assumed, because oxygen is consumed by respiring cells. This may be an unsuspected artefact even when partial pressures of oxygen are thought to be high enough to give full oxygen sensitization, particularly when the cells are in, or below, liquid medium that is not vigorously stirred or shaken. The effectiveness with which respiring cells deplete surrounding media of oxygen in closed vessels has indeed deliberately been used by some experimenters to render suspensions anoxic. But it has not been completely established whether the cells made anoxic by that technique will respond to radiation in just the same way as those from which oxygen has been removed by gas displacement.

Oxygen is also consumed in radiation chemical reactions, so a constant partial pressure can be maintained only if the rate at which it is supplied to the irradiated system is at least sufficient to balance the rate of its consumption, which in its turn will be governed by the dose rate. Adequate equilibration may not be achieved even when a gas mixture is vigorously bubbled through a suspension of cells, especially if the partial pressure of oxygen is rather low, as in an experiment to determine the *K* value (Alper 1976a).

The effect of oxygen consumption within cells by radiation chemical action was illustrated in extreme form first by Dewey & Boag (1959) and later in more extensive experiments by Epp, Weiss & Santomasso (1968). When bacteria were held in an environment in which the P_{O_2} was rather low, and exposed to radiation delivered in very short pulses, survival curves were of the type described by fig. 5.4. The terminal regions were parallel to survival curves on bacteria that were anoxic from the start of irradiation.

The magnitude of the constants *m* and *K* of equation 6.1 are important for some aspects of radiotherapy as well as radiobiology, so the precautions necessary for accuracy in their measurement merit attention. Direct determination of *m* requires that irradiations be done both under rigorous anoxia and with sufficient oxygen present so that a further increase in *P* will not detectably increase sensitivity. The necessity for these conditions are probably more widely recognized than are the pitfalls in estimating *K*. It follows from equation 6.1 that, when $K = P$, $r = (m + 1)/2$, and it is not uncommon to find *K* incorrectly defined as that partial pressure at which *r* is equal to $(m + 1)/2$. But this will be true only if equation 6.1 is a valid description of change in sensitivity with oxygen pressure. The constant *K* can therefore not properly be determined from a series of measurements unless these show that the results are fitted by the equation. For that purpose it is convenient to use some function of relative sensitivity that will be linearly related to *P*, if the equation is valid. Marked departure from linearity may then be detected by visual inspection of a graph of the observations, and, if a linear relationship holds, errors on the estimates of the parameters are easily calculated. Two transformations of equation 6.1 have been used.

One of them requires that sensitivity S_P at partial pressure *P* be related to the maximum sensitivity, S_{max}, attained when the partial pressure is high enough so that the sensitivity cannot be detectably increased by adding more oxygen (Alper, Moore & Smith 1967b). If we denote the ratio S_{max}/S_P by *R*, it can be shown (see appendix) that

$$\frac{1}{R-1} = \frac{m}{K(m-1)} \cdot P + \frac{1}{m-1} \qquad (6.3)$$

K can be determined from the slope of the line, and the intercept on the axis $P = 0$ gives $1/(m - 1)$.

Analysis of results by that method has the advantage that all sensitivities are measured relative to the maximum, achieved with high partial pressures of oxygen, and that determination is less subject to technical error than when rigorous anoxia is called for. The method could be particularly useful if the test material were of a nature not to tolerate extreme anoxia, since the maximum o.e.r., or the constant m, can be estimated from the intercept of the line with the axis $P = 0$; it need therefore not be measured directly. If it can be measured, the two estimates are a useful check on each other.

A disadvantage of the method is that the relative errors in determining

Fig. 6.3. To demonstrate a misinterpretation by Alper, Moore & Smith (1967) attributable to the large error in determining $1/(R - 1)$ when R was near to 1, in the determination of K for *S. flexneri* in suspension. Line a: for X-rays at 13 to 31 grays per minute. Line b': as fitted for electrons at 250 to 450 grays per minute. The slope of the line was determined mainly by the points at low oxygen partial pressures, for which the 95% confidence intervals (shown by vertical lines) were small.
Line b: probably a correct fit, with the terminal slope parallel to a. The initial lesser slope of line b was due to oxygen depletion at the high dose rate, despite continuous gas bubbling. Oxygen was replaced during intervals between sample removals, so evidence of depletion was obscured.
Line c was obtained when each survival curve at the high dose rate was based on exposures without interruption.

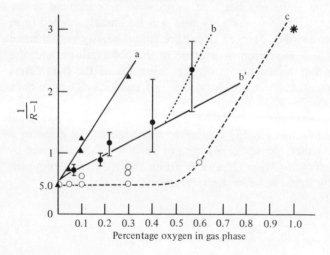

R increase with increasing P, so that a departure from linearity may be obscured by the magnitude of the errors in R at higher P. Figure 6.3 shows an example in which this artefact led to a wrong conclusion.

An alternative transformation yielding a linear relationship is given by rearranging equation 6.1 (Alper, 1976a):

$$\frac{P}{K} = \frac{r-1}{m-r} \tag{6.4}$$

If the hyperbolic relationship is obeyed, a plot of $(r-1)/(m-r)$ against P should give a straight line passing through the origin, with slope K (fig. 6.4). The relative errors in $(r-1)/(m-r)$ are least in the region $P = K$, so this method of analysis complements that provided by equation 6.3.

The presence of an unsuspected constant amount of oxygen, additional to what is deliberately introduced into the experimental system, cannot cause the data to fail to fit equation 6.1. This artefact would result only in a shift from what should be the true position of the axis $P = 0$ (fig. 6.2). In such a case, the true value, K_t, will be related to the estimated value, K, by

$$K_t = K - P' \tag{6.5}$$

Fig. 6.4. Results of Cullen & Lansley (1974) plotted according to equation 6.4.

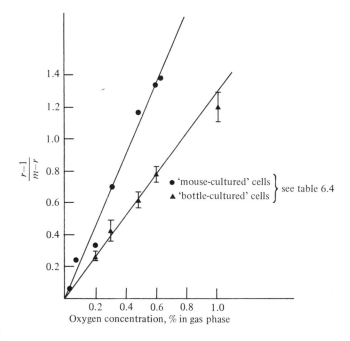

Oxygen concentration, % in gas phase

where P' is the partial pressure of the unsuspected oxygen content (see appendix). The true value, m_t, will be related to the estimated values, m and K, by

$$m_t = \frac{m(K - P')}{K - mP'} \qquad (6.6)$$

On the other hand, the data will not fit equation 6.1 satisfactorily if the amounts of contaminating oxygen vary with the partial pressure of oxygen in the system, as would be the case if absorbed oxygen were gradually released from the materials of the apparatus (Chapman *et al.*, 1970).

It is easier to detect an artefact resulting from the consumption of oxygen before and during irradiation by cell respiration, or by radiation chemical action during irradiation. The effective partial pressure of oxygen in the system at any measured level will depend on the balance between rates of consumption and supply, so depletion will be more important at low partial pressures. The transformations given in equations 6.3 and 6.4 betray this type of artefact readily, by virtue of the changing slope, which will be less at low than high value of P (e.g. Moore, Pritchard & Smith, 1972; fig. 6.3). Provided the artefact is recognized, and determinations of relative sensitivity are made also at oxygen pressures sufficiently high to render the oxygen depletion negligible, the value of K may still be determined from the slope of the curve at its steepest.

Theoretical derivation of equation 6.1

The equation may be derived theoretically on the assumption that, in the absence of oxygen, a radical induced in a target will react with one or other molecule in its immediate environment, the nature of the reacting species determining whether the damage is metionically fixed or restituted. If we designate the postulated species as F (damage-fixing) and S respectively and their rates of reaction with the radical species as k_f and k_s, the relative proportions of fixed lesions and restituted molecules will be determined by the concentrations $[F]$ and $[S]$ in the environment of the radical, and by k_f and k_s. If oxygen is now added, the reaction $R^{\cdot} + O_2 \xrightarrow{k_O}$ (fixed lesion) will act as an additional pathway to damage fixation.

In terms of these symbols, it may be shown (see appendix) that r, the radiosensitivity relative to that observed with no oxygen present, is given by

$$r = \frac{\dfrac{k_f[F] + k_s[S]}{k_f[F]} \cdot [O_2] + \dfrac{k_f[F] + k_s[S]}{k_O}}{[O_2] + \dfrac{k_f[F] + k_s[S]}{k_O}} \qquad (6.7)$$

where $[O_2]$ is the concentration of oxygen in the environment of the radical. Equation 6.7 may be written

$$r = \frac{m[O_2] + K}{[O_2] + K} \tag{6.8}$$

$$\text{if } m = \frac{k_f[F] + k_s[S]}{k_f[F]} \tag{6.9}$$

$$\text{and } K = \frac{k_f[F] + k_s[S]}{k_O} \tag{6.10}$$

K is in units of oxygen concentration.

In practice $[O_2]$ cannot be known unless the immediate environment of $R^\cdot$ is known, since concentration will depend on the solubility of oxygen in that environment as well as on its partial pressure.

The solubility of oxygen in cellular components has indeed been shown to vary very widely at a given P_{O_2}. The fluorescence of certain dyes is quenched by oxygen in proportion to its concentration, and Longmuir *et al.* (1977) used this phenomenon to compare oxygen solubility co-efficients in small volumes of non-respiring liver cells. Values of the coefficient varied from 0 to 125. It is not meaningful, therefore, to use 'oxygen concentration' as a variable to express experimental results referring to cells. For that reason, equation 6.1 was written with P, the partial pressure of oxygen, as the variable. In that form, K will be in units of oxygen pressure (or 'tension').

'Oxygen-independent damage'

The theoretical equation 6.7 differs somewhat from that derived by Howard-Flanders (1958). His derivation involves three possible reactions of a radical induced in a target structure, one of them leading to lethal injury even in the absence of oxygen. This was called 'oxygen-independent damage'. No proposals were made as to the type of radical reactions envisaged as leading to damage fixation. As used by Powers, however, in many publications on the radiobiology of the spores of *Bacillus megaterium*, the phrase has a different connotation. To describe survival curves, Powers and his co-workers have used equations of the form 5.2, and have used separate exponents, K_I and K_{II}, for 'oxygen-independent damage' and damage done in the 'immediate presence' of oxygen, (e.g. Powers, 1962). Survival curves for anoxic and well-oxygenated spores would be expressed respectively as $\ln f = -K_I D$, $\ln f = -(K_I + K_{II})D$. Thus the concept 'oxygen enhancement ratio', designated here by m, does not find a place in Powers' analysis. That constant would be expressed by $1 + K_{II}/K_I$, if Powers' terminology were used.

Dry bacterial spores, as used by Powers and his colleagues, and also dry seeds, are suitable materials for the direct observation of radiation-induced radicals, which may then be correlated with information on lethal damage. From observations by microwave spectroscopy on the radicals induced in spores of *Bacillus megaterium* Powers (1962) equated 'oxygen-independent' with non-radical damage which, he suggested, represents direct injury to the targets for cell killing.

That a contribution to lethal damage should be oxygen-independent, or relatively so, is a salient feature of my own model, which postulates at least two types of cellular target of a chemically different nature. The model is described briefly in the next section and discussed more fully in Chapter 15.

Variability in the constants m and K

It is important to note that these constants are specific not only to the cell line under investigation, but also to the experimental conditions. One or both of the constants may vary with the method of irradiation, the physiological state of the cells and even with the post-irradiation conditions of culture. This is illustrated by tables 6.2, 6.3 and 6.4.

Table 6.2 shows how m, the maximum enhancing effect of oxygen present *during* irradiation, may depend on *subsequent* methods of culture. The tabulated results with *Escherichia coli* B were the starting point for my deduction that there are at least two sites of primary energy deposition, such that damage fixation by oxygen is much more important in one site (type O damage) than in the other (type N damage). Differential repair or bypass of type N damage would then result in an increase in the observed value of m (Chapter 15).

The results of Cullen & Lansley (1974) and of Cullen (1976) with mouse

Table 6.2. *Dependence of o.e.r. on post-irradiation culture conditions,* Escherichia coli *B*

Medium	Incubation temperature immediately after irradiation (°C)	D_0, grays		o.e.r.
		Anoxic	Oxygenated	
A [a]	45	120 ± 3.0	32.5 ± 1.0	3.67 ± 0.15
B [a]	45	90 ± 3.4	27.1 ± 0.6	3.32 ± 0.13
A	37	87.2 ± 1.7	30.1 ± 1.3	2.90 ± 0.14
A + NaCl, 4g/1	37	63.8 ± 1.3	23.4 ± 0.4	2.73 ± 0.08
B	37	24.7 ± 0.8	12.2 ± 0.5	2.02 ± 0.21
A	19	21.1 ± 1.1	11.8 ± 0.5	1.79 ± 0.11
B	19	9.8 ± 0.4	5.7 ± 0.3	1.63 ± 0.11

After Alper (1961a).
[a]Medium B contained NaCl, 4g/1, and had a higher peptone content than A.

ascites tumour cells (table 6.4) may illustrate the dependence of K on the concentration of intrinsic damage-fixing and restituting metionic reactants, in the immediate environment of the target radical, as suggested by equation 6.10. As mentioned in Chapter 4, there is experimental evidence of hydrogen donation to radicals by sulphydryl compounds in chemical systems. If there is a realistic basis for the derivation of those equations, we might expect that both $[F]$ and $[S]$ could vary with the state of the cells at the time of irradiation. The –SH groups may reasonably be assumed to be involved in restitutory metionic reactions, and the dependence of K on $[-SH]$ seems to be supported by further results of Dr Cullen (personal communication). She has found the concentration

Table 6.3. *Dependence of m and K for* Shigella flexneri *on conditions of irradiation, X-rays and fast electrons*

Method of irradiation	Dose rate, (Gy min^{-1})	K^a (P_{0_2} in mm Hg)	95% confidence interval, K	m^a	m (Alper, Moore & Bewley, 1967, direct measurement)
Bacteria on cellophane carriers					
5-MeV electrons	250–450	0.80	0.646–1.05	2.60	2.79
250 kVp X-rays	26–32	1.85	1.47–2.49	3.34	3.41
Bacteria in suspension					
250-kVp	13–32	1.47	1.20–1.90	2.98	2.99

After Alper, Moore & Smith (1967).
[a]From regression line, equation 6.3.

Table 6.4. *Values of m and K, Ehrlich ascites tumour cells, grown* in vivo *(MC) or in* vitro *(BC) before irradiation.*

	K^a (P_{0_2} mm Hg)	95% confidence interval, K	m^a	95% confidence interval, m	Reference
MC, attached	3.27	2.96–3.65	2.79	2.61–3.01	Cullen & Lansley, 1974
BC, attached	6.16	5.47–6.99	3.04	2.93–3.19	Cullen & Lansley, 1974
MC in suspension	1.98	1.60–2.58	2.78	2.56–3.06	Cullen, 1976
BC in suspension	4.03	3.50–4.71	2.82	2.69–2.96	Cullen, 1976

[a]From regression line, equation 6.3.

of –SH groups to be less in mouse-cultured than bottle-cultured cells by a factor of about two. This could be causally related to the smaller value of K observed with the mouse-cultured cells.

The use of equation 6.1 to test proposed mechanisms of cell killing

If a set of results on variation in radiosensitivity with oxygen content fails to be fitted by equation 6.1, and the possibility of experimental artefact has been excluded, this would indicate a mechanism of oxygen sensitization other than any predicated in a theoretical derivation of the equation. With some sets of results on the killing of cells an analysis of the data does show that they cannot be fitted satisfactorily by equation 6.1; but in none of these cases have the authors themselves made such an analysis, so possible reasons for the failure have not been sought. But it is reasonable to conclude that the equation should be generally applicable when cell death is under consideration, because the results were found to conform adequately in experiments with biologically very different cells: bacteria (Alper, Moore & Smith, 1967b), algae (Bryant, 1973) and mammalian cells (Cullen & Lansley, 1974). The same equation should therefore apply, and with the same constants, to results on any lesion postulated to be the cause of cell death.

One such hypothesis is that death of eukaryotic cells is attributable entirely to the deposition of energy within, and resulting damage to, the chromosomes. Variation in radiosensitivity with oxygen content was investigated for induction of chromosomal aberrations in mammalian cells (Deschner & Gray, 1959; Hawes, Howard & Gray, 1966) and Tradescantia pollen tubes (Evans & Neary, 1959). In those instances it was concluded that equation 6.1 fitted the results. However, simultaneous tests for the killing of cells in identical conditions were not made, so it is not known whether the constants of the equation would be the same for both end-points.

An analagous hypothesis is that cell death ensues as a result of a double-strand break in the DNA (the nuclear DNA, in eukaryotic cells). Double-strand breaks are postulated to be caused by two single-strand breaks on opposite strands within a short distance of each other, so the mode by which single-strand break induction is modified should be an indication also of what happens with double-strand breaks. A technique that has been widely used gives information on the total of DNA strand breaks. This was devised by McGrath & Williams (1966). The DNA of cells to be irradiated is labelled by feeding a radioactive precursor. After irradiation the cells are layered on to the top of a sucrose gradient, made very alkaline so that the cells lyse and the DNA is released without being subject to shearing forces. The tubes containing the gradients are centrifuged, after which the layers (called fractions) are separately collected and their radioactivity measured. The DNA sediments in accordance with its size, so the patterns of sedimentation of broken and intact DNA will differ. This makes it possible to relate the yield of strand breaks to some variable, like radiation dose or a modifying agent like oxygen. Sensitivity both to induction of strand breaks, in Chinese hamster cells, and to killing were examined at a wide range of partial pressures of oxygen by Chapman *et al.* (1974). Figure 6.5 has been constructed according to equation 6.4 from raw data very kindly supplied by Dr J. D. Chapman. The curves do not appear to coincide: the oxygen concentration required for a given value of the function $(r - 1)/(m - r)$ is always less in the case of strand breakage. It would also appear that the data for the latter end-point are not fitted by a straight line. But these differences may not be large enough to be attributable to anything more than the experimental errors involved in making such observations (Dr J. D. Chapman, personal communication).

It should be noted that some results obtained by the technique of McGrath & Williams may need to be re-assessed, because several problems associated with the method have been reported. These are mentioned in Chapter 16.

Comparisons of change in sensitivity with oxygen concentration for the same two end-points in bacteria were made by Johansen, Gulbrandsen

Fig. 6.5. Results of Chapman *et al.* (1974) plotted according to equation 6.4, from raw data supplied by Dr J. D. Chapman. Abscissa: concentration of dissolved oxygen in water.

○ Values for the killing of Chinese hamster V79 cells at 20 °C; several determinations at each oxygen concentration

● at 0°C, one determination at each concentration

□
■ } as above: yield of single-strand breaks

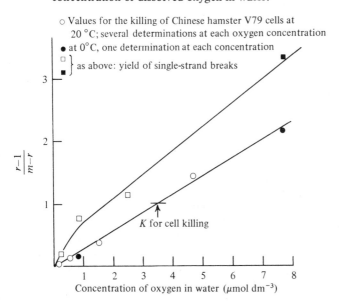

& Pettersen (1974). Certain small particles capable of reproducing themselves are to be found in bacteria, these 'episomes' often consisting of
circular DNA. A break in one strand causes the loss of the circular form,
so the broken and intact particles are easy to separate by centrifugation.
Johansen *et al.* infected several variants of *Escherichia coli* K12 with a
bacteriophage named λ, which has that circular form. They measured
change in sensitivity as a function of P_{O_2} for the killing of uninfected
bacteria as well as for induction of strand breaks in the intracellular
phage DNA, which could plausibly be assumed to reflect the induction
of breaks in the bacterial DNA. Results of those experiments gave very
different values of K for the two end-points, the value for cell killing being
greater by a factor of sixteen.

Mechanism for the enhancement of cell killing by oxygen

At the time I put forward the hypothesis underlying the theoretical derivation of equation 6.1 (Alper, 1956), the favoured view of the
mechanism for the 'oxygen effect', and indeed for radiobiological damage
in general, was their mediation through oxidizing species deriving from
the radiolysis of water. With oxygen present, it was reasoned, the primary
radical H would react with an O_2 molecule to give HO_2, so that the oxidative yield would be greatly increased (see e.g. Gray, 1954; Bacq &
Alexander, 1955). Since so much material is present in the cell that is
able to react with H, its lifetime would be expected to be very short
indeed. The HO_2 hypothesis therefore implied that oxygen had to be
present in solution at the time of irradiation if it were to act as a radiosensitizer by an indirect mechanism. But the damage-fixation hypothesis
suggested the possibility that the organic radical R might remain available
for reaction long enough to enable observations of an effect of oxygen
within a measurable time after a very short pulse of radiation. Some
evidence for this possibility had been provided by Gordy, Ard & Shields
(1955) from studies on radiation-induced radicals, for which they used
microwave spectroscopy. With various dried materials of biological
origin, irradiated *in vacuo*, they observed that the subsequent admission
of air resulted in either alteration or decay of the ESR signals. Subsequently, analogous observations were made when dry seeds or bacterial
spores were irradiated (e.g. Conger & Randolph, 1959; Ehret *et al.*, 1960).

However, the lifetimes of radicals are generally very much shorter in
wet than in dry material. Howard-Flanders & Moore (1957) attempted to
measure the lifetimes of the postulated radicals induced in vegetative
(i.e. wet) bacterial cells, as a test of the hypothesis that the oxygen would
interact directly with the ionized targets. Their experimental arrangements did not permit them to observe lifetimes shorter than about 20 ms,
which proved to be not short enough.

Resolution of the time interval between cessation of radiation and re-action with oxygen in vegetative cells was made possible by the 'gas-explosion' technique devised by Michael *et al.* (1973). Bacteria were irradiated in an oxygen-free space by single pulses of electrons, of dura-tion $2\mu s$. Oxygen was then released into the space by 'explosion' from a reservoir, the opening to which was controlled electrically, so that the oxygen could be admitted in as little as 0.1 ms after the electron pulse. From the theory leading to equation 6.1 it would be predicted that there should be an exponential decay with time in the number of radicals re-maining able to react with oxygen and the results of Michael *et al.* agreed with that prediction at least to a first approximation. They estimated that after 0.5 ms about half of the reactive radicals were still available for reaction. Subsequently some minor problems in the technique were solved, enabling Michael and his colleagues to delineate the decay curve more accurately (Michael *et al.*, 1978). They found that it could not be described by a single exponential function; but a combination of two exponential functions was adequate (fig. 6.6). From this they concluded that, in the bacteria they used, two reactive species are produced, giving resolvable decay curves from which half-lifetimes of respectively 0.4 and 4 ms were computed. A less direct experimental method gave very much lower estimates for the lifetimes of reactive species in bacteria. In the experiments of Epp *et al.* (1968), referred to on page 55, the irradiated bacteria were in an environment containing oxygen throughout, and the authors reasoned that the same 'anoxic tails' to the curves could not have been observed at all if the species had remained reactive for times long enough to allow oxygen to diffuse to the target sites. Subsequently Epp *et al.* (1973) used double pulses of irradiation, separated by times from 1 ms upwards, to calculate the diffusion time. The estimates gave an upper limit of 10^{-4} seconds for the lifetime of any reactive species such that combination with oxygen would kill the cell. The discrepancy between that conclusion and the measurements of Michael *et al.* may reflect radio-biological differences between anoxic cells not previously irradiated, as used by those authors, and the test bacteria in the experiments of Epp *et al.*, which were survivors of large doses and must necessarily contain combination products of oxygen with a variety of radical species. Such differences were indeed suggested by Kessaris, Weiss & Epp (1973) in their analysis of the calculations of diffusion times of oxygen into bacteria.

In wet bacterial spores reactive species induced by radiation retain for much longer times their ability to experience either a damaging reaction with oxygen or an alternative restitutory metionic reaction. *Bacillus mega-terium* spores in anoxic suspension were exposed to a beam of electrons for 1s, at various times after which buffer containing dissolved oxygen was added to them. The results indicated that there were two species susceptible to the fixing of damage by oxygen, with the unexpectedly long half-lives of 9 and 120 s (Stratford *et al.*, 1977). With wet spores of the same species rendered anoxic by exposure to single pulses of radiation of 3 ns duration, Weiss & Santomasso (1977) estimated the half-life of an oxygen-dependent species to be of the order of 10 s.

The gas-explosion method was used also to observe the kinetics of sensitization by oxygen of mammalian cells (Watts, Maughan & Michael, 1978). For technical reasons, the timing curve could not be as clearly analysed as was possible with bacteria; but the results showed that cell killing depended in part at least on a species capable of interacting with oxygen after the delivery of a pulse of radiation; the half-lifetime was apparently 1 ms.

All those results are in conflict with any hypothesis attributing oxygen sensitization to radiolysis products of water, since they are much too short-lived. However, one such hypothesis has again been actively canvassed since the discovery of the enzyme superoxide dismutase (SOD) which 'dismutes' superoxide (0_2^-) ions, a natural byproduct of intracellular chemical reactions. The enzyme is considered to catalyse the

Fig. 6.6. To show times up to which incipient lesions in *Serratia marcescens* remain reactive, so that damage can be fixed by contact with oxygen. Each point was determined from the slope of a full survival curve. The calculated half-lifetimes of the more and less rapidly decaying species are about 0.4 and 4 ms respectively. The solid line is drawn according to the equation $1/D_0 = 0.0173$ $e^{-t/0.55} + 0.014e^{-t/6.0} + 0.018$. (Figure provided by Dr B. D. Michael.)

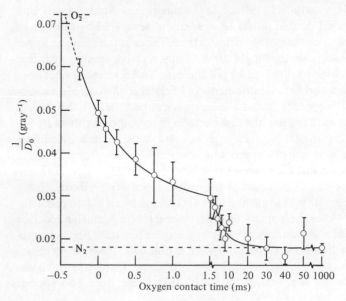

reaction

$$2H^+ + 2O_2^- \rightarrow O_2 + H_2O_2 \quad \text{(McCord \& Fridovich, 1969)}$$

It has been reasoned, therefore, that the radiosensitizing action of oxygen may be mediated through the formation of the superoxide ion, which may be generated in water by the attachment of a 'hydrated electron' (e_{aq}^-) to an O_2 atom; or by the dissociation of HO_2 into $H^+ + O_2^-$.

If this were correct, it might be expected that increasing the intracellular concentration of superoxide dismutase in a given population of cells would result in a reduction in radiosensitization by oxygen. Experiments to test that prediction have yielded both negative and positive results (Goscin & Fridovich, 1973; Oberley *et al.*, 1976). Important observations of Petkau & Chelack (1974) were made with a strain of mycoplasma – a very small micro-organism that is parasitic on cells, but it is not a virus and can form colonies. The organisms do not contain intrinsic superoxide dismutase, and Petkau & Chelack could affect survival after irradiation in the presence of oxygen by adding superoxide dismutase to the suspension of organisms. Significant results were that the 'protective' effect of the enzyme was much greater when the irradiated organisms were held for 4 hours after irradiation before plating, and that the addition of the enzyme 10 minutes after irradiation was nearly as effective as when it was present during irradiation. The latter result, in particular, argues against the agency of the superoxide ion formed *during irradiation* in radiosensitization by oxygen. Petkau & Chelack reasoned that the increased effectiveness of SOD, when plating was delayed, was attributable to its acting as a check in the steps of autoxidation of lipids initiated by the radiation, a chain reaction during which O_2^- is formed and then initiates further oxidation. The implication of lipid peroxidation has an important bearing on the site of damage that is susceptible to radiosensitization by oxygen, as will be discussed in later chapters. In any event, it is difficult to reconcile the hypothesis that oxygen acts via the radical O_2^- with the observation that oxygen can exert its effect for as long even as 0.1 ms after a pulse of radiation, since this would demand an implausibly long lifetime for the precursors which might react with oxygen molecules to yield O_2^-. Sensitization by oxygen at very much longer times after a pulse of radiation, as in bacterial spores, robs the O_2^- hypothesis of credibility.

Appendix: derivation of some of the equations of chapter 6

Equation 6.3

$$\frac{S_{max}}{S_N} = m \qquad \frac{S_p}{S_N} = \frac{mP + K}{P + K}$$

$$\therefore \frac{S_{max}}{S_p} = R = m \div \frac{S_p}{S_N}$$

$$= \frac{m(P + K)}{mP + K}$$

$$R - 1 = \frac{K(m - 1)}{mP + K}$$

$$\frac{1}{R - 1} = \frac{m}{K(m - 1)} \cdot P + \frac{1}{m - 1}$$

Equations 6.5 and 6.6

If radiosensitivites are measured relative to the maximum, $R (= S_{max}/S_p)$ they will be the same whether or not the measured partial pressure is true, P_t, or false, P, where $P = P_t - P'$; and the graph of $1/(R - 1)$ plotted as a function of P will have the same slope whether the origin is the true one, or a false origin giving $P = 0$ when $P_t = P'$. Let the true and false parameters be respectively m_t, K_t and m, K. From equation 6.3, the slope to $1/(R - 1)$ plotted against P is $m/K(m - 1)$ or $m_t/K_t(m_t - 1)$

But

$$m = m_t \cdot \frac{P' + K_t}{m_t P' + K_t} \qquad m - 1 = \frac{K_t(m_t - 1)}{m_t P + K_t}$$

and

$$\frac{m}{m - 1} = \frac{m_t(P' + K_t)}{K_t(m_t - 1)}$$

$$\therefore \frac{m}{K(m - 1)} \equiv \frac{m_t}{K_t(m_t - 1)}$$

$$\frac{m_t(P' + K_t)}{K_t(m_t - 1)K} \equiv \frac{m_t}{K_t(m_t - 1)}$$

i.e.

$$\frac{P + K_t}{K} = 1$$

or

$$K_t = K - P$$

whence equation 6.6.

Derivation of equation 6.7

Assumptions:

(1) That a radical R, formed in a target structure will result in cell death if it reacts to give a fixed lesion L.

(2) That the accumulation of fixed lesions is directly proportional to dose. After dose D the total number of fixed lesions will be λD, the magnitude of λ depending on the fraction of lesions fixed. For different conditions of fixation, λ will have different values, say λ_1, λ_2 etc. Survival will be given by surviving fraction $f = e^{-\lambda_1 D}$ or $e^{-\lambda_2 D}$ etc., and the relative sensitivities will be in the ratio $\lambda_1 : \lambda_2$ etc.

(3) That reaction $R^{\cdot}$ with O_2 will give L: $R^{\cdot} + O_2 \xrightarrow{k_o} L$.

(4) That other damage-fixing reactants designated F are present, in concentration $[F]$. $R^{\cdot} + F \xrightarrow{k_f} L$.

(5) That restitution occurs through reaction of $R^{\cdot}$ with S, present in concentration $[S]$. $R^{\cdot} + S \xrightarrow{k_f} R_s$.

(6) That at a given time $t = 0$ during irradiation, concentration $[R^{\cdot}]_0 \ll [F]$ or $[S]$ or $[O_2]$, so that $[R^{\cdot}]$ will decrease by a first-order process. Then, at a time t,

$$[R^{\cdot}]_t = [R^{\cdot}]_0 \{e^{-(k_o[O_2] + k_f[F] + k_s[S])t}\}$$

If $[L]_t$ and $[R_s]_t$ are the concentration of fixed damage and restituted radicals at time t

$$[R]_t = [R^{\cdot}]_0 - \{[L] + [R_s]\}$$

i.e. $[R^{\cdot}]_0 - [R^{\cdot}]_t = [L] + [R_s]$

$$= [R^{\cdot}]_0 \{1 - e^{-(k_o[O_2] + k_f[F] + k_s[S])t}\}$$

When all radical reactions are complete, i.e. t is large,

$$[R]_0 = [L] + [R_s]$$

The fraction of fixed lesions

$$\frac{[L]}{[L] + [R_s]} = \frac{k_o[O_2] + k_f[F]}{k_o[O_2] + k_f[F] + k_s[S]}$$

The fraction of fixed lesions in the absence of oxygen is

$$\frac{k_f[F]}{k_f[F] + k_s[S]}$$

By assumption (2), $\lambda_{[O_2]}$ and λ_{anoxic} will be in the ratio

$$\frac{k_o[O_2] + k_f[F]}{k_o[O_2] + k_f[F] + k_s[S]} : \frac{k_f[F]}{k_f[F] + k_s[S]}$$

which gives relative sensitivity r

$$r = \cfrac{\dfrac{k_s[r] + k_s[S]}{k_f[F]}[O_2] + \dfrac{k_f[F] + k_s[S]}{k_o}}{[O_2] + \dfrac{k_f[F] + k_s[S]}{k_0}}$$

7

Radiosensitization of hypoxic cells by other compounds

The damage fixation hypothesis for sensitization by oxygen carries the corollary that other agents might likewise have chemical properties enabling them to fix damage by adding to radiation-induced radicals in target structure. This would prevent a restitutory metionic reaction, such as recombination with a free electron, which would restore the target to a functioning condition (Alper, 1956). It is an integral part of that hypothesis that all restitutory metionic reactions will be prevented if oxygen is present in sufficiently high concentration in the immediate environment of the target at the time the radical is formed (see derivation of equation 6.1, appendix to Chapter 6). Since the hypothesis is supported by experimental evidence, it is not to be expected that compounds which sensitize hypoxic cells by radiation–chemical, i.e. metionic, reactions with the target radicals should enhance radiation damage to cells already experiencing full sensitization by oxygen. It has been suggested also that sensitization might occur through attack on cell targets by radicals formed of the sensitizer molecules (Fisher et al., 1978).

A great many compounds have been reported to sensitize only hypoxic cells, or to sensitize oxygenated cells to a much lesser extent. The action of some of these, or a part thereof, may be through effects on some intermediate biochemical pathway. Some drugs in that category may be regarded as 'hypoxic cell sensitizers' if the relevant biochemical action is more probable in those cells when they are held hypoxic for a period, as they are bound to be before irradiation in anoxic conditions. Conversely, compounds known to affect cellular biochemistry may still act at the metionic level.

Agents may act as sensitizers specifically of hypoxic cells if toxic products are formed only in the absence of oxygen; for example, Cramp (1967) found this to be the mode of sensitization of bacteria by copper compounds. Sometimes toxic products by radiation are very short-lived, so it is difficult to be sure that an agent is acting metionically unless techniques are used for observing responses of cells within short times of

71

contact. The sensitizing action of iodoacetamide through generation of short-lived toxic products was revealed only in a rapid-mix experiment by Dewey & Michael (1965). They set up arrangements for a solution to be injected into, and mixed with, a suspension of bacteria at times down to 10 ms before or after a pulse of radiation lasting about 2 μs. The time resolution of such an arrangement was improved with the development of the 'fast-flow, rapid-mix' technique by Adams, Cooke & Michael (1968). Liquids are driven by electrically operated plungers through two tubes and come together at preselected times before or after a short pulse of radiation. If a suspension of cells is passed through one tube, and a sensitizing (or protective) compound through the other, it can be determined how long before irradiation the compound must be in contact with the cells, to exert its effect; and also whether radiosensitivity is modified by post-irradiation contact. Mixing can be timed to occur a few milliseconds before or after a pulse of radiation, so this technique makes it possible to decide whether the mechanism of action of a particular compound is more likely to be through a biochemical pathway, through metionic reaction, or through the formation of short-lived toxic products of radiation.

Nitric oxide and nitroxyl compounds

Reasoning that nitric oxide, like oxygen, should have the chemical properties requisite for damage fixation by direct addition to radiation-induced radicals, Howard-Flanders (1957) observed that bacteria (*Shigella flexneri*) were indeed sensitized when nitric oxide was bubbled through anoxic suspensions. Partial pressures of O_2 and NO that gave roughly the same concentration in water produced about the same level of sensitization (Howard-Flanders & Jockey, 1960). The equivalence of NO and molecular oxygen as radiosensitizers was confirmed by Kihlman (1958) for the production of chromosomal aberrations in bean roots and by Gray, Green & Hawes (1958) for lethal effects on mouse ascites tumour cells. Unlike oxygen, however, the presence of NO during and after irradiation was protective for very dry test systems (plant seeds: Sparrman, Ehrenberg & Ehrenberg, 1969; bacillus spores: Powers, Kaleta & Webb, 1959).

Another difference in the radiobiological action of O_2 and NO, on *Shigella flexneri*, was noted by Dale, Davies & Russell (1961). Exposure of the bacteria to NO for varying periods before irradiation was not in itself toxic, provided the suspensions were kept free of oxygen, but when the NO was flushed out and replaced by nitrogen before irradiation the bacteria were considerably more sensitive than if they were irradiated in a nitrogen atmosphere without previous exposure to NO.

Despite the development during the 1960s and 1970s of techniques for observing fast processes in radiation chemistry and biology these observations on similarities and dissimilarities in the action of O_2 and NO have not been followed up; yet their elucidation might well contribute to our understanding of the phenomenon of sensitization in general.

Demonstration of the radiosensitizing properties of NO led Emmerson & Howard-Flanders (1964, 1965) to test certain substituted nitroxides which are water-soluble stable free radicals. Some compounds when tested for their ability to sensitize anoxic *Escherichia coli* B/r were almost as effective as oxygen, provided their solubility in water was high enough to allow of their use in sufficiently high concentration. The first compound shown to have these properties was di-t-butyl nitroxide (DTBN). An analogous one, triacetoneamine-N-oxyl (TAN) also proved very effective, maximum sensitization of the bacteria being by a factor of about 2.8 when the concentration was 10^{-3} mol dm^{-3} (Emmerson, 1967). However, TAN was less effective in sensitizing anoxic mammalian cells *in vitro*: the maximum sensitization achieved by Parker, Skarsgard & Emmerson (1969) was by a factor of about 1.5.

A phenomenon that has implications for the mechanism by which oxygen modifies radiosensitivity was demonstrated in experiments of Emmerson (1968). Using the repair-deficient mutant *Escherichia coli* K12 AB2463 with which, in the conditions of his experiments, the o.e.r. was 1.9, he found that when they were irradiated anoxically, with TAN present, they were more sensitive than when fully aerobic. The presence of TAN made no difference to the sensitivity of the oxygenated cells. The phenomenon seems to be associated to some extent with the property that reduces the o.e.r. of a mutant to less than that for the wild-type parent. It was not observed with another rather less radiosensitive mutant, K12 AB3058, for which the o.e.r. was even higher than for the parent strain.

Johansen (1974) also used the sensitive mutant *E. coli* K12 AB2463 in investigations with the nitroxyl compound tetramethyl piperidinol-N-oxyl (TMPN). He examined the relationship between degree of sensitization and the partial pressure of oxygen, expressed in terms of its concentration when dissolved in water. His results (fig. 7.1) demonstrated that this repair-deficient mutant, when anoxic, was subject to considerable sensitization by the TMPN, but was protected against it by the presence of oxygen, to a degree which depended on the P_{O_2}.

Observations of Sapora, Fielden & Loverock (1977) with yet another nitroxyl compound, nor-pseudopelletierine N-oxyl (NPPN), suggest that 'super-oxic' sensitization of certain bacterial mutants for which o.e.r.'s are low may be a general effect of such compounds. From their results with repair deficient mutants of *E. coli* K12 and *E. coli* B and B/r, and also the wild-type parents, it may be seen that there were reasonably good correlations between the values of D_0 with and without oxygen present, as well as with and without NPPN present (fig. 7.2). A deviant is the strain JG 112, deficient in a DNA polymerase, for which the o.e.r. was 4.3. The figure shows that there was full sensitization by NPPN, to the extent conferred by oxygen, with all strains for which the o.e.r. was less than about 2.2; values of D_0 for anoxic cells irradiated in the presence of NPPN were significantly less than for oxygenated cells in the case of two mutants for which the o.e.r.'s were 2.4 and 1.3. Their common deficiency was inability to repair UV damage by recombination of intact pieces of DNA.

n-ethylmaleimide and other sulphydryl-binding agents

The early trials of nitric oxide and of compounds containing free radicals of the nitroxyl group were suggested by the damage-fixation hypothesis for the oxygen effect (Alexander & Charlesby, 1955; Alper, 1956; Howard-Flanders, 1958), but hypoxic cell sensitizers belonging to different groups of compounds have been selected on the basis of quite other considerations. Among the earliest to be reported upon was n-ethylmaleimide (NEM), tried by Bridges (1960) because of its ability to bind sulphydryl groups. Various hypotheses have been canvassed over the

Fig. 7.1. Results of Johansen (1974) with the recombination-deficient strain *E. coli* K12 AB2463.
Ordinate: radiosensitivity, defined as reciprocal of dose to give surviving fraction of 1%. O abscissa: concentration of oxygen; △ abscissa: concentration of TMPN; ● abscissa: concentration of oxygen with TMPN present in concentration 5.10^{-3} mol dm^{-3}.

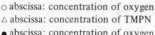
o abscissa: concentration of oxygen
△ abscissa: concentration of TMPN
● abscissa: concentration of oxygen with TMPN present in concentration 5×10^{-3} mol dm^{-3}

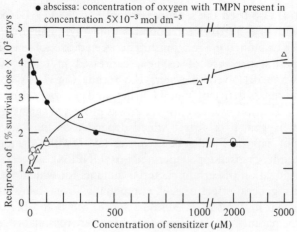

years attributing an important role to these groups in protection against cellular radiation damage.

Research on radioprotective agents, pursued with vigour in the 1940s and 1950s, had shown that aminothiols, in particular, were very effective. Bridges reasoned that intracellular free –SH groups might act to protect cells against radiation damage, so that they would be sensitized if the –SH action were blocked by an appropriate compound. Bridges' first experiments were with *Escherichia coli* B/r, and the NEM sensitized both aerobic and anoxic bacteria, though the latter were more effectively sensitized. Full sensitization was seen only after incubation of the bacteria with the drug for 5 minutes before irradiation, which suggests that the sensitization may not all have been directly at the metionic level. On the other hand, the sensitizing action of NEM required the presence of the drug during irradiation. When it was washed out of the suspensions beforehand it failed to sensitize the bacteria, although the –SH content of the cells did not increase for some time after the removal of NEM (Dewey, 1965). Some strains of bacteria (*Pseudomonas, Serratia*) responded to the NEM as if it were a true metionic sensitizer, in that it sensitized only hypoxic cells. Its mode of action, at least on *Serratia*, was confirmed as being at that level by the rapid-mix experiments of Adams *et al.* (1968); they found that NEM sensitized the hypoxic bacteria if it was present only 4 ms before a pulse of irradiation, a time presumed to be much too short to permit the binding of a significant fraction of intracellular –SH. NEM therefore

Fig. 7.2. Results of Sapora, Fielden & Loverock (1977) on values of D_0 for survival of variants of *E. coli* of the K12 and B strains. Results plotted to show relationships between values of D_0 for aerobic bacteria and for anoxia bacteria with and without NPPN (10^{-3} mol dm^{-3}) present.

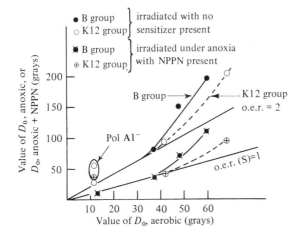

seems to sensitize hypoxic bacteria at least partly through its 'electron-affinic' or oxidative properties, as many hypoxic cell sensitizers are presumed to do, but action through biochemical pathways certainly contributes to some of the radiobiological action of NEM.

Such action, not confined to hypoxic cells, is illustrated by experiments of Sinclair (1973) with Chinese hamster cells; his observations related only to aerobic cells. His object was to test whether NEM would affect cells differently when they were in different stages of the cell cycle, since the radiation response had been found to be very variable as the cells proceeded from one mitosis to the next (Chapter 12). Sinclair (1969) had observed differential protection through the cycle by an–SH compound, the effect being greatest in those stages in which the survival curves had the lowest extrapolation numbers. The effect of the NEM was indeed in the opposite sense, the shoulders to survival curves being greatly reduced when they would normally have been large.

Since Bridges' review (1969) of the whole field of sensitization by sulphydryl-binding agents, others have been selected as being likely to sensitize specifically hypoxic cells, the rationale being that protection by anoxia might be attributable to the restitutory metionic reactions of target radicals with non-protein –SH groups. An agent found to be very effective in oxidizing cellular glutathione in mammalian cells is diazenecarboxylic acid (N,N-dimethylamide) or 'diamide', and this has attracted attention as a hypoxic cell sensitizer since Harris & Power (1973) reported on its effects on three lines of Chinese hamster cells as well as on the bacterium *Pseudomonas fluorescens*. The compound sensitized the bacteria only when they were hypoxic, and was slightly more effective than NEM used in the same concentration (5×10^{-4} mol dm^{-3}). A striking effect on the mammalian cell survival curve was a marked reduction in extrapolation number, but only for anoxically irradiated cells. When they were treated before and during irradiation by diamide in concentration 2×10^{-4} mol dm^{-3} survival of V79 cells was exponential, whereas the normal survival curve had an extrapolation number of about four. The value of D_0, however, was not much reduced. The concentration of diamide required to change the survival curve in that manner was far in excess of that needed to oxidize all the intracellular glutathione, so the assumption on which the trial was based was not validated. Reduction in extrapolation numbers of survival curves is not normally seen with nitroxyl sensitizers or with 'electron-affinic' sensitizers in similar experiments.

Evidence for action by diamide at the metionic as well as the biochemical level came from experiments of Watts Whillans & Adams (1975). They used Chinese hamster cells of the V79 line and also two strains of bacteria: *Serratia marcescens*, for which survival was exponential, with D_0 under anoxia about 50 Gy; and a resistant strain, *Micrococcus sodonensis*, with which the extrapolation number was about 100, and anoxic D_0 about 400 Gy. Under ^{60}Co-γ irradiation sensitization was conferred on those two strains in the dose-multiplying mode, that is to say, the extrapolation number to the survival curve for the *Micrococcus* was unaffected. The survival of well-oxygenated bacteria was unchanged by the diamide. But its effects on the radiation response of the Chinese hamster cells were qualitatively different from those on bacteria. They agreed with those of Harris & Power (1973), except that the 40 minutes' pre-treatment time used by Watts *et al.* resulted in a reduction in the extrapolation number of the survival curve also for oxygenated cells.

Further qualitative differences were revealed in the fast-flow rapid-mix experiments of Watts *et al.* Again the compound acted differently on bacteria and mammalian cells, and, with the latter, it differed in its action from that of other hypoxic cell sensitizers that have been used in rapid-mix experiments. These could not be done with *Micrococcus sodonensis*, because the dose per pulse available was not large enough. With *Serratia*, diamide in concentration 3×10^{-3} mol dm^{-3} was enough to confer maximum sensitization in steady-state experiments, and, when mixed with the bacteria before irradiation to give that concentration, the anoxic cells were sensitized to the same extent whether they were mixed with the diamide 4 or 40 ms before irradiation, although the sensitivity was less than in the steady-state experiments. But when diamide was mixed with mammalian cells 40 ms before irradiation, the longest time that could be used in the experiments of Watts *et al.*, there was no effect at all of the compound.

Thus there are similarities in the action of NEM and diamide, as far as the experimental evidence goes. Given sufficient time of action on mammalian cells, both compounds reduce the shoulders of survival curves, even when the cells are well oxygenated; and both compounds sensitize bacteria, or at least *Serratia*, within a time of contact which is unlikely to be long enough for biochemical action. The mechanism of action of diamide, at least, differs from that of sensitizers classed as electron-affinic, which mostly act in a dose-multiplying mode also on mammalian cells, and which exert some sensitizing action on hypoxic cells after times of contact as short as 5 ms (Adams *et al.*, 1975; Whillans & Hunt, 1978). It should be noted, however, that these properties are not necessarily to be associated with electron affinity, since Wold & Brustad (1974) and Whillans & Neta (1975) found that the electron affinity of diamide was higher than that of several-radiosensitizers commonly allocated to the 'electron-affinic' class.

Electron-affinic compounds

'Electron affinity' as a basis for effectiveness in sensitization was proposed by Adams & Dewey (1963), on the grounds that cell killing could be attributable to attachment at a critical site of the hydrated electron, e_{aq}^-, formed in the radiolysis of water (see Chapter 6, p. 67). It was reasoned that longer life, and therefore greater probability of effectiveness, would be conferred upon the electron if it were captured by a molecule acting as electron acceptor, which would then, as negative radical-ion, carry the potentially damaging electron to the target. It was envisaged that the ideal carrier molecule should combine high electron affinity with a structure such that the electron could be readily detached. On that basis, several compounds with the required properties were tested for their ability to sensitize hypoxic bacteria (*Serratia marcescens*), with some degree of success.

It was subsequently thought unlikely that cell killing could be attributable to a significant extent primarily to the action of the hydrated electron (see, e.g., Adams, 1972); nevertheless, that approach led to the identification of a large number of so-called electron-affinic sensitizers, comprising several groups of compounds. Some of these are as effective in sensitizing higher cells as bacteria, and almost as effective as oxygen. Their modes of action have aroused considerable interest because of their potential clinical usefulness.

Selective sensitization of hypoxic cells is a topic of interest in radiotherapy because of the likelihood that some tumours contain hypoxic cells, the comparative radioresistance of which makes the tumours impossible to sterilize without causing intolerable damage to the normal tissues in the radiation field. For this reason, research into sensitizers specifically of hypoxic cells is currently vigorous, especially in respect of compounds that can be used *in vivo*. Begg, Sheldon & Foster (1974) reported that the dose required to sterilize C3H mammary tumours in mice was reduced if the animals were treated with metronidazole (a 5-nitroimidazole) before irradiation; and Denekamp & Harris (1975a) found even more effective sensitization of a different tumour line when they used a 2-nitroimidazole, designated at that time as Ro-07-0582, later named misonidazole. The potential clinical importance of the nitroaromatic compounds is emphasized in a general review on radiosensitizers by Adams (1977), and their properties are separately reviewed by Wardman (1977). Only some results that might reflect on mechanisms of action of cell sensitizers, particularly at the metionic level, will be dealt with here. Some points especially relevant to their use *in vivo* are discussed in Chapter 17.

Sensitizers selected for possible action at cell membranes

It is well established that the activity of anaesthetic drugs is associated with their binding to cell membranes. Shenoy, Singh & Gopal-Ayengar (1974) and George *et al.* (1975) therefore tested procaine hydrochloride as a possible radiosensitizing drug. *E. coli* B/r irradiated while in contact with the procaine were indeed sensitized, and only when they were anoxic. There was considerable sensitization also of anoxically irradiated bacteria treated with procaine immediately after irradiation. The deduction that sensitization was not at a DNA site was confirmed by lack of any sensitization to germicidal UV. There was evidence also of radiosensitization of Yoshida ascites tumour cells, the effect increasing with time of incubation after irradiation. Viability was assessed by exclusion of dye by the cells, a less sensitive test than colony-forming ability.

The compound cetylpyridinium chloride (CPy Cl) has a group that is highly electron-affinic and a hydrocarbon chain that directs selective distribution in lipids. Redpath & Patterson (1976) found that its presence sensitized oxygenated as well as hypoxic *Serratia marcescens*, the latter to a somewhat greater extent. It may be seen from their results that the values of o.e.r. in the absence and presence of CPy Cl were respectively about 2.7 and 2.0. No sensitization was observed with either of two other compounds, one of them containing the same electron-affinic group, but no long hydrocarbon side chain, while the other contained the same hydrocarbon moiety as CPy Cl but not the pyridinium group.

The high electron affinity of the pyridinium group prompted the investigation of some bipyridinium and nitropyridinium compounds by Anderson & Patel (1977) and Anderson, Patel & Smithen (1978). Two of the four bipyridinium compounds tested increased the sensitivity of anoxic *Serratia marcescens* by a maximum factor of about two, compared with the o.e.r. of 3.4 in those conditions. An unusual property of methyl-viologen, which had been shown not to cross cell membranes, was the mode of increase in effectiveness with concentration. At about 10^{-4} mol dm^{-3}, sensitivity was increased by a factor of two. This was maintained until the concentration was about 10^{-3} mol dm^{-3}, when further increases rendered the cells as sensitive as they were under aerobic irradiation. The authors suggested that this mode of concentration-dependence might be attributable to formation under irradiation of radical cations that could cross the outer membrane, the cell wall and the inner membrane of the bacterium; they would then be trapped in the cell, the dicationic form having been restored by the intermittent oxygenation of the cell suspension which had been used to regenerate reduction radiation products of that compound. On the basis of that model the authors inferred that the other bipyridinium compounds sensitized the anoxic bacteria by action at a site associated with the outer membrane; while the methyl-viologen acted also internally because charge delocalization had allowed it to cross the cell membranes and cell wall.

The biphasic form of concentration dependence observed with methyl-viologen was seen also by Anderson *et al.* (1978) when they used two especially synthesized nitropyridinium compounds which, they assumed, would be able to cross the cell membrane, unlike the dicationic bypyri-dinium compounds. One of the nitropyridinium compounds, when used in concentration at about the limit of solubility, caused sensitization beyond that achieved by oxygen alone. No results were given of their effects on aerobically irradiated bacteria.

Comparisons of the efficacy and modes of action of hypoxic cell sensitizers

At the time of writing it is common practice to compare hypoxic cell sensitizers one with another in terms of the extent to which they increase lethal effects of radiation on hypoxic mammalian cells *in vitro*: and a Chinese hamster cell line, V79, has often been the test system chosen. A conventional measure of effect is the enhancement ratio (abbreviated ER) – i.e. the ratio of doses required to reduce survival to a given level without and with the sensitizer present. When the compound acts in the dose-multiplying mode, the ER is given by the ratio of D_0 values, or the inverse of the ratio of the survival curve slopes. However, the term oxygen enhancement ratio (o.e.r.) is almost always used in comparing 'full oxygen sensitivity' with anoxic sensitivity; whereas the term ER has come to connote a quantity that varies with the concentra-tion of the sensitizer, i.e. it corresponds with the '*r*' of equation 6.1.

It is not common practice to compare survival curves for irradiation under oxygen and anoxia with sensitizer present in both cases. This comparison gives 'o.e.r. in the presence of sensitizer', which I shall abbreviate o.e.r.(S). This is a useful parameter for some aspects of mechanisms of action (see fig. 7.2).

Sensitizers may be regarded as the more efficacious the closer they can come, when used in tolerable concentration, to rendering hypoxic cells as sensitive as oxygen can make them, i.e. to giving an o.e.r.(S) of 1. Alternatively, their efficacy can be judged in terms of the concentration required to give an ER of some selected value with a given test organism. Adams *et al.* (1976) used an ER of 1.6 for V79 Chinese hamster cells as the standard value in comparing the efficacies of a series of nitro-aromatic compounds with different one-electron reduction potentials. If a larger value had been used, not all of the compounds cited could have been included. With some, the maximum concentration that could be used, and therefore the maximum ER attainable, was limited by toxicity or solubility; with one compound, p-nitroacetophenone (PNAP), the ER for Chinese hamster cells increased with concentration to reach a maximum value of 1.7, not exceeded even at higher concentrations. (Adams *et al.*, 1971a; Chapman, Webb & Borsa, 1971).

The results of the comparisons gave a roughly linear relationship between one-electron potential and the logarithm of the concentration needed to give an ER of 1.6, with concentrations ranging from about 2×10^{-5} to 4×10^{-3} mol dm^{-3}. (Adams *et al.*, 1976). The scatter of the points about the line was sufficient (over a factor of about ten in concentration) to accommodate some marked differences between sensitizers in the mode of increase in ER with concentration. The data for oxygen were shown to be consistent with the general trend; but enough information is available to suggest that the correlation does not necessarily signify uniformity in the mechanisms of action of that group of sensitizers with each other or with oxygen.

One method of getting evidence on that point is to confer a degree of sensitization on cells by one compound, then to test whether the introduction of another increases sensitivity in a manner to be expected if the two compounds act by the same mechanism. If one of them is used in concentration that gives its maximum effect, and this is less than maximal sensitization by oxygen, the achievement of further sensitization by a second compound may plausibly be regarded as evidence that the two sensitizers operate at least in part by different mechanisms.

Such 'additivity' experiments have been done with representatives of various classes of sensitizer, used in pairs. Bacteria as well as mammalian cells have been used in such tests. The mode of action of diamide, at least on mammalian cells, was shown by the rapid-mix experiments of Watts *et al.* (1975) to differ from that of other hypoxic cell sensitizers, and this was confirmed by experiments of those authors, as well as of Harris & Power (1973), in which it was shown that nitro-compounds used together with diamide would give greater sensitivity than could be achieved by diamide alone.

Evidence of similarities or differences in the mechanisms of action of nitroxyl compounds and those designated as electron-affinic was sought in several sets of experiments in which bacterial spores or vegetative bacteria were used as test organisms. In all these, the compound p-nitroacetophenone (PNAP) was chosen as representative of the electron-affinic class. Fielden, Ewing & Roberts (1974) found that NPPN and PNAP were additive in sensitizing bacterial spores; Sapora *et al.* (1977) failed to find 'super-oxic' sensitization of *E. coli* K12 AB 2463 by PNAP; and Johansen *et al.* (1977) achieved greater sensitization of *E. coli* K12 AB1886 and *E. coli* B/r with a combination of PNAP and NPPN than could be achieved with either compound acting alone, while with the mutant AB 2463 the addition of PNAP failed to protect against 'super-oxic' sensitization by NPPN, as Johansen (1974) had found oxygen to do. When used in combination with NPPN, PNAP increased the sensitivity of AB2463 even more. Johansen *et al.* also used NPPN in combination with misonidazole, which when used alone sensitizes mammalian cells as well as bacteria to an extent almost equal to maximum sensitization by oxygen. The additive effect of misonidazole with NPPN was even greater than of PNAP. The conclusion from all these experiments was that the mechanisms of sensitization by the stable free radical compounds and the electron-affinic (nitro-aromatic) compounds must differ.

When the electron-affinic compound PNAP was used in concentration sufficient to confer its maximal degree of sensitization on *Serratia marcescens* no change in sensitivity occurred through the addition of bipyridinium compounds in concentrations that would sensitize the bacteria to the same extent as the PNAP alone. Additivity experiments with misonidazole gave the same result (Anderson & Patel, 1977). The authors concluded that those compounds, like the viologens, sensitized *Serratia* by acting at a membrane site. Anderson *et al.* (1978) likewise failed to find additivity of sensitization by nitropyridinium compounds and PNAP.

How oxygen-mimetic are hypoxic cell sensitizers?

Attention was drawn in Chapter 6 (p. 59) to evidence of considerable variation in the solubility of oxygen within cell structures. Any chemical modifier of radiation damage must be expected to have its own distinctive pattern of solubility, so precise 'oxygen mimicry' by chemical sensitizers is not to be expected unless radiation-induced cell death can be attributed to the occurrence of only one kind of lesion located in only one structure. Current evidence suggests that such a view is hardly tenable. Nevertheless, some broad features of sensitization by oxygen may be matched to a greater or lesser degree by other sensitizing compounds.

According to the simple model outlined at the beginning of this chapter a compound that mimics the sensitizing action of oxygen should fix damage by reacting with target radicals, and, if present in low concentration, it should compete with restitutory metionic reactions. Dependence of sensitizing ability on concentration should therefore be described by an equation of the form of equation 6.1. Other criteria are that action should be in the dose-multiplying mode when the experimental conditions are such that oxygen acts by that mode; and that sensitization should be discernible when contact with the cells is only as short as is required for the compound to diffuse to the site at which it must react to fix damage.

Dose-multiplication is a more exacting criterion when survival curves are shouldered. It is satisfied for almost all radiosensitizers classed as electron-affinic, but with some reservations, since some of them have been found to be toxic to hypoxic cells, given periods of contact of some hours. Whitmore, Gulyas & Varghese (1978) found that prolonged incubation of hypoxic mammalian cells with misonidazole caused a reduction in the extrapolation number to the survival curve. The criterion of action within short contact times has also been satisfied for PNAP and misonidazole, which Adams *et al.* (1975) reported to be sensitizing, though not to the maximum extent, if they were mixed with mammalian cells a few milliseconds before a pulse of radiation. But the first criterion mentioned, concentration dependence of the type demonstrating competitive reaction, seems to be only partially satisfied by such nitro-aromatic compounds as have been tested (Adams, personal communication). As shown in Chapter 6, rearrangement of equation 6.1 gives the relationship $(r - 1)/(m - r)$ proportional to oxygen concentration (equation 6.4, p. 57). With sensitizers, the parameters analogous with r and m are respectively ER and ER_{max} where ER_{max} is the maximum enhancement ratio that can be achieved. Results with misonidazole as a sensitizer of *Serratia marcescens* and of V79 hamster cells, kindly supplied by Dr I. J. Stratford, have been used to construct fig. 7.3. $(ER - 1)/(ER_{max} - ER)$ is plotted as a function of sensitizer concentration for both classes of cell. The data points for *Serratia* may be defined by a straight line, if allowance is made for experimental error; but those for the hamster cells would be difficult so to define.

It may be that some sensitizers are oxygen-mimetic with respect only

Fig. 7.3. The ratio $(ER - 1)/(ER_{max} - ER)$ for misonidazole, as a sensitizer of hypoxic Chinese hamster V79 cells (○) and *Serratia marcescens* (●) (Raw data, from experiments by Dr. J. C. Asquith and I. J. Stratford, provided by Dr Stratford.)

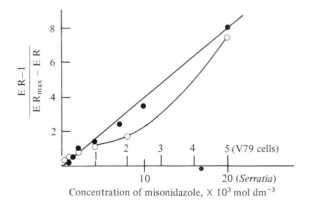

Concentration of misonidazole, $\times 10^3$ mol dm^{-3}

to one 'component' of oxygen sensitization. For example, Michael *et al.* (1978) observed that in *Serratia marcescens* a pulse of radiation created two reactive species with which oxygen would react to produce a lethal lesion. These species were resolved by their different lifetimes: that is to say, the species of shorter half-life was the more rapidly restituted to give a non-lethal product. Two 'components' of the oxygen effect were brought to light also by Shenoy *et al.* (1975) in experiments with mammalian cells. They used the fast-flow rapid-mix technique to gain information on the relationship between pre-irradiation time of contact of the cells with oxygen and the degree of sensitization. An anoxic suspension of cells was mixed with buffer containing oxygen at times between 4 and 40 ms before the radiation pulse. Full survival curves showed the oxygen to have acted in the dose-multiplying mode, whatever the P_{O_2}. Maximum sensitization (o.e.r. ≈ 2.8) was attained only when mixing occurred 40 ms before irradiation, but with all values of P_{O_2} tested (from 1% to 50% oxygen in the gas phase, in the final mixture) the cells were 1.7 times as sensitive as anoxic cells when mixing had occurred only 4 ms before irradiation, that being the shortest time that could be achieved.

Of several interpretations they discussed, Shenoy *et al.* favoured a model involving incipiently lethal lesions formed in different sites, such that different times are required for oxygen to diffuse to them. Alternatively, it was suggested that the 40 ms contact with oxygen required for its full sensitizing effect might be a measure of the time it takes to reach its final concentration at either of two reactive sites with differing lifetimes. Only the longer-lived species would be available to react with oxygen admitted shortly before the pulse of radiation. If this were so, the so-called 'fast component' seen in the results of Shenoy *et al.* refers to interaction with a comparatively long-lived species, perhaps analogous with the long-lived species detected by Michael *et al.* (1978) in their experiments with *Serratia marcescens*.

Adams *et al.* (1975) reported that PNAP appeared to mimic oxygen in respect of the 'fast component', which was no longer evident when PNAP was used in concentration sufficient to give its maximum ER. From experiments in steady-state conditions McNally & de Ronde (1978) concluded that misonidazole as well as PNAP reacted with that component of damage in mammalian cells. When misonidazole and oxygen were introduced into the cell suspension together there was some additivity in sensitization, but only provided that both were in concentrations lower than needed to give an ER of less than 1.6, which was the maximum that could be achieved with PNAP. With higher concentrations of O_2 or misonidazole there was no further additivity although sensitization was less than the maximum that could be conferred by either agent acting separately. Since yet higher concentrations of either agent would have increased sensitivity, it was concluded that misonidazole could not be acting like oxygen overall.

Michael *et al.* (1978) compared mechanisms of action of some radio-sensitizers with each other and with oxygen by a modification of the gas-explosion technique. The bacteria to be irradiated were held in contact with the sensitizer, or with oxygen at low P_{0_2}, during the pulse of radiation which was followed at measured times by the rapid admission of 100% oxygen. All three sensitizers tested, TAN, misonidazole and metronidazole, as well as oxygen at low partial pressure, behaved as if they reacted with the species that would otherwise have persisted the longer, so the kinetics observed were those relating to the shorter-lived species. The conclusion was that TAN behaved like oxygen, in respect of interaction with that species, but that misonidazole did not; results with metronidazole were inconclusive. The authors' interpretation was that radiation engendered radicals with which misonidazole would react to fix damage, but which, in its absence, would give rise to secondary radicals that could either be restituted or would result in fixed damage if they interacted with molecules of TAN or oxygen. As noted by McNally & de Ronde (1978) a scheme of that nature could account also for their results. All these observations together lead to the conclusion that the mechanism of radiosensitization by oxygen is mimicked only in part, perhaps only to a minor extent, by the nitro-aromatic compounds PNAP and misonidazole.

In contrast, all the criteria set out for oxygen-mimicry have been satisfied in experiments with nitroxyl compounds. Brustad (1968) reported that TAN sensitized bacteria within milliseconds after contact, in rapid-mix experiments. The results of Parker *et al.* (1969) show that it was dose-multiplying, when used with mammalian cells; and Emmerson, Fielden & Johansen (1971) found that dependence of sensitizing ability on concentration was of the predicted type, with three different compounds. However, Johansen *et al.* (1977) commented that the mechanism by which oxygen sensitized bacteria must be different, because it could protect certain radiosensitive mutants against 'superoxic' sensitization by those compounds.

It has often been emphasized that oxygen is bifunctional since it is highly electron-affinic and also has an unpaired electron. Oxygen is often protective, but never sensitizing, when model protein or nucleic acid targets are irradiated in aqueous suspension (table 4.1), unless reducing substances have been added that are specifically protective in the absence of oxygen. Oxygen is well known to combine readily with e_{aq}^- and hydrogen atoms produced in the radiolysis of water, so if such species contribute to the inactivation of DNA within the cell, as they do outside it (e.g. Ebert & Alper, 1954; Dewey & Stein, 1970) protection by oxygen may perhaps become evident in cells which are deficient in capacity to repair lesions so induced. This speculation is elaborated further in Chapter 15. If it is valid, protection by oxygen could be expected in the combined circumstances of complete fixation of one or more other types of damage by chemical compounds and deficient repair of some DNA lesions. In that case nitroxyl compounds and oxygen may act similarly as sensitizers, although the former fail to act protectively against damage by reducing species.

The suggestion that oxygen might protect against damage to DNA, or at least some form thereof, is in direct contradiction to the assumption usually made that the damage fixed by oxygen is primarily in DNA. Some tenuous evidence was quoted above in support of the view that the primary lesions involved in sensitization by oxidizing agents are associated with cell membranes. At the time of writing the great majority of radiation chemical studies on radiosensitizers have focussed on their reactions with DNA or constituent molecules thereof. It may well be that analogous experiments with model membrane systems will help to elucidate the processes of radiosensitization.

8

Chemical protection

The search for chemical means of protecting man against the hazards of ionizing radiation was a major brief for many of the national radiobiology research institutions soon after the end of World War II. From the outset, a good deal of effort was devoted to testing the proficiency of possible protective agents in animals. Observations at the cellular level were confined for the most part to experiments with micro-organisms, since viable counting techniques with mammalian cells were not developed until 1955 (by Puck & Marcus). Most of the information on chemical protective agents therefore still derives from experiments with bacteria. In parallel with the surge of interest in mammalian cell radiobiology, dating from 1956, there was a growing tendency to frame research work within a radiotherapy context; protection against radiation hazards, as a practical end-point, received correspondingly less attention. Perhaps for that reason, and in contrast with work on radiosensitizers, there is surprisingly little to report on the protection of mammalian cells by chemical agents. Interest in the topic has recently revived, in the context of possible clinical applications. It was suggested many years ago that differential protective effects on oxygenated and hypoxic cells might have some clinical applications, in the sense that hypoxic cells in tumours would be less well protected than those in normal tissues by an appropriate agent. This possibility is currently receiving some attention (Chapter 17).

Chemical protection and hypoxia

Hypoxia is the most effective means of protection against ionizing radiation and many agents that were effective in raising LD_{50} doses in animals proved to act by reducing tissue oxygen tension by one mechanism or another (see e.g. van der Meer & van Bekkum, 1959; van den Brenk & Moore, 1959). Even when agents were tested for protective action *in vitro*, results were confusing and confused, because unsuspected oxygen effects were operating. For example, when fairly concentrated

87

suspensions of bacteria were kept in closed vessels, their respiration rate affected the partial pressure of oxygen in the suspending medium. Depletion of oxygen would be speeded up by a substance that increased the rate of respiration so that substance might be deemed 'protective'. Although the aminothiols, cysteine and cysteamine (otherwise known as β-mercaptoethylamine) were already regarded as well-established chemical protectors, Gray (1956) reported that those compounds, being readily oxidizable, rapidly reduced the partial pressure of oxygen in buffer at pH 6 or higher in closed vessels. Such observations induced scepticism as to whether chemical protection was a real phenomenon, apart from the induction of hypoxia. This might be in the medium surrounding the cells, or, perhaps, in the micro-environment of the cellular target, by inhibition of the access of oxygen thereto. It was suggested by some that chemical protection could be proven only by demonstrating its occurrence in rigorously anoxic cells. Insofar as methods for rendering cell suspensions anoxic were themselves often of dubious efficacy, it was not easy to find convincing proof that cells could be protected by chemical agents. It did not become clear until about 1960 that true chemical protection had been difficult to establish with confidence because protective agents are almost always less effective when cells are anoxic during irradiation.

At the Second International Congress of Radiation Research, 1962, this aspect of chemical protection was extensively discussed. Those interested in the historical development of the topic will find it useful to consult rapporteurs' reports of papers presented on three aspects: chemical systems (Alexander, 1963); protection at the cellular level (Wright, 1963); and chemical protection of mammals (Scott, 1963). The greater effectiveness of protective agents when cells are oxygenated was first noted by Alper (1962) and Bridges (1962a) from experiments with bacteria, and the same differential effect was seen also in mammalian cells, although it is difficult to find a clear-cut demonstration before the paper of Chapman *et al.* (1973). The operation of this differential effect also *in vivo* (see Chapter 17) may be deduced from experiments in which mice were kept hypoxic for a short time during which they were subjected to whole-body irradiation. They were protected by cysteamine, but to a much smaller extent than when they breathed air during irradiation (Wright, 1962).

Table 8.1 is set out to illustrate the greater efficacy of protection afforded to aerobic than anoxic cells *in vitro* by protective agents of several classes of compound. Data on protection by the aminothiols, cysteine and cysteamine, have been excluded from the table, because of the special difficulties in interpreting the results, emphasized by the report of Gray (1956) quoted above. Nevertheless the same

Table 8.1. *Some protection ratios, aerobic and anoxic cells (sparsely ionizing radiation)*

Test cells	Compound	Concentration ($mol\,dm^{-3}$)	Protection ratios		Reference
			aerobic	anoxic	
Pseudomonas	dimethyl sulphoxide	1.0	4.0	1.9	Bridges (1962a)
E. coli B/r	dimethyl sulphoxide	1.3	3.0	1.0	Alper (1963a)
Chinese hamster	dimethyl sulphoxide	3.0	2.6	1.3	Chapman *et al.* (1973)
B. megaterium (spores)	glycerol	6.5	1.9	1.3	Webb & Powers (1963)
E. coli B/r	glycerol	1.0	3.1	1.7	Alper (1963a)
E. coli B/r	glycerol	6.0	3.2	2.4	Alexander, Lett & Dean (1965)
S. flexneri	thiourea	0.3	3.5	2.2	Cramp (1966)
S. flexneri	guanylthiourea	0.3	2.3	1.4	
S. flexneri	thiosemicarbazide	0.1	2.4	1.7	

differential effect operates when those compounds are used, and it is unlikely that the more effective protection of fully oxygenated cells by the aminothiols can all be attributed to oxygen depletion. This point is discussed in the following section.

The aminothiols

Cysteine was first demonstrated to be a protective agent *in vivo* (Patt *et al.*, 1949). At that time, radiobiological phenomena, including the oxygen effect, were widely believed to be mediated almost entirely by the radiolysis products of water. Sulphydryl-containing enzymes in aqueous solution had been found to be oxidized by doses much lower than were required for other observable effects on biological molecules, so it was thought that an increase in the content of reducing substances in the blood would act protectively. Another aminothiol, β-mercapto-ethylamine, or cysteamine, was soon found in the early experiments to protect bacteria and animals. Many hypotheses were advanced as to the mechanisms of protection by these compounds. Undoubtedly some mis-interpretations are attributable to unsuspected involvement with the oxygen effect.

Protection of bacteria (*Escherichia coli* B/r) by cysteine was first de-monstrated by Kohn & Gunter (1959). To achieve maximum protection they found it necessary to introduce the cysteine into the bacterial suspensions some 10 to 15 minutes before irradiation, and to keep the

Fig. 8.1. Rapid-mix experiments, showing difference required in time of incubation of mammalian cells with cysteamine and dimethyl sulphoxide to give detectable protection (after Whillans & Hunt, 1978).

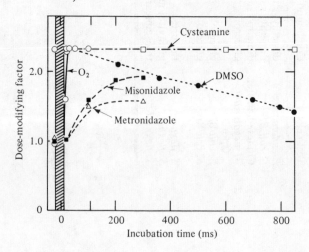

suspensions anoxic during that time to avoid oxidation of the cysteine. The requirement for pre-irradiation incubation of cells with cysteine or cysteamine has largely been overlooked when hypotheses have been proposed for the mechanism of their protective action. Rapid-mix studies of Whillans & Hunt (1978) have confirmed this requirement in the pro- tection of mammalian cells by cysteamine. Their arrangements per- mitted them to observe results of mixing cells with chemical modifying agents at times from 5 to 1000 ms before a pulse of radiation. No protec- tive effect of cysteamine was seen even when mixing took place 1 s before irradiation, whereas dimethyl sulphoxide exerted some protection from the earliest time of mixing onwards (fig. 8.1).

The biochemical change wrought in cells by the aminothiols is made evident also by the radiation response of cells after the compounds have been washed out of the suspending media. In my own experiments, reported heretofore only in an abstract (Alper, 1963a), I found that this treatment, when applied to *Escherichia coli* B/r, left them more resistant to radiation after the cysteine had been thoroughly washed out of the suspension (fig. 8.2). Analogous effects with mammalian cells *in vitro* were noted by Vergroesen, Budke & Vos (1967) in tests with cysteamine. As

Fig. 8.2. To show residual protective effect of cysteine after it was washed out of a suspension of *E. coli* B/r.

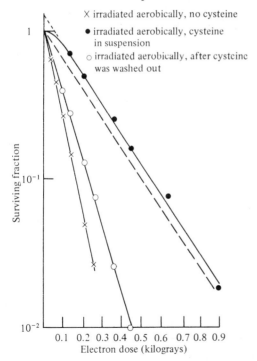

X irradiated aerobically, no cysteine

● irradiated aerobically, cysteine in suspension

○ irradiated aerobically, after cysteine was washed out

Surviving fraction

10^{-1}

10^{-2}

0.1 0.2 0.3 0.4 0.5 0.6 0.7 0.8 0.9
Electron dose (kilograys)

noted below, such after-effects of treatment by protective agents are not seen with alcohols or dimethyl sulphoxide.

The necessity for biochemical interaction before irradiation, and the effects of this interaction after removal of the compounds, together make it improbable that the aminothiols can be protecting cells *per se* by metionic reactions.

However, in respect of greater efficacy of protection afforded to cells irradiated under full oxygenation, cysteine and cysteamine seem to act like other protective agents. While autoxidation of the compounds, with consequent reduction in intracellular P_{O_2}, might partially account for this effect, it is unlikely to be the only explanation. In my own experiments, it was established by various controls that the phenomenon could not be ascribed to oxygen depletion in the suspending medium; for example, there was no protection at all of yeast cells, although they were handled in just the same way as the various strains of bacteria (table 8.2). Of course there could be no control on intracellular P_{O_2}, but it is implausible that the differential effects illustrated by the table can be ascribed to any significant extent to intracellular reduction in P_{O_2}, *E. coli* B/r were most effectively protected when they were irradiated on the surface of agar gel containing cysteine, so that oxygen from the air could diffuse freely into them (fig. 8.3). Evidence comes also from the experiments with *E. coli* B, with which the extent of sensitization by the oxygen present during irradiation depends on the post-irradiation

Table 8.2. *'Protection ratios' (ratios of doses to give the same effect) for cysteine: influence of oxygen*

Strain	Protection ratio		Oxygen enhancement ratio	
	Cells aerobic	Cells anoxic	normal	with cysteine
E. coli B/r	2.4	1.2	3.0	1.5
Salmonella Typhimurium	2.5	2.0	2.5	2.0
Yeast: single cells	1.0	1.0	3.6	3.6
Budding cells	1.0	1.0	1.7	1.7
E. coli B Values of D_0 for anoxic irradiation, depending on conditions of culture $\begin{cases} Gy \\ 120 \\ 87 \\ 28 \\ 21 \end{cases}$	2.4 2.2 1.6 1.4	1.2 1.3 1.3 1.3	3.7 2.9 2.0 1.8	1.7 1.5 1.6 1.7

After Alper (1963a).

plating conditions (Alper, 1961a). When cysteine was present its protective effect on the aerobically irradiated bacteria was greatest when plating conditions yielded the highest value of D_0; so that the o.e.r. with cysteine present was approximately constant, whereas the o.e.r. itself was not.

A favoured hypothesis for the mechanism of protection by aminothiols is hydrogen donation from the free –SH, as a reaction competing with damage fixation (e.g. Adams, 1972). As noted in Chapter 4, sulphydryl compounds can indeed be protective when proteins or viruses are irradiated dry or in the frozen state (Ginoza & Norman, 1957; Braams, 1960) or in suspension in the presence of high concentrations of scavengers of the radiolysis products of water (Howard-Flanders, 1960). But the rates of reaction of model radicals are considerably greater with oxygen than with the –SH group (see Adams, 1972, for summary). It is to be expected, therefore, that protection of extracellularly irradiated

Fig. 8.3. To show large protective effect of cysteine on bacteria exposed to air during irradiation. The bacteria were on cellophane carriers resting on the surface of agar incorporating or not incorporating cysteine.

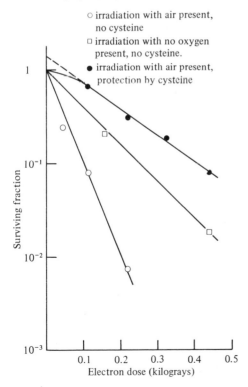

○ irradiation with air present, no cysteine

□ irradiation with no oxygen present, no cysteine.

● irradiation with air present, protection by cysteine

entities should be more effective when oxygen has been excluded. This was indeed the case in the experiments of Howard-Flanders (1960) and of Howard-Flanders, Levin & Theriot (1963: fig. 8.4) on the protection afforded by cysteamine to bacteriophages. These were in suspensions containing broth, which was assumed to scavenge all radiation products that might otherwise inactivate the phage by indirect action. However, it is difficult to tell whether the protective action of the cysteamine was, in fact, by virtue of hydrogen donation to radicals directly formed in the phage DNA, or of scavenger action additional to that provided by the organic material in the broth. As shown in Chapter 4 (p.23), there is evidence of indirect action on sub-cellular entities irradiated in aqueous suspension even in the presence of protein in high concentration. Results of Hotz & Müller (1962) suggest that cysteamine may indeed act as an effective scavenger when no oxygen is present in an irradiated suspension. Bacteriophage T1 was irradiated in broth suspension or in protein-free buffer. With increasing cysteamine concentration the inactivation dose for the phage in protein-free suspension increased more rapidly than for the suspension in broth, and with concentrations above 10^{-2} mol dm^{-3} there was no difference: evidently the cysteamine alone was even more effective than the components of the broth.

Thus it is not clear whether the protection afforded by cysteamine to bacteriophage in aqueous suspension can be unequivocally attributed to hydrogen donation to, and therefore restitution of, radicals formed by direct action. But if it is, this mechanism plays a much more important role when oxygen is absent during irradiation, as it does when dry test systems are used (e.g. Ormerod & Alexander, 1963; Hotz, 1966). In

Fig. 8.4. Achievement of anoxic protection of bacteriophage T2 in suspension by introducing mercaptoethanol. Protection is considerably more marked in anoxic than oxygenated suspensions (from Howard-Flanders, Levin & Theriot, 1963).

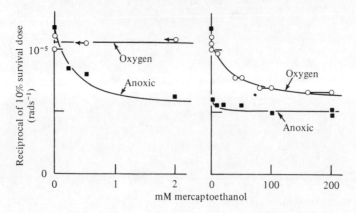

contrast, –SH compounds are more effective in protecting oxygenated than anoxic cells against killing by radiation. This important difference has not been commented on when hydrogen donation by –SH compounds has been accepted as the mechanism by which they protect cells. It has been argued that the difference is accounted for by the intrinsic –SH content of cells even when no protective agent with a thiol group has been added. But this explanation could not account for the comparatively low protection ratios afforded to hypoxic cells by aminothiols, unless it could be shown that protection of oxygenated cells, as well as of model systems, reaches a maximum with increasing concentration of the –SH, then declines. This was not observed in the experiments of Howard-Flanders *et al.* (1963) with bacteriophage nor of Chapman *et al.* (1973) with mammalian cells.

Thiourea and analogous compounds

Thiourea is an effective protective agent for bacteria (Bridges, 1962b) and mammalian cells (Vergroesen *et al.*, 1967). Because of the widely held belief that chemical protection at the metionic level is at its most effective when a free –SH group is involved, it has been assumed that this is also the mode of action of thiourea. If this were the case, it would have to react in a form in which a free thiol is available. This supposition was specifically tested by Cramp (1966) in experiments in which he examined the protective ability of a series of analogous compounds, such that the existence of a tautomer with a free –SH group was considerably less likely even than for thiourea itself. Some of Cramp's results on the protection of *Shigella flexneri* by thiourea and analogous compounds were given in table 8.1. There was little difference in their efficacy, despite the improbability that an –SH group could be involved when analogues were used.

From this evidence, and the several points made above about the protection of cells by aminothiols, it must be doubted whether these very effective protective agents do indeed act directly by hydrogen donation to radicals induced in target structures.

Alcohols

The earliest reports of chemical protection of bacteria came from workers at the Oak Ridge National Laboratory. Among many compounds they tested, the alcohols formed an important group. As mentioned in the introduction to this chapter, it is difficult to interpret the data from work in which the oxygen content of irradiated suspensions was not carefully controlled, which was not the case in experiments done before the late 1950s. Marcovich (1957) tested the ability of a group of alcohols, including glycerol, to protect *E. coli* K12 irradiated while oxygen was bubbled through the suspensions. The efficacy of protection varied with concentration in much the same way for all, a maximum being reached with concentrations of about 1.5 mol dm^{-3}. A similar result was reported by Dewey (1963), with respect to the protection of *Serratia marcescens* by glycerol, ethylene glycol and ethyl alcohol, although his results suggest that maximum protection had not been attained even with concentrations of 2 mol dm^{-3}. In contrast, Cramp (1969) did not find ethyl alcohol and glycerol to behave similarly with respect to the concentration-dependence of their ability to protect *Shigella flexneri*. The alcohols he used were those for which data were available on rates of reaction with hydroxyl (OH) radicals and his results are discussed in that context below.

Of the alcohols, only glycerol has been used to protect eukaryotic cells against killing by radiation, perhaps because of its low toxicity. Erikson & Szybalski (1961) tested the protection it afforded to mammalian cells that had been rendered more sensitive to irradiation by the incorporation of a thymine analogue, 5-bromodeoxyuridine, into the DNA. Glycerol protected both the normal and the sensitized cells, though their results suggested somewhat more effective protection of the latter.

Dimethyl sulphoxide
The radioprotective action of this compound was discovered first by Ashwood-Smith (1961) who had used it as a solvent for drugs to be injected into mice that were subsequently irradiated. A few experiments with bacteria have been reported, and also a few with mammalian cells *in vitro*. The concentrations required to give substantial protection ratios are much higher than with many other chemical protective agents, but the maximum protection ratios are also higher, and high ratios can be achieved with well-oxygenated mammalian cells as well as bacteria because the compound is not toxic even when used in concentration as high as 3 mol dm^{-3}.

Vos & van Kaalen (1962) compared protection of mammalian cells by glycerol and dimethyl sulphoxide at 77 K and at room temperature, and found that protection factors were the same. This result has a bearing on the widely canvassed association of chemical protection with the scavenging of OH radicals (see below).

General comments
Even before techniques for rapid mixing were developed some information was available as to how long before irradiation protective chemicals had to be in contact with cells to be effective, and also whether there was any effect after they were washed out of the cell suspensions.

Evidence was given above that rather long times of contact with amino-thiols are needed, of the order of minutes; and also of residual protective action after they have been washed out of suspensions. In contrast other protective agents, like the alcohols, or dimethyl sulphoxide, have no effect when they are left in contact with cells for a period, but washed out before irradiation (e.g. Bridges, 1962a). Only a few protective compounds have been investigated in this context by rapid-mix techniques. Brustad & Singsaas (1971) observed that glycerol at 1 mol dm^{-3} exerted protective action on *E. coli* strain B if it was mixed with the bacterial suspension within milliseconds before a pulse of radiation. Maximum protection was achieved within a contact time of about 200 ms. Glycerol may be taken as representative of the alcohols. The results of Whillans & Hunt (1978) with cysteamine and dimethyl sulphoxide as protective agents for mammalian cells were shown in fig. 8.1.

Although dependence of protection ratios on concentration of the agent has been studied by several authors, few have investigated the applicability of an equation that would suggest competitive reactions, like equation 6.1. Dewey (1960) fitted an equation of this type to his results on the protection of *Serratia marcescens* by glycerol; and the results of Bridges (1962a) on protection of *Pseudomonas* by dimethyl sulphoxide were also so fitted. Bridges (1962b) used several compounds as protectors of *Pseudomonas* irradiated under anoxia, so maximal protection ratios were rather too small for accurate curve-fitting. However, the form of the concentration curves appeared to be of the type mentioned except with thiourea.

The hydroxyl radical-scavenging hypothesis

In chapter 6 it was mentioned briefly that the sensitizing action of oxygen has been attributed by some authors to interactions with radiolysis products of water. One view of the mechanism of action of protective agents likewise embodies the concept that a large fraction of lethal damage is attributable to damaging interaction between the target and a radiolysis product of water. Such damage, it is reasoned, would be reduced if the damaging species were selectively scavenged by agents introduced into the cell. Johansen & Howard-Flanders (1965) and later Sannner & Pihl (1969), using bacteria as test cells, graded several protective compounds in order of their efficacy, and also in order of their rates of reaction with some of the species engendered by radiation in water. They established a rough correlation between protective effectiveness and rate of reaction only with the hydroxyl (OH) radical. From this correlation and their experimental observations on protection ratios they concluded that only 30–40% of the lethally damaging events in cells could be attributable to direct action, the rest being attributed to damage inflicted by OH radicals which would be scavenged by the addition of an appropriate compound. Roots & Okada (1972) analysed in similar fashion the protective action of some alcohols and –SH compounds against the induction of single-strand breaks in the DNA of mouse leukaemia cells and came to roughly the same conclusions. They attributed 70% of the effect to the actions of OH radicals. Their observations were not correlated with dose–effect curves for cell killing: they used a dose of 100 Gy throughout, which is very much higher than that at which mammalian cell survival *in vitro* can normally be observed. The force of the correlation they claimed must in any case be weakened by the inclusion of –SH compounds, since, according to the evidence quoted above, their mechanism of action is unlikely to be the same as that of the alcohols.

Correlation between efficacy of protection and reaction rates of the agent with OH radicals was studied in more detail by Cramp (1969) who used only alcohols. Excluding β-mercaptoethanol, which has an SH group, the correlation as tested by such a closely related chemical group should be good, if the hypothesis is valid. The alcohols he used, and rate constants from NSRDS-NB 559 (National Bureau of Standards), are given in table 8.3. Cramp found that ethanol was less protective than it should have been, if it acted by OH radical-scavenging, while 1,4-butanediol was no more effective than glycerol or ethanediol, although its rate constant was more than twice as high.

Chapman *et al.* (1973) accepted the OH radical-scavenging hypothesis as having been established by work from their group on reactions of OH with DNA in aqueous suspension, and of the effects of adding a

Table 8.3. *Protection of* Shigella flexneri *by ethanol, β-mercaptoethanol and other alcohols*

Alcohol	Rate of reaction with OH $(dm^3 mol^{-1} sec^{-1} \cdot 10^9)$	Protection ratios	
		aerobic	anoxic
Ethanol	1.9	1.9	1.2
Ethanediol	1.1	2.7	1.4
Glycerol	1.8	2.7	1.4
Erythritol [a]	2.0	2.7	1.4
1,4-butanediol	3.4	2.5	1.4
β-mercaptoethanol	8.5	3.3 [b]	2.0 [b]

Protection data from Cramp (1969); rates of reaction from NSRDS-NB559 (1977).
[a] Measured for pH 9. All others for pH 7.
[b] Concentration 0.3 mol dm^{-3}; ratios for all other compounds at 1 mol dm^{-3}.

radiosensitizer or dimethyl sulphoxide to the suspension. On this assumption, they interpreted their results on protection of Chinese hamster fibroblasts by dimethyl sulphoxide as a method of measuring the contribution of OH radicals to the overall lethal effect of radiation.

Like other authors previously mentioned, they experimented with anoxic as well as well-oxygenated cells (table 8.1), and examined the dependence of protection ratios on protector concentration for both conditions. Protection ratios increased with increasing concentration until a plateau was reached. With dimethyl sulphoxide maximum effectiveness was achieved with about 3 mol dm^{-3}. The authors assumed that OH radicals were completely scavenged, with that concentration, and that radiobiological effects should then all be attributed to 'direct action', i.e. direct absorption of energy in the target, presumed to be cellular DNA for that as well as for indirect action. On the basis of those assumptions, it was calculated that, to be effective, an OH radical would have to be formed within 0.8 to 2 nm from the DNA (Chapman *et al.*, 1976). The presence of oxygen has not been found to affect reactions of OH radicals with the molecules with which they interact, so the hypothesis that chemical protective agents act by scavenging OH radicals requires additional assumptions to accommodate the greater effectiveness of protection on aerobic than anoxic cells. The ancillary hypothesis introduced into the model of Chapman *et al.* (1973) is that the interaction of OH with DNA causes formation of a radical which then reacts just as if it had been engendered by 'direct action', i.e. by an energy absorption event within the DNA. Thus they postulate the reactions

$$\text{OH} + \text{DNA} \longrightarrow \text{DNA radicals}$$

$$\text{DNA radicals} + \text{S} \left[\begin{array}{l} \longrightarrow \text{DNA} + \text{S} \\ \longrightarrow \text{DNA}^+ + \text{S}^- \end{array} \right\} \text{chemical fixation}$$

where S is a radiosensitizing substance, either oxygen or a chemical sensitizer.

A numerical point of interest arises from the results of Chapman *et al.* (1973, 1976) seen in the light of their model. With as much dimethyl sulphoxide present as they considered necessary to scavenge all OH radicals, they observed an o.e.r. of 1.6, which they therefore associated with the operation of direct action, presumed also to be on DNA. The o.e.r. in the absence of protector was about 2.8, which must therefore represent a weighted average of the o.e.r.'s associated with the radicals engendered by direct action and by interaction with OH radicals. Adopting the estimates variously given for the fraction of the action attributable to the latter, 60–70%, the o.e.r. associated with the postulated DNA + OH radicals must lie between 3.3 and 3.6, more than twice as great as that associated by Chapman and his colleagues with DNA radicals directly engendered. The model proposed by Chapman *et al.* (1973) does not distinguish between the DNA radicals formed by the two pathways they postulate; but the inference must be that the metionic reactions with the two postulated kinds of radical are very different.

The plausibility of the OH radical hypothesis for lethal damage to cells – and therefore for chemical protection – cannot be profitably discussed further without consideration also of other radiobiological and radiation chemical phenomena; for example, the yield of OH radicals decreases, as LET increases, whereas cells are more effectively killed. Many phenomena require to be taken into account also in discussing the additional assumption that sensitization or protection are exclusively to be accounted for in terms of incipient lesions in cellular DNA. Consideration of these fundamental issues is therefore deferred to later chapters.

9

Quality of radiation

Ionizing and non-ionizing radiations

Electromagnetic radiations are characterized by their frequency or wavelength, but they may also be described in terms of the energy of the photons, given by $h\nu$, where h is Planck's constant and ν the frequency. The ways in which energy is transferred from the photons to the atoms with which they interact depends on their energy. If it is more than 10 electron-volts (about 1.6×10^{-18} J), orbital electrons may be ejected from their parent atoms, so ion pairs will be formed. The corresponding wavelength for photons that are just sufficiently energetic is about 125 nm. The transfer of quanta smaller than about 10 eV may cause orbital electrons to be raised to levels of higher energy, and the atoms in which this occurs are said to be in an excited state. If the photon energy is insufficient to cause ionization, i.e. if the wavelength is more than, say, 130 nm, only excitations will occur.

Although the electromagnetic spectrum is continuous, ionizing and non-ionizing radiations are clearly separated in their practical applications. Ultraviolet light of wavelengths less than about 180 nm (energy greater than about 7 eV) are so readily absorbed by all matter, including air, that objects can be irradiated only if there is vacuum between them and the source. This difficulty pertains also to very 'soft' X-rays, in addition to which their production by X-ray tubes is limited by the nature of the materials that can be used. The least energetic X-rays with which radiobiological experiments have been performed are of wavelength 0.83 nm, the maximum energy of the photons being about 2.4×10^{-16} J or 1.5 keV.

Thus ultraviolet light, as that term is commonly understood, is non-ionizing radiation, which transfers energy to atoms almost wholly by excitations. The commonest source of UV is a low-pressure mercury discharge lamp, emitting most of its energy at 254 nm. This is just within the range of wavelengths (250–270 nm) specifically absorbed by nucleic acids and these are the most effective in the usable UV spectrum in

101

killing viruses and bacteria; so such lamps are often referred to as 'germicidal'. When radiation by UV is referred to in radiobiological literature without specification of the wavelength it may usually be assumed that 'germicidal' UV is meant.

The absorption of UV at any given wavelength depends on the characteristics of the molecules on which it impinges, whereas with biological materials there is in effect no selective absorption of ionizing radiation. Thus 'UV dose' is normally specified in terms of incident energy per unit area, and this cannot easily be converted into 'energy absorbed per unit mass', which is the method for specifying a dose of ionizing radiation. There are difficulties involved in making such a conversion for UV, even if it is done by taking into account a known mass absorption coefficient for a specific cell component, like DNA. This would still not necessarily be relevant to the particular kind of biological damage under test, and in any case there would be considerable variation between classes of cell in respect of the UV actually incident on such a component after it passed through others, which would 'shadow' the component concerned. Only very approximate comparisons can therefore be made of the relative effectiveness of exciations and ionizations in killing cells.

An attempt at such a comparison was made by Lea & Haines (1940), who based their conclusions on observations with three species of bacteria: *Serratia marcescens*, *Escherichia coli* and spores of *Bacillus mesentericus*. From the survival parameters they calculated the numbers of ionizations and of UV quanta required to inflict averages of one lethal event: the ratios were respectively about 700, 500 and 100 for the three species. To convert incident UV dose into quanta they adopted a figure for the absorption coefficient of 'bacterial protoplasm'. This is much lower than the absorption coefficient for pure nucleic acid, at the wavelength they used (254 nm), so the ratios calculated by Lea & Haines must be a lower limit. It is evident from these approximate calculations that excitations are biologically very much less effective than ionizations, so the effects of energy loss through excitation can be ignored when the biological effects of ionizing radiation are analysed.

The various kinds of ionizing radiations

With electromagnetic radiation that is energetic enough to cause ionization, i.e. X- or γ-rays, the first processes leading to observable biological effects are interactions between photons and atoms. These result in the emission of fast-moving electrons, which in their turn are energetic enough to create ion pairs, the negative ion (the electron) perhaps again having enough energy to act as an ionizing particle. Thus the energy imparted by the photon may ultimately be degraded through the ejection from atoms of a series of electrons of successively lower energy. It is customary to refer to electrons of rather low energy as δ-rays.

If a beam of energetic electrons is used as radiation source, the processes of energy loss are much the same as for electromagnetic radiation. The effects of more densely ionizing radiation can be observed if sources are available from which atomic nuclei can be ejected at high velocity; these sources may be natural emitters of α-particles (helium nuclei) or beams of protons, deuterons or of nuclei of higher atomic number accelerated by machines to sufficiently high velocities. The mode of energy degradation of such particles involves fewer steps than is the case with high-energy photons: the atomic nuclei make collisions with atoms in their path, creating pairs of ions; some of the electrons i.e. the δ-rays, will have sufficient energy to be capable of causing ionization in their turn.

Only very thin samples of biological material can be used in irradiations by α-particles from natural radioactive sources, because their ranges in tissue are of the order of only tens of μm. Beams of helium or heavier ions accelerated to much higher energies by machines will penetrate considerably further in biological tissue; but qualities that are significantly different from that of γ-rays or fast electrons can still be attained within only a small fraction of their path, the 'Bragg peak', which is the region of greatest density of ionization (fig. 9.1). Thus if the quality is to be reasonably uniform through the irradiated object it is still necessary to use rather thin objects, like single layers of cells.

Fig. 9.1. 'Bragg peaks'; peaks in the ionization caused in air by α-particles from polonium and radium C' (From: *Radiations from Radioactive Substances*, by E. Rutherford, J. Chadwick and C. D. Ellis, Cambridge University Press, 1930.)

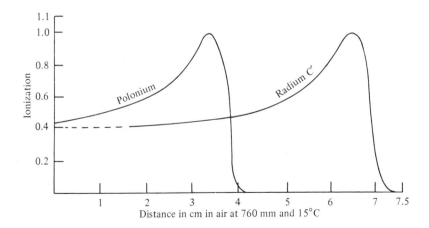

Distance in cm in air at 760 mm and 15°C

For observations on thicker samples, or on animal or plant tissues *in vivo*, it is customary to use a penetrating radiation that will give rise to densely ionizing particles as a result of interaction with atomic nuclei within the medium. Energetic neutrons and negative π-mesons have this property, and neutrons, in particular, have been extensively used in experiments designed to compare the effects of radiation of different quality. Like electromagnetic radiation, neutrons are 'ionizing' only by virtue of the products of their interactions with atoms. These may be elastic collisions with hydrogen atoms, which they 'knock on' to give fast-moving protons. If they collide with heavier nuclei, they may be captured, with the emission of γ-rays; and the capture may result in nuclear instability, with the consequence that protons, deuterons or α-particles may be emitted, the α-particles being the most frequent. The energy imparted to the heavier nuclei in these interactions may also be sufficient for them to collide with macromolecules and so to cause biological effects.

Thus the energy deposited by a neutron beam will be divided mainly between knock-on protons and the more densely ionizing α-particles and heavy recoil nuclei, in a manner that depends on the energies of the neutrons in the beam, and also on the nature of the material through which they pass. In this respect there is no great variation from one type of biological material to another, except that if there is a high fat content, there will be more hydrogen present, and this will result in relatively greater production of knock-on protons. If animals are irradiated, allowance must be made for the considerably lower hydrogen content of bone.

The 'dose' received by any irradiated object is defined as energy

Fig. 9.2. Kerma in the ambient medium and the absorbed dose for a parallel beam of neutrons perpendicularly incident on a slab of material (schematic). (After Report 26, International Commission on Radiation Units and Measurements, 1977.)

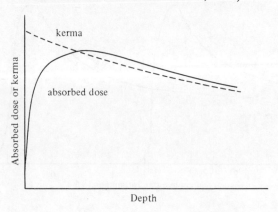

absorbed per unit mass (p. 6), and this is usually adequate also as a definition of dose delivered by a beam of neutrons. However, when a small object is irradiated, the neutrons may lose energy in the production of charged particles the range of which is greater than the dimensions of the object, so the energy lost by the neutrons is not the same as that absorbed by the object. To allow for this difficulty (which may indeed arise also when small objects are exposed to other types of radiation) a unit called the kerma was introduced, defined as the sum of initial kinetic energies of the charged particles liberated within a given volume, divided by the mass of that volume. Figure 9.2 is a schematic diagram showing variation of both kerma and absorbed dose with depth in a hypothetical homogeneous medium when this is irradiated by a parallel beam of neutrons.

Unlike neutrons, which carry no charge, negative π-mesons (or pions) are themselves ionizing particles and therefore lose energy as they traverse matter. When a pion has been sufficiently slowed down, it will interact with an atomic nucleus – mainly hydrogen, oxygen, nitrogen or carbon, in biological tissue. Those nuclei which have captured pions become unstable, and disintegrate, emitting neutrons, protons and stripped ions of higher atomic number, like α-particles, with sufficient energy to act as ionizing particles. Interactions with atomic nuclei occur for the most part within a narrow layer of tissue, at a depth that depends on the initial energy of the pions. The dose within that layer is large, compared with that delivered outside it; this property of pion beams has been suggested as a potential advantage for the treatment of cancerous tumours, the object of the radiotherapist being always to irradiate the normal tissues as little as possible while treating the tumour adequately. An analogous advantage is possible with beams of very energetic heavy ions, which deposit a large fraction of their energy in the Bragg peak (fig. 9.1). Radiobiological investigations with beams of negative pions and very high-energy ions have been made in the specific context of their possible use in radiotherapy, rather than as a method of investigating radiobiological mechanisms, for which purpose they are indeed not very suitable.

The specification of radiation quality

It is evident that the processes of energy loss by all types of ionizing radiation are complex, so 'quality' is impossible to specify in any simple way. Before 1950 it was customary to use the description 'density of ionization', defined, for X- or γ-rays or energetic electrons, by the average initial kinetic energy of the primary electrons, divided by the average energy required to create an ion pair and divided also by their average range. This gives a quantity 'number of ion pairs per unit path length'. However, it is only in gases that proper measurements can be made of the quantities involved in the creation of ion pairs. Since radiobiological phenomena involve the interaction of radiation with condensed media, the need was recognized for a description of radiation quality that did not refer specifically to ionization. The term Linear Energy Transfer, proposed by Zirkle, Marchbank & Kuck (1952) followed on an earlier suggestion of Zirkle (1940) that radiation quality be designated by the 'linear energy absorption' of the medium from the ionizing particle.

Linear Energy Transfer, abbreviated LET, is defined as the energy lost by a charged particle per unit length of the medium it traverses. Values of LET are usually given in KeV per micrometre. The SI units are J/m, and 1 KeV/μm corresponds with 1.602×10^{-10} J/m. Some energy transfers may be large enough to result in the emission of electrons that can themselves act as ionizing particles, so it is necessary, in quoting a value for LET, to stipulate the maximum energy loss per event that will be taken into account in defining the quality of the radiation in a particular context. For example, it may be considered that energy losses of greater than 100 eV or 1.6×10^{-17} J will not be relevant to the purpose for which the description of quality is required. In that case, the relevant Linear Energy Transfer will be designated L_{100}. In general, the symbol L_{Δ} implies that only energy losses less than Δ units in value are being taken into account. For some purposes it is thought desirable to specify the LET in terms of total energy loss, and the symbol used is LET_{∞}. This may be regarded as representing the manner in which the ionizing particles are slowed down or 'stopped' in the medium, and L_{∞} is often now referred to as the 'stopping power' – i.e. the stopping power of the medium for the particle under consideration. If the quality is specified in terms of stopping power, the appropriate units involve the density of the absorbing material, ρ, formerly in cm^{-3}. The dimensions are energy per unit length/ρ, and stopping power has been expressed conventionally as MeV/unit length^2/density, or MeV cm^2 g^{-1}. The SI units are J m^2 kg^{-1}, and 1 MeV cm^2 g^{-1} corresponds with 1.602×10^{-11} J m^2 kg^{-1}. The older units will be used in the ensuing discussion.

Katz (1970) proposed an alternative method of specifying the quality of beams of ions, based on the conclusion that their effects on biological macromolecules as well as on physical detectors are attributable mainly to the δ-rays ejected from their tracks as they pass through matter. In his view, events in the track core are of much less importance than the distribution of the energy deposited by the δ-rays. Their total energy varies as $Z^{*2}\beta^{-2}$, where Z^* is the effective charge of an atomic nucleus (ion) of atomic number Z, and β its velocity relative to that of light. The energy density varies inversely with the square of the distance from the path of the moving particle, up to approximately the range r_m of the δ-rays of maximum energy. According to Katz, therefore, the response of macromolecules through which the ions pass should vary with $Z^{*2}\beta^{-2}$, provided their average radii are considerably less than r_m. Because the ions pick up electrons as they slow down, Z^* may be less than Z. For atomic nuclei up to $Z = 18$, calculations of $(Z^*)^2$ as a function of β are to be found in Butts & Katz (1967).

On the assumption that fast-moving ions exert their effects through the energy deposited by δ-rays, rather than in the track core, Butts & Katz (1967) made predictions about the inactivation of small macromolecules (enzymes and viruses) by beams of ions, and these agreed well with experimental results. The use of the quantity $Z^{*2}\beta^{-2}$ in analysing effects on cells cannot be justified on the grounds of the target dimensions, since these are not known. However, as shown below, some apparently discrepant results of irradiation with ions become the less so when $Z^{*2}\beta^{-2}$ rather than LET_{∞} is used as the quality-specifying variable.

Electromagnetic radiation, neutrons and π-mesons cannot meaningfully be described as having one specific quality, however this is specified, because of the variety of energies of the electrons or other charged particles that are together responsible for energy transfers to the medium. For such radiations, LET spectra were calculated by several authors. An example is shown in fig. 9.3. The LET for radiations other than monoenergetic beams of atomic nuclei has often been quoted as an 'average LET', usually calculated either as 'track-average' or 'absorbed dose-average' LET (alternatively, 'energy average').

These averages are calculated by considering a theoretical division of LET into small intervals, so that the contributions pertaining to each

Fig. 9.3. Dose distribution for the irradiation of water by neutrons from different sources (after Edwards & Dennis, 1975).

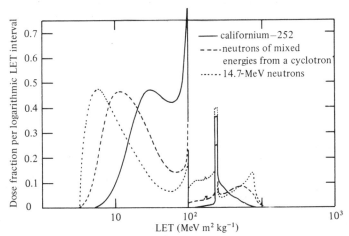

interval can be summed, and an average taken. For any ionizing particle traversing the medium, the LET will vary along its track, and an average track LET for that particle may be calculated by estimating the proportion of the total track length that is within each LET interval. Thus equal statistical weight is assigned to each unit of track length. To calculate the track average LET as a whole, the contributions of all the ionizing particles must be summed and averaged.

The absorbed dose-average LET takes into account the energy deposited by an ionizing particle in the medium it traverses. The average LET is calculated by estimating the proportion of the energy that is deposited within each LET interval, so equal statistical weight is assigned to each unit of energy deposited. The absorbed dose-average LET as a whole is again calculated by summing and averaging the contributions from all the ionizing particles.

The two methods can yield very different values of 'average LET', the differences depending on the kind of radiation, and also on the cut-off energy chosen for specifying LET. Table 9.1 shows some calculated averages (International Commission on Radiation Units, Report No. 16, 1970). The table explains why different authors assign different specific values to the 'LET' of sparsely ionizing radiation like ^{60}Co γ-rays. The value quoted will depend on the choice of track-average or absorbed dose-average LET, and on the choice of cut-off energy.

The use of one or other 'average LET' is particularly unsatisfactory as a means of specifying the quality of a beam of fast neutrons. The knock-on protons contribute a major fraction of total track length, but

Table 9.1. *Track-average and absorbed dose-average values of LET in water irradiated with various radiations*

Radiation	Cut-off energy Δ (eV)	$\bar{L}_{\Delta(track)}$ (keV/μm)	$\bar{L}_{\Delta(dose)}$ (keV/μm)
^{60}Co γ-rays	Unrestricted	0.24	0.31
	10 000	0.23	0.48
	1 000	0.23	2.8
	100	0.23	6.9
22-MV X-rays	100	0.19	6.0
2-MeV electrons (whole track)	100	0.20	6.1
200-kV X-rays	100	1.7	9.4
^{3}H β-rays	100	4.7	11.5
50-kV X-rays	100	6.3	13.1
5.3-MeV α-rays (whole track)	100	43	63

the much more densely ionizing α-particles and heavy recoils contribute proportionately much more to the energy deposited. As shown below, there is considerable variability in the response of cells as a function of L_∞, even among closely related mammalian cell lines. It follows that the overall response to neutron irradiation will depend on the relative importance, in cell killing, of the different products of neutron interactions, so no method of averaging quality will have useful predictive value.

Microdosimetry

The unsatisfactory features of 'average LET' as a specification of quality have led to its replacement by a presentation, usually in graphical form, of the distribution of a physical property of the radiation in question: for example, fig. 9.3, showing calculated dose as a function of stopping power of a medium of unit density. Graphs are now quite commonly constructed from actual measurements of the distribution of the sizes of energy deposits in small volumes. The techniques of 'microdosimetry' have been developed in the context of biological effects of radiation, so the volumes within which the measurements are made are designed to simulate small volumes of tissue, of the order of a micrometre in diameter. The most frequently used measuring instrument is a specific type of ionization chamber known as a proportional counter, the important feature of which is that the response is proportional to the quantity of energy deposited in a single event. By this means a record is made of the occurrence of single events which occur at random (stochastically); whereas methods for measuring absorbed dose yield averages for all events occurring within a given mass.

The counters customarily used in measuring event size distributions are constructed with walls of the same atomic composition as biological tissue, and the volumes contain tissue-equivalent gas. The small volume of tissue is simulated by having the gas at as low a pressure (and therefore density) as is practicable, so that the number of collisions an ionizing particle makes in its passage over a distance of, say, 100 nm would be about the same as if it were traversing about 1 μm of tissue.

Linear Energy Transfer, defined as the energy transferred to the medium by an ionizing particle per unit length of its path, takes no account of the stochastic nature of the energy-deposition events. Microdosimetric measurements, however, are customarily made in terms of a quantity known as 'lineal energy', for which the symbol y is used. This is defined as 'the quotient ϵ by $\bar{d}$, where ϵ is the energy imparted to the matter in a volume during an energy deposition event and $\bar{d}$ is the mean chord length in the volume of interest' (International Commission on Radiation Units, Report 19, 1971). For any given type of radiation the sizes of the events will be distributed according to their probability, the function of which is denoted by $F(y)$ or $D(y)$. Results may be plotted with $y \times D(y)$ as ordinate, and y as abscissa, which gives a visual presentation of distribution of event size. Some authors plot $y \times yD(y)$, i.e. $y^2D(y)$ on the ordinate scale, which then presents the distribution of energy contributions per interval of y. Figure 9.4 is an example of such methods of visual presentation.

Radiolysis products of water as a function of LET_∞

Brief mention was made in previous chapters of hypotheses concerning the involvement of primary radiolysis products of water in the killing of cells, radiation effects only at 'low LET' having then been under discussion. To consider the validity of such hypotheses in the light of variation in biological effects with radiation quality, it is neces-

Fig. 9.4. To illustrate different distributions of $y D(y)$ as a function of y in KeV/μm (after Kellerer & Rossi, 1972).

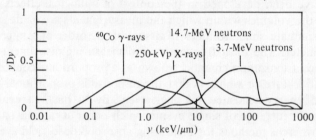

Fig. 9.5. Yields of products in the radiolysis of water by radiations of different LET (adapted from Kuppermann, 1974).

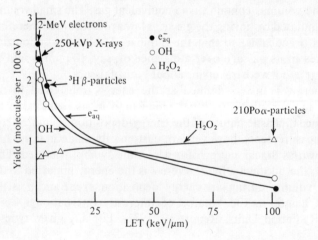

sary to take into account the variation in the yields of primary products as the density of ionization increases. Figure 9.5 has been adapted from Kuppermann (1974) to show experimental results for the variation with LET_∞ of the primary yields of e_{aq}^-, OH and H_2O_2 as measured in pure (de-oxygenated) water. The ordinate is in 'G-values', these being the units commonly employed in radiation chemistry: the G-value is the number of molecules or radicals formed or destroyed per 100 eV absorbed. The yields of e_{aq}^- and OH decrease rapidly at first with increasing LET_∞, then more slowly; the yield of H_2O_2 rises slightly as LET_∞ increases.

Relative biological effectiveness of monoenergetic charged particles: 'track segment method'

Comparisons between effects of radiation of different quality are often stated in terms of their Relative Biological Effectiveness or RBE, and this can, of course, be measured even when no single number can justifiably be used to express radiation quality. RBE is conventionally defined as the ratio of doses required to give the same biological effect. Some term is clearly required to denote a closer spacing of energy-loss events with one type of radiation than another. In older terminology, the phrase 'greater density of ionization' was used; according to the conventions used in microdosimetric contexts, radiations of different quality are compared in respect of greater or lesser tendencies towards high values of y. Since the comparative phrases 'higher LET' or 'lower LET' are generally understood in radiobiological literature in that context, I shall keep to the term 'LET', the quotation marks being used when the term is to be considered as qualitative.

As we have seen, quality can with reasonable validity be specified by one of the variables LET_∞ or $Z^{*2}\beta^{-2}$ when monoenergetic beams of atomic nuclei are used in the irradiation of very thin samples. If the RBE is constant, that is, independent of the level of effect (or dose), it may be represented without constraint as a function of the variable chosen to signify quality. Biological macromolecules irradiated extracellularly are almost always inactivated exponentially (Chapter 4), so RBE values are synonymous with ratios of values of D_0 (the dose required to give an average of one inactivating event per macromolecule). As was shown by fig. 4.4 there is a decrease in RBE values for loss of reproductive or biochemical functions as $Z^{*2}\beta^{-2}$ increases, when 'model targets' are irradiated extracellularly. This conforms with the prediction of target theory for single-hit detectors (Chapter 4).

In contrast, RBE values for cell killing almost always increase with increasing 'LET', until a maximum value is attained. It was formerly believed that vegetative bacteria were to be excluded from that general statement; RBE values were quoted (and sometimes still are) as constant over a range of values L_∞, then decreasing (e.g. International Commission on Radiation Units, Report 16, 1970). But with *Shigella flexneri* and *Escherichia coli* B/r we found that the effectiveness of charged particle irradiation increased to a maximum at $Z^{*2}\beta^{-2}$ approximately 240, then decreased (Alper, Moore & Bewley, 1967a; Moore, unpublished fig. 9.6). The extent to which effectiveness increased, and the value of LET (or $Z^{*2}\beta^{-2}$) at which the maximum occurred, are well matched in results of Munson *et al.* (1967) for strains of *E. coli* that are known to be proficient in repairing damage to DNA inflicted by ultraviolet light. With repair-deficient strains, in contrast, those authors found that there was no increase in effectiveness with increasing LET_∞. This difference in response to change in radiation quality between repair-proficient and repair-deficient strains, noted also when fast neutrons are used as source of 'high LET' radiation (Alper & Moore, 1967; Alper, 1971) has a bearing on the mechanisms of radiation action (Chapter 15).

The value of $Z^{*2}\beta^{-2}$ at which effectiveness is maximal for repair-proficient bacteria is considerably lower than had previously been observed for yeast, mammalian cells or bacterial spores, so the increase in effectiveness for repair-proficient vegetative bacteria with increasing $Z^{*2}\beta^{-2}$ may have been missed by other investigators (e.g. Brustad, 1961). In the killing of dry bacterial spores and haploid yeast the values of

Fig. 9.6. The effectiveness of deuterons and helium ions in killing *Shigella flexneri*, compared with that of 7-MeV electrons (relative effectiveness = ratio of D_0 value with electrons to that with the ion beam). Vertical and horizontal bars show 95% confidence intervals. (After Alper, Moore & Bewley, 1967a).

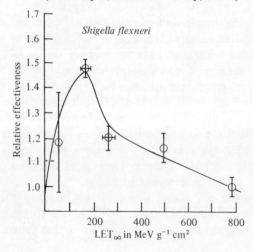

$Z^{*2}\beta^{-2}$ for the most effective radiations were respectively about 2000 (Powers, 1965) and 1300 (Manney, Brustad & Tobias, 1963).

In all experiments quoted above, except for those of Munson *et al.*, values of RBE could be plotted as single-valued functions of stopping power, because survival curves were all exponential, or were of the same shape for all qualities of radiation. Thus, for any two qualities, the ratio of doses to give the same effect was independent of the level of effect. In contrast, an alga (*Chlamydomonas reinhardii*) and some lines of mammalian cells that have been exposed to monoenergetic beams of ions are characterized by shouldered survival curves that change markedly in shape as the quality of the radiation is changed (e.g. fig. 9.7 from the work of Deering & Rice, 1962).

In most cases cell survival curves with radiations of different qualities have been described by the authors in terms of equation 5.4, and have therefore been defined by the values of n (the extrapolation number) and D_0 (the dose to give an average of one lethal event per cell, in the terminal exponential region of the survival curve). Where n is not constant, it decreases with increasing 'LET'; values of D_0 decrease to a minimum, then increase.

Results of experiments with radiations of different quality are often

Fig. 9.7. Survival curves for HeLa cells irradiated by X-rays (dotted lines), 105-MeV carbon ions (●) and 130-MeV oxygen ions (○). The RBE of the carbon ions, compared with X-rays, decreases with increasing dose. (Adapted from Deering & Rice, 1962).

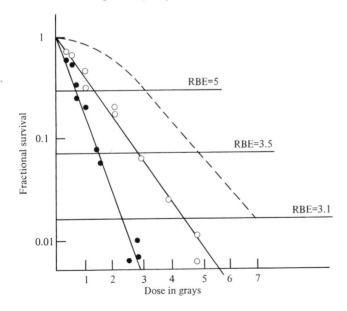

shown with RBE plotted as a function of LET_∞. If the shapes of the survival curves vary, RBE varies with the level of effect at which it is measured, so measurements of RBE must be made at some arbitrary level (fig. 9.7). For the purpose of gaining insight into the mechanism of radiation action there is no particular merit in choosing this method of analysing results: it is preferable to compare the effectiveness of radiations of different quality in terms of a parameter with radiobiological significance, like the value of D_0. When survival curves are defined by the parameters D_0 and n, with n decreasing as LET_∞ increases, ratios of values of D_0 give limiting high dose values of the RBE as commonly defined.

Most authors have given results in terms of D_0, or an approximation thereto. However, Cox *et al.* (1977b) used the parameters of equation 5.12 to describe their survival curves for Chinese hamster cells taken at different values of LET_∞; all survival curves for cultured human fibroblasts were exponential.

Of several sets of experiments with mammalian cells exposed to monoenergetic beams of ions, two were on precisely the same cell line (kidney cells of human origin, named T1), irradiated in roughly similar conditions (Barendsen *et al.*, 1966; Todd, 1964). The main difference lay in the use by Todd of beams of charged particles for which the energy per atomic mass unit was constant, so that the velocity was constant; whereas Barendsen *et al.* used beams of deuterons and helium ions, achieving changes in quality by slowing these down so that the same particles had different kinetic energies in the plane of the cells. Todd

Fig. 9.8. Variation in effectiveness of radiation in killing T1 (human kidney) cells, as a function of LET.

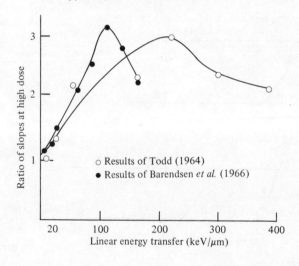

o Results of Todd (1964)

● Results of Barendsen *et al.* (1966)

Ratio of slopes at high dose

Linear energy transfer (keV/μm)

reported his results on the basis that equation 5.8 was applicable to the survival curves, and gave values of D_0, $_1D_0$ and n for each survival curve. Barendsen *et al.* did not regard that equation as a good description of their survival curves, but reported parameters analogous with D_0 and $_1D_0$, viz. values of 'D_{37} in the high dose region' and 'D_{37} in the low dose region'. Figures 9.8 and 9.9 show the two sets of results together presented with D_0 ratios and 'D_{37}, high dose' ratios as ordinate and value of LET_∞ as abscissa in one case and $Z^{*2}\beta^{-2}$ in the other. Although it is conventional to plot stopping power (or $Z^{*2}\beta^{-2}$) on a logarithmic scale, there is no rationale for this, other than convenience when a very large range of LET_∞ is used. A logarithmic plot that includes results at 'low LET' must necessarily have the scale compressed in a region of great interest, namely that at which effectiveness is maximal; so linear scales for LET_∞ and $Z^{*2}\beta^{-2}$ have been used here.

Figures 9.8 and 9.9 show that the LET_∞ for peak effectiveness (in terms of D_0 ratios or their equivalent) was about twice as high in the experiments of Todd as in those of Barendsen *et al.*; whereas the use of $Z^{*2}\beta^{-2}$ as quality-specifying variable brings the results closer together, though there is still a difference of some 50% in the values of $Z^{*2}\beta^{-2}$ at which maximum effectiveness was observed. It may be that the better agreement between the two sets of results, when $Z^{*2}\beta^{-2}$ is used, is a justification of the view that even in cells the effectiveness of heavy ion bombardment depends more on the distribution of energy deposition by secondary electrons than on track-core events (cf. Dennis, 1977).

In models set up to account for the influence of radiation quality, importance may attach to the LET_∞ for maximum effectiveness as well as to the magnitude of the relevant RBE. The results of several sets of observations on cell killing are therefore summarized in table 9.2, which

Fig. 9.9. As fig. 9.8, with $Z^{*2}\beta^{-2}$ as abscissa.

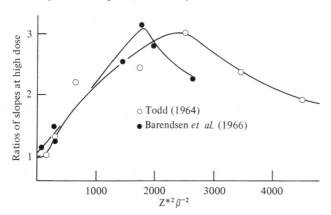

Table 9.2. *Results from irradiation of asynchronous populations of several mammalian cell lines by beams of ions*

Cell line	n	Photon value D_0 (Gy)	Minimum D_0 (Gy)	Maximum ratio, values of D_0	$LET_\infty(L)$ (MeV cm² g⁻¹) for $n \to 1$	$LET_\infty(L)$ (MeV cm² g⁻¹) for minimum value, D_0	Reference
HeLa	4	1.3	0.55	2.4	1890 > L > 430	1900	Deering & Rice (1962)
Human kidney, T1	6	2.0	0.66	3.0	2200 > L > 1650	2200	Todd (1964)
Human kidney, T1	(a)	1.8 (b)	0.57	3.2 (b)	610 > L > 260 (c)	1100 (c)	Barendsen et al. (1966)
Chinese hamster, CH2B₂	5	1.7	1.00	1.7	3510 > L > 1890 (d)	2000	Skarsgard et al. (1967)
Chinese hamster, ovary	no data		0.48		6560 > L > 1910	1233	Bird & Burki (1975)
Chinese hamster, V79	(e)	(e)	(e)	3.5 (f), 3.8 (f)	> 2000 (c)	900 (c), 1600 (c)	Cox et al. (1977)
Human skin fibroblasts	1	1.26	0.32	4.0	all	900 (c)	

Notes:
(a) No value of *n* fitted: survival curve continuously bending.
(b) Parameter used in author's 'D_{37}, high dose'.
(c) Quality specified in keV μm⁻¹. Value multiplied by ten to give MeV cm² g⁻¹.
(d) $n = 1$ for asynchronous population; but survival curves for synchronized cells in early and late S were shouldered (Skarsgard, 1974).
(e) Parameters fitted by authors to equation $\ln f = -\alpha D - \beta D^2$.
(f) Calculated for $\ln f = -5$ from fitted parameters. Two maximum values of RBE at the 10% survival level were given.

includes values of extrapolation numbers when these have been given by the authors.

In the contexts of radiotherapy and radiobiological protection, cell-survival curve parameters are most relevant in respect of their initial regions (Chapters 14, 17), and these are often characterized by a non-zero slope, even for sparsely ionizing radiation (see: T. Alper, ed: *Cell Survival after Low Doses of Radiation*, 1975). The use of radiations of 'high LET' is becoming increasingly prominent in radiotherapy, so precise information on changes in initial slopes with quality is important. Of the work done with monoenergetic beams of ions, Todd (1964) and Barendson *et al.* (1966) have given such information; the T1 cell line was used in both sets of experiments. Their results have been plotted in fig. 9.10 with 'D_1' (Todd) or 'D_{37}, low dose' (Barendsen *et al.*) as dependent variable. The latter authors did not give estimates of their errors. Vertical bars are used to show 95% confidence intervals on Todd's data, calculated by taking twice his estimates of standard errors. If it is assumed that the confidence intervals for the determinations of Barendsen *et al.* are of the same order of magnitude, the overall results must be regarded as substantially the same. Results of Cox *et al.* (1977b) on

Fig. 9.10. The increase in initial slope (decrease in $_1D_0$) with increasing $Z^{*2} \beta^{-2}$. Vertical bars are 95% confidence limits on Todd's estimates, for which he gave standard errors.

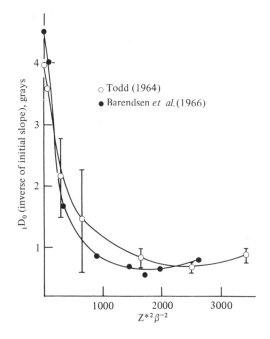

Chinese hamster V79 cells are similarly plotted in fig. 9.11, but with LET_∞ as abscissa. Information necessary for conversion to $Z^{*2}\beta^{-2}$ was not given, but an approximate conversion suggests that the initial slope increased more sharply in the case of the V79 cells.

Biological damage by fast neutrons

Beams of fast neutrons have been extensively used for investigations on test systems *in vitro* as well as *in vivo*. Some beams are monoenergetic, others comprise neutrons of a spectrum of energies, as for example when high-energy deuterons impinge on a beryllium target. In either case, the ionizing particles engendered by the neutron beam are of mixed quality, as illustrated by fig. 9.3; so the effectiveness will be a weighted average of the effectiveness of all the particles that deposit energy within the cells under test. Contributions to the damage may be regarded broadly as deriving from the effects of knock-on protons on the one hand, and α-particles and heavy recoil atomic nuclei on the other. With thin samples, the relative contributions of the two groups of particles can be estimated from experiments in which the cells are exposed directly to the neutron beam as nearly as possible without intervening material. The knock-on protons arising within the cells will, in general, move forward and will have sufficiently long ranges to deposit very little energy within the cells. In these conditions, 'without build-up', most of the biological effect of the neutron beam is attributable to the densely ionizing α-particles and recoil nuclei. If now the cells

Fig. 9.11. The increase in initial slope (increase in coefficient α) with increasing LET for asynchronous populations of Chinese hamster V79 cells (After Cox *et al.*, 1977).

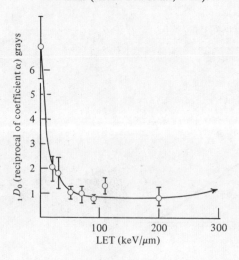

are exposed to the neutron beam with sufficient intervening tissue-like material to give proton equilibrium within them, the results are attributable to both classes of particles combined, so the contribution from the knock-on protons can be calculated. Unfortunately, the results are complicated to some extent by the unavoidable accompaniment of γ-radiation whenever a neutron beam is used.

The technique is exemplified by experiments of Broerse, Barendsen & van Kersen (1968) and of Bewley, McNally & Page (1974). The former authors used monoenergetic neutrons from a generator accelerating deuterons onto a tritium target, yielding neutrons of energy 14 MeV, the latter the neutron beam from the Medical Research Council's medical cyclotron. That beam consists of neutrons of various energies, with a mean of 7.5 MeV. Different mammalian cells were used. Values for knock-on protons of o.e.r., and of RBE at arbitrary levels of survival, were calculated from measurements with X-rays and with neutrons used without build-up. Values of o.e.r. are given in Chapter 10.

Qualitatively the results in the two sets of experiments were similar. The relative effectiveness of the densely ionizing particles was greater than the overall effectiveness of the neutron beam. Values of RBE of the α-particles and heavy recoils, compared with X-rays, were found by Broerse *et al.* and Bewley *et al.* respectively to be 3.1 and 6.9. It was thought possible that the difference in that respect, in the two sets of results, might have been attributable to knock-on protons, in the 'no build-up' conditions used by Broerse *et al.* On the other hand, as shown in the previous section, there is considerable variability in the mode of change of response of cells of different lines with change in LET_∞, so there is no reason to expect the same parameters for different cell lines in experiments such as these.

Indeed, the many comparisons of effects of neutrons with those of X- or γ-rays display no consistent pattern of change in survival curve parameters, even with a similar group like mammalian cells. While with these there appears to be no exception to the rule that neutrons are the more effective form of radiation, the increased RBE is, in some instances, due almost entirely to a reduction in extrapolation number, as, for instance, in the survival curves for stem cells in the crypts of Lieberkühn (Withers, Brennan & Elkind, 1970; fig. 9.12). The values of D_0 were the same, within experimental error, so the RBE would approach unity at very high dose. On the other hand, at least two sets of results have shown neutron survival curves with extrapolation numbers equal to those found with X-rays (Hornsey & Silini, 1961; Nias & Gilbert, 1975). The equality of the extrapolation numbers was subsequently confirmed by Hornsey & Silini (1962) by means of observations on post-irradiation recovery (Chapter 13). Thus while smaller extrapolation numbers are commonly observed in neutron survival curves for eukaryotic cells, that is not an invariable rule. In contrast, survival curves for bacteria more commonly retain the same shape when radiations of different quality are used. Strains of enterobacteria often yield exponential survival curves when they have been grown in nutrient broth, but shouldered ones when glucose has been added. Shekhtman, Plokhoi & Filippova (1958), using *Escherichia coli* communis, exposed cells grown with and without glucose to α-rays as well as X-rays. Extrapolation numbers of ten were observed with the glucose-grown cells, for both kinds of radiation, whereas those grown without glucose survived exponentially. In analogous experiments we grew *E. coli* B/r and *Shigella flexneri* in nutrient broth and with added

Fig. 9.12. Single-dose survival curves for the jejunal cells of intestinal mucosa of mice exposed to 200-kVp X-rays or 14-MeV neutrons. Vertical bars show 95% confidence intervals (after Withers, Brennan & Elkind, 1970).

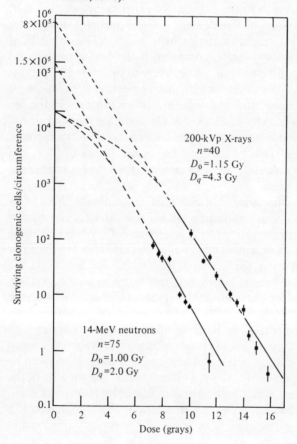

Fig. 9.13. To show equality of extrapolation numbers, survival curves for *S. flexneri* grown before irradiation in nutrient medium with added glucose.

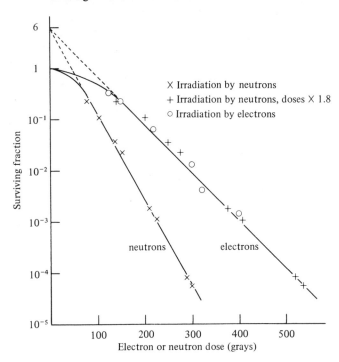

Table 9.3. *Survival curve parameters and ratios of values of D_0 for two bacterial strains grown with and without glucose before irradiation*

Parameter	Escherichia coli B/r		Shigella flexneri Y6R	
	without glucose	with glucose	without glucose	with glucose
D_0 electrons, grays	83.2	113	50.4	46.6
95% confidence limits	76.4–91.4	108–118	0–54.2	43.4–50.3
D_0 neutrons, grays	50.2	65.4	27.3	26.1
95% confidence limits	46.6–54.3	63.1–68.0	25.7–29.1	24.8–27.5
Common extrapolation no., electrons and neutrons	2.46	5.72	0.86	6.00
95% confidence limits	1.50–4.03	4.50–7.29	0.65–1.13	4.07–8.89
Ratio of D_0 values	1.66	1.73	1.85	1.79
95% confidence limits	1.55–1.77	1.68–1.78	1.73–1.97	1.71–1.88

glucose. Both strains yielded survival curves with large shoulders only for the glucose-grown cells, with extrapolation numbers equal after neutron and X-irradiation. RBEs of the neutrons were independent of the pre-irradiation growth medium (fig. 9.13; table 9.3).

As mentioned above, repair-deficient bacterial mutants show lesser increases than their wild-type parents in the effectiveness of 'high LET' radiations, the RBEs of neutrons for very sensitive strains being about unity (Alper & Moore, 1967; Alper, 1971; Redpath, 1978). Radiosensitive mutants of many other classes of cell have been selected, including yeasts, algae and slime moulds, but observations comparing effects of neutrons (or other 'high LET' radiation) and X-rays do not seem to have been made with these mutants and related to comparable observations with the wild-type parents. The nearest approach has been a comparison of the RBEs of neutrons for sub-lines of sensitive mutant of a mouse lymphoma line first selected by Alexander (1961). With two sub-lines of differing sensitivity to X-rays, Dr P. Coultas (personal communication) found that the RBEs were 1.3 and 2.2 respectively for the more and less sensitive sub-lines (see also Chapter 16, table 16.3).

If correlation between repair proficiency and the magnitude of RBE of 'high LET' radiation is manifested with cells of different classes, the phenomenon clearly has a bearing on the mechanisms of radiation-induced cell killing. It is to be hoped that more information on this point will become available.

10

Variation in sensitization and protection with radiation quality

A differential protective effect of hypoxia, depending on radiation quality, was first noted by Thoday & Read (1949), in their experiments on induction of chromosomal aberrations in bean roots by α- particles. These were emitted by radon gas dissolved in the water surrounding the bean roots and taken up with it. There was considerably less difference in the aberration yield when dissolved oxygen had been removed from the water than in comparable observations with X-rays. Read (1952) made analogous observations with inhibition of root growth as a test of damage, this being associated with the killing of proliferating cells in the meristem. Those first observations were soon followed by several in which fast neutrons were used as the source of 'high LET' radiation. Values of the o.e.r. were always lower, irrespective of the kinds of cells used, and again this was observed for induction of chromosome aberrations as well as for cell killing.

A few series of experiments have been done by the track segment method, described in Chapter 9 (p. 111), to examine in detail variation with radiation quality of the o.e.r. for cell killing. The several classes of cell used have been biochemically and radiobiologically very dissimilar. Manney, Brustad & Tobias (1963) exposed yeast to beams of ions from the Berkeley heavy-ion accelerator. Helium ions and deuterons from the Medical Research Council's cyclotron were used to irradiate oxygenated and anoxic mammalian cells (Barendsen *et al.*, 1966), *Shigella flexneri* (Alper, Moore & Bewley, 1967) and an alga, *Chlamydomonas reinhardii* (Bryant, 1973). In those three sets of experiments methods of exposing the cells and of achieving anoxia were all the same. As shown in fig. 10.1, the results are qualitatively similar, but they differ in detail. Also shown in the figure are the results of Moore (unpublished) on change in o.e.r. with change in quality for another test of damage to bacteria, inhibition of the ability to induce the enzyme galactosidase. The o.e.r. at 'low LET' for this type of damage is higher than for killing (Pauly, 1959; Moore, 1965; Forage, 1971), and, as the figure shows, this difference was

123

maintained with all qualities of radiation tested. Comparisons of Todd (1964) and Barendsen *et al.* (1966) of the effectiveness of radiations of different qualities in killing mammalian cells of the same line were discussed in Chapter 9. Discrepancies were shown to be reduced when $Z^{*2}\beta^{-2}$ rather than LET was used as quality-defining variable. Curtis (1970) showed that this method of analysing the results would partially resolve apparent discrepancies also in their measurements of o.e.r. with different qualities of radiation. Although differences between the results of Barendsen *et al.* and Todd on values of o.e.r. have been commented upon (e.g. Bewley, 1968), the confidence limits on both sets of measurements are wide enough to make it difficult to be sure that they differ significantly when $Z^{*2}\beta^{-2}$ is used to define the quality.

Reduction in the sensitizing action of oxygen with increasing density of ionization is clearly a general phenomenon. With enterobacteria, however, values of o.e.r. with X-rays of 250 or 300 kVp were significantly greater than with 7-MeV electrons. This was observed with bacteria irradiated while resting on a surface, but not when they were in suspension. Furthermore, the effect was seen only with repair-proficient bacteria (Forage & Alper, 1970; table 10.1). The results could not be accounted for by lack of 'build-up' when the bacteria were irradiated on surfaces (unpublished observations).

Fig. 10.1. Change in o.e.r. with quality of radiation, for experiments using beams of deuterons or helium ions from the Medical Research Council's cyclotron at Hammersmith Hospital. Bars represent 95% confidence intervals.

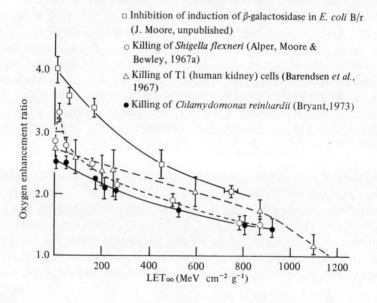

□ Inhibition of induction of β-galactosidase in *E. coli* B/r (J. Moore, unpublished)

○ Killing of *Shigella flexneri* (Alper, Moore & Bewley, 1967a)

△ Killing of T1 (human kidney) cells (Barendsen *et al.*, 1967)

● Killing of *Chlamydomonas reinhardii* (Bryant, 1973)

Table 10.1. *To show greater o.e.r. for more densely ionizing radiation, with repair-proficient but not deficient bacteria. Effect seen only with bacteria irradiated while exposed to ambient gas*

Bacterial strains	Oxygen enhancement ratios				
	bacteria in suspension		bacteria resting on surface		
	5-MeV electrons	250-kVp X-rays	6-7 MeV electrons	250-kVp X-rays	~300-keV electrons + X-rays (a)
Repair-proficient					
S. flexneri	2.91 ± 0.06		2.59 ± 0.10	3.41 ± 0.07	3.55 ± 0.13
E. coli B–H		3.18 ± 0.07	3.04 ± 0.08	4.05 ± 0.15	3.85 ± 0.05
K12 AB1157			3.13 ± 0.07		4.51 ± 0.18
B$_{s-12}$ (resistant component)			2.0		3.5
Repair-deficient					
E. coli B$_{s-1}$		1.70 ± 0.03	1.47 ± 0.06	1.55 ± 0.07	1.59 ± 0.09
B$_{s-12}$ (sensitive component)		1.11 ± 0.09	1.00 ± 0.06	1.00 ± 0.07	1.00 ± 0.06

After Forage & Alper (1970).
(a) A block of Perspex 28.5 mm thick was placed in the 7-MeV electron beam. The bacteria were therefore irradiated by electrons of residual energy about 300 keV plus X-rays engendered within the block.

An exception to the general phenomenon of reduction in o.e.r., with increasing density of ionization, occurred in experiments of Mortimer, Brustad & Cormack (1965). They exposed diploid yeast to monoenergetic beams of ions from the Berkeley heavy-ion linear accelerator and measured values of o.e.r. for the induction of mutations at three known loci, also for cell killing and for the induction of colour change in the colonies. Only with the last test did the o.e.r. approach one at 'high LET', and there was a suggestion that the o.e.r. increased at still higher values. Results of Mortimer *et al.* for cell killing and the induction of the three specific mutations are reproduced in table 10.2.

Oxygen enhancement ratios, fast neutrons

Many o.e.r.'s for fast-neutron irradiation have been compared with those for radiation of 'low LET', with test systems *in vivo* and *in vitro*. It was suggested that an 'effective' value of the LET for a neutron beam might be estimated from a knowledge of the LET_∞ which would give an equal value of the o.e.r. (Bewley, 1972) but when this is done the RBEs do not match. This is because the o.e.r. for neutrons depends on the contributions to damage from the different regions of the neutron 'LET spectrum'. As shown by the experiments of Broerse, Barendsen & van Kersen (1968), and Bewley, McNally & Page (1974), the α-particles and heavy recoils contribute more to the biological damage to mammalian cells than do the knock-on protons, so the o.e.r. is accordingly weighted towards the low values (near 1) observed for those densely ionizing

Table 10.2. *Oxygen enhancement ratios for the killing of yeast cells and induction of mutations at three loci by beams of ions*

		Oxygen enhancement ratios		
			Reversion of requirement for tryptophan	
Ions and LET MeV cm^2 g^{-1} (mean of range)	Lethal effect	Reversion of requirement for histidine	reverse mutation	dominant suppressor
D 45	1.9	2.25	2.22	2.71
He 185	1.6	2.11	2.57	2.39
Li 427	1.6	1.85	1.96	1.83
B 1280	1.6	1.75	2.22	1.93
C 2000	1.3	1.61	1.48	1.83
Ne 5920	1.2	2.35	1.72	1.63
X-rays, 50 kVp (unfiltered)	1.6	2.77	1.92	2.52

After Mortimer, Brustad & Cormack (1965).

particles (table 10.3). With bacteria, on the other hand, the RBE of those particles is rather low (fig. 9.6) so the o.e.r. for fast neutrons is determined to a greater extent by that which pertains to the knock-on protons. Figure 10.2 gives comparisons between oxygen enhancement ratios with fast neutrons and with 'low LET' radiation, for various strains of bacteria and several mammalian cell lines. The ratio of o.e.r.'s, neutrons and sparsely ionizing radiation, is always less for bacteria.

Chemical protection and the OH radical-scavenging hypothesis
Confusion between true chemical protection and induced hypoxia in some systems may well be responsible for the statement sometimes made that protective agents, like anoxia, decrease progressively in effectiveness as radiation becomes more densely ionizing. Since o.e.r's do decrease with increasing LET, any protection attributable to oxygen depletion will likewise become relatively less effective. Comparisons of the efficacy of protection with different qualities of radiation have not been nearly as extensive as measurements of oxygen enhancement ratios. As it happens, protection of well-oxygenated cells by glycerol has been studied in two track segment experiments, and in one experiment in which protection against 25-MeV helium ions and 250-kVp X-rays were compared. The results, normalized for comparison, are graphed in fig. 10.3. This shows that the protection factor may be constant over a wide range of LET_∞, or may even increase. Cysteine, likewise, has been found to protect cells as effectively against radiation of 'high' as 'low' LET (plant cells, Biebl, 1963; bacteria, Alper, Bewley & Fowler, 1962; fig. 10.4). As mentioned in Chapter 8, it has come to be quite widely accepted that OH radicals contribute to a major extent to the lethal effects of radiation, and that chemical protective agents act by specific scavenging of OH. Figure 9.5 showed how the yield of those radicals decreases with increasing LET_∞. Such measurements are of the radicals

Table 10.3. *Values of o.e.r. for photons and neutrons*

	Human kidney (T1) cells (a)	Rat fibrosarcoma cells (b)
Mean neutron energy	14 MeV	7.5 MeV
Measured o.e.r., 250-kVp X-rays.	2.5	2.25
Measured o.e.r., neutrons with build-up	1.4–1.6	1.43
Measured o.e.r., α-particles and heavy recoils (no build-up)	1.1	1.0
Calculated o.e.r., knock-on protons	2.1	2.13

(a) After Broerse *et al.* (1968).
(b) After Bewley *et al.* (1974).

diffusing out of the tracks of ionizing particles: and those would be the products which would be scavenged by chemical protective agents, on the hypothesis mentioned.

In fig. 10.5 results of Alper *et al.* (1967a) are related to the yields of OH radicals as summarized by Kuppermann (1974; fig. 9.5). Results plotted are for the killing of anoxic *Shigella flexneri* and protection afforded to aerobic and anoxic bacteria by glycerol at 1 mol dm^{-3}. Fortunately the problems of matching values of 'LET' adopted in the two reports are lessened by Kuppermann's having shown specific yields for irradiation by energetic electrons, 220-kVp X-rays and polonium α-particles, the latter being of about the same energy as the least energetic helium ions used by Alper *et al.*

Fig. 10.2. Comparisons of oxygen enhancement ratios for the killing of cells by fast neutrons and photons. Mammalian cells: (1) Rhabdomyosarcoma cells (Barendsen, 1968); (2) RIB5 rat tumour cells. (Bewley, McNally & Page 1974); (3) Mouse lymphocytic leukaemia (Berry, Bewley & Parnell, 1965); (4) T1 human kidney cells (Barendsen *et al.*, 1966); (5) Mouse ascites tumour assayed *in vitro* (Cullen, 1976); (6) Mouse ascites tumour assayed *in vivo* (Hornsey & Silini, 1961).
Bacteria: [7] *E. coli* B$_{s-1}$; [8] *E. coli* K 12
AB2463; [9] *E. coli* B;
[10] *S. flexneri*; [11] *E. coli* B-H
[12] *E. coli* K 12 AB1157;
[7] to [12] Alper & Moore (1967); [13] *E. coli* B$_{s-1}$; [14] *E. coli* B/r.
[13] and [14] Cramp & Bryant (1975).

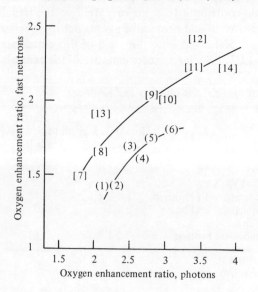

Fig. 10.3. To show variation with $Z^{*2} \beta^{-2}$ of pretection by glycerol. Results normalized to 'protection factor 100%', for maximum factor observed. Horizontal lines across the graph show the protection of *Shigella* against 7-MeV electrons (factor 2.51) and of haploid yeast against 50-kVp X-rays (factor 2.44).

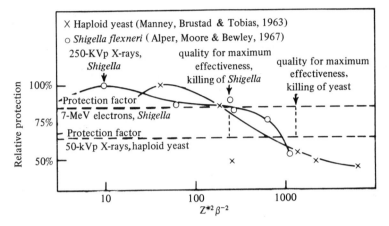

Fig. 10.4. Protection of *E. coli* B/r by cysteine against α-particles of energy 27 and 5.2 MeV (after Alper, Bewley & Fowler, 1962).

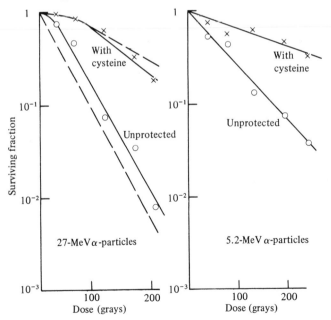

If we ignore the lesser protection of the bacteria against electrons than 250-kVp X-rays, some degree of correlation could be claimed between the yield of OH and protection by glycerol in the presence of oxygen; but there is none either for the protection or killing of anoxic cells. Reactions of OH radicals are, of course, unaffected by the presence of oxygen.

Chemical sensitization of hypoxic cells

Information available at the time of writing comes from a few experiments in which neutron beams were used as the source of radiation of 'high LET', plus investigations of Chapman *et al.* (1978) made with beams of heavy ions from the 'BEVELAC' heavy-ion accelerator at Berkeley. The latter have been the nearest approach to date to track segment experiments. Chapman *et al.* irradiated Chinese hamster V79 cells in suspension in the presence or absence of radiosensitizing drugs. To achieve radiation of 'high LET' with ion beams passing through those comparatively thick samples it was necessary to spread the Bragg peaks by the use of special filters, so no attempt was made to specify the quality of

Fig. 10.5. Effectiveness of the killing of *Shigella flexneri*, and of protection by glycerol at 1 mol dm^{-3} (results of Alper *et al.*, 1967a) related to yield of OH radicals (from summary of Kuppermann, 1974).

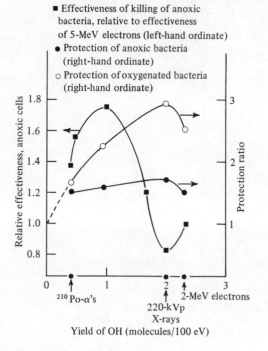

■ Effectiveness of killing of anoxic
   bacteria, relative to effectiveness
   of 5-MeV electrons (left-hand ordinate)
● Protection of anoxic bacteria
   (right-hand ordinate)
○ Protection of oxygenated bacteria
   (right-hand ordinate)

the radiations. The authors gave results obtained with four radiosensitiz-ing drugs. Those with metronidazole and misonidazole have been selected for presentation in fig. 106*a* and *b* because the same ones were used also in experiments with neutron beams. (Hall *et al.*, 1975; NcNally, 1976).

The rationale for the method chosen in plotting fig. 10.6. is the expecta-tion that the degree of sensitization should be smoothly correlated with sensitization by oxygen. Drug enhancement ratios have therefore been plotted as ordinate and oxygen enhancement ratios as abscissa. The simplest correlation that could be expected would be one defined by a straight line joining the origin (o.e.r. = 1, E.R. = 1) with the point for maximum o.e.r. and corresponding E.R. for the drugs concerned. The observations of the experimenters who used neutron beams do fall reasonably well on those lines. McNally's calculations included 95% confidence intervals for the ratios, so the fit is particularly satisfactory. In contrast, most of the data points from the results of Chapman *et al.* lie below the lines joining the origin and the point for the most sparsely ionizing radiation: that is to say, the enhancement ratios with radiations in the higher range of LET values were smaller, or the values of o.e.r. (S) were larger, than might have been expected on the basis of the simple correlation assumed by a straight-line relationship. There could be a real difference between the effects of sensitizers used with neutron beams and with radiations that give the same values of o.e.r. but have less complicated distributions of radiation quality. As noted above, values of o.e.r. with beams of fast neutrons are weighted downwards

Fig. 10.6. (*a*) Enhancement ratios for metronidazole compared with oxygen enhancement ratios for radiations of different quality. (*b*) As *a*, for misonidazole.

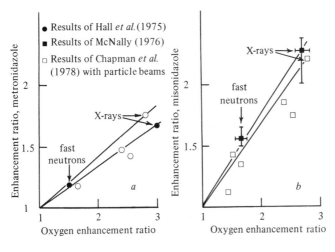

by the contributions from the α-particles and heavy recoils, and sensitization may occur only at the loci of events due to the less densely ionizing knock-on protons. However, the evidence presently available is too slender to enable a decision on this point to be made.

11

Models for variation in effectiveness with radiation quality

Sources of radiation of high LET were initially limited to preparations of naturally occurring radioactive substances that emitted α-particles, and, after the discovery of neutrons, to the rather weak sources provided by mixtures of radium and beryllium, neutrons being produced in the reactions between α-particles and beryllium nuclei. By the 1940s experiments in which these sources were used had shown that viruses, enzymes and bacteria were less effectively inactivated by densely than sparsely ionizing radiation; whereas aberrations in the chromosomes of higher cells were more effectively induced. Before the view became prevalent that radiobiological effects were mediated by the radiolysis products of water, results were interpreted rather in terms of ionizations directly in target structures. A decrease in what is now called the RBE of radiation of high LET was symptomatic of single-hit action; whereas an increase was interpreted by Lea & Catcheside (1942), for example, as a demonstration that several ionizations within the target were needed to bring about the effect under consideration – in that instance, a chromatid aberration. The books of Lea (1946) and Timoféeff-Ressovsky & Zimmer (1947) did not take into account the sensitizing effect of oxygen on cellular damage by X-rays, and their publication preceded the observation of Thoday & Read (1949) that the effect was considerably reduced when α-particles were the damaging agents. Thus there was no occasion before 1949 for the formulation of a hypothesis for LET effects that would accommodate also a reduction in o.e.r. with an increase in the density of ionization.

The apparent need for such accommodation arose at about the time when radiobiological thinking tended to be dominated by the concept that damage to cells was mediated by the radiolysis products of water. In its simplest form that hypothesis was in any case inadequate to account for increasing effectiveness on oxygenated cells, with increasing LET, since the yield of the primary decomposition products of water was known to decrease (p. 110, fig. 9.5). One suggestion was that the damaging action of formed hydrogen peroxide could account simultaneously for increased RBE and decreased o.e.r. values with radiation of 'high LET'. When water is exposed to radiation of low LET, hydrogen peroxide is formed, but it is concomitantly decomposed. An equilibrium is achieved which depends on the oxygen content of the water: it is negligibly low in completely deaerated water. With radiation of high LET, however, the equilibrium yield of H_2O_2 is much less dependent on oxygen content, presumably because of interactions not involving molecular oxygen. It was therefore envisaged that the damaging role of the active radicals formed in high yield at low LET was gradually taken over by the higher concentration of H_2O_2 formed in the tracks of more densely ionizing particles; and that o.e.r. was consequently reduced (see, e.g. Gray, 1954). But that model failed to find support in experimental evidence. For example, RBE values for the killing of cells were seen to increase within a range of LET values within which the yield of H_2O_2 in aerated water decreased (see Alper, 1956).

As was mentioned in Chapter 6, the late 1950s saw a strong tendency to return to the concept of direct action of radiation on cellular targets and in effect, therefore, to the idea of Lea & Catcheside (1942), that the manifestation of certain types of damage required the co-operation of several energy deposits within the target itself; in fact, those authors calculated that the breakage of a chromatid required the deposition of about 600 eV (about 20 ionizations). This calculation appeared to find support in experimental observations of Catcheside & Lea (1943). My own proposal to account for the reduced o.e.r. at high LET was that interaction between radicals formed close together would serve to fix damage, so that reaction with oxygen would become increasingly less important as the spacing between events became less. These concepts were the basis for calculations by Howard-Flanders (1958) of the number of events necessary to produce various radiobiological effects, from data on the dependence of effectiveness on LET_∞.

Surprisingly, that 'interacting-radicals' hypothesis, apparently able to account for reduction in o.e.r. with increasing LET, was never subjected to arithmetical test until long after Neary (1965) had challenged it. Such a test may be made if data are available on o.e.r.'s and ratios of slopes of survival curves for two qualities of radiation. As shown by Alper & Bryant (1974), the interacting-radicals hypothesis leads to the prediction that the ratio of slopes of survival curves cannot be greater than $(1 - m_x^{-1})/(1 - m_y^{-1})$, where m_x and m_y are the o.e.r.'s respectively at the lower and higher values of LET. Tables 11.1 and 11.2 demonstrate the failure of that prediction.

Probably because the quantitative implications have not been taken into account, it is to this day widely regarded as axiomatic that physico-chemical interaction between radicals formed close together, in a

Table 11.1. *Results of Barendsen et al. (1966) with T1 kidney cells, used to test the interacting-radicals' hypothesis. [Baseline 1.49 MeV deuterons: $D_0 = 1.60$ Gy, $m_x = 2.6$, $(1 - 1/m_x) = 0.615$]*

Radiation	o.e.r. (m_y)	$1 - 1/m_y$	D_0 value, grays (in high dose region)	Ratios of final slopes	
				Maximum theoretical value $\dfrac{1 - 1/m_x}{1 - 1/m_y}$	Observed value
Deuterons, 3.0 MeV	2.40	0.592	1.45	1.035	1.10
He ions 25 MeV	2.40	0.590	1.20	1.035	1.33
He ions 8.3 MeV	2.05	0.512	0.84	1.195	1.90
He ions 5.1 MeV	1.70	0.411	0.70	1.490	2.29

After Alper & Bryant (1976).

Table 11.2. *Results for MRC cyclotron neutrons and low LET radiation compared, as a test of the 'interacting-radicals' hypothesis*

Type of cell used	o.e.r., X-rays (m_x)	o.e.r., fast neutrons, MRC cyclotron (m_y)	Ratios of final slopes		Reference
			maximum theoretical value $\dfrac{1 - 1/m_y}{1 - 1/m_x}$	observed value	
Lymphocytic leukaemia	2.5	1.7	1.46	2.10	Berry et al. (1965)
T₁ human kidney	2.7	1.6	1.70	2.25	Barendsen et al. (1966)
Tetraploid Ehrlich ascites tumour	3.1	1.8	1.52	2.10	Hornsey & Silini (1961)
Mouse ascites tumour *in vitro*	2.8	1.8	1.45	2.20	McNally & Bewley (1969)
RIB5 rat tumour	2.25	1.43	1.84	2.09	Bewley et al. (1974)
Green alga: *Chlamydomonas*	2.21	1.50	1.65	2.08	Byrant (1973)

After Alper & Bryant (1974).

particle track, should concomitantly account for increasing biological effectiveness and decreasing o.e.r. with increasing density of ionization, and this assumption forms the basis of the models that are most often quoted (see Dennis, 1977, for a review). For example, Katz *et al.* (1971) distinguish between 'gamma kill' and 'ion kill', the latter being attributable to intratrack interactions. The assumption that intratrack and intertrack interactions differ qualitatively has been challenged only by Neary (1965); but that aspect of his 'theory of RBE' does not appear to have made as much impact as his deduction that there is substantial spatial separation between the loci of energy deposits that interact to cause a chromosomal aberration. Perhaps because of that deduction, similar to one relevant to cell killing, and based on microdosimetric considerations by Kellerer & Rossi (1972), a fundamental difference between Neary's model and others currently under discussion seems not to have been widely recognized.

Neary's 'Theory of RBE'

The basis for Neary's theory was the re-investigation by Neary, Savage & Evans (1964) of the effectiveness of monochromatic X-rays of very low energy in inducing chromatid aberrations in mature pollen of *Tradescantia*. They were repeating experiments of Catcheside & Lea (1943), the results of which had seemed to confirm the previous deduction of Lea & Catcheside (1942) mentioned above. An important aspect of that confirmation had been the apparent inefficiency of X-rays of energy 1.5 keV in producing chromatid aberrations in *Tradescantia* pollen. The photoelectrons, of maximum energy 1 keV, were apparently not energetic enough to deposit within a chromatid the 600 eV thought to be needed for the electron to break it as it passed through. But Neary *et al.* found that X-rays of that low energy were quite efficient at producing chromatid breaks, provided care was taken to avoid artefacts easily occasioned with radiation of such low penetrating power. The yield of aberrations produced by 3-keV X-rays (maximum electron energy 2.5 keV) was linearly related to the dose. Their conclusions referred mainly to mechanisms for the production of aberrations of different types, but the general 'theory of RBE' based by Neary (1965) on those conclusions was thought by him (Neary, 1967) to be applicable to the killing of cells by radiation.

Essentially, Neary departed from the traditional (and still current) view of dependence of effectiveness on LET in his conclusion that the primary events leading to observable biological effects of radiation (in the case he was discussing, chromosome aberrations) were due to single-energy depositions at high as well as at low LET. Production of chromosomal aberrations was due to 'interaction' of two chromosome regions in which these energy deposits has occurred, and the production of several energy deposits within a rather short region would increase the probability that the necessary interaction could occur. Neary did not specify what he meant by 'interaction', except in a negative way, that is, the events were presumed by him not to 'co-operate', but it is clear from his text that he had in mind interaction at a biochemical level, certainly post-metionically, for intratrack as well as intertrack interactions.

The 'Theory of dual radiation action'

This theory, discussed in full by Kellerer & Rossi (1972), rests basically on an analysis of how the RBE of neutrons compared with X-rays decreases with increasing neutron dose (see fig. 9.11). Using a variety of end-points, including mutation induction and effects on organized tissues (not all of them obviously correlated with cell death), they concluded that the relationship could be uniformly expressed as

$$RBE \propto (\text{neutron dose})^{-1/2}$$

within the limits of accuracy of the biological data, although some exceptions were later conceded by Rossi & Kellerer (1974). Since RBE is defined by D_x/D_n, the ratio of doses of X-rays and neutrons required to give the same effect, that relationship gives $D_x/D_n = \text{constant} \times D_n^{-0.5}$ i.e. $D_x = K \times D_n^{0.5}$ or $D_x^2 = K \times D_n$.

On the basis of evidence for single-hit killing by radiation of high LET, Keller & Rossi reasoned that inactivation of a cellular target should require the passage of only one densely ionizing particle, so that the 'high LET' components of neutron irradiation would inflict lethal lesions in direct proportion to the dose. From this it would follow that, with radiation of 'low LET', lesions accumulate only as the square of the dose, i.e. a single lesion requires energy deposition by two independent electrons within the target volume. The sub-lesions they inflict must interact if a lethal lesion is to result. In essence, that is the reasoning behind the derivation of equation 5.12, $\ln f = -(\alpha D + \beta D^2)$. In summary, Kellerer & Rossi regard the linear term for dose-dependence in equation 5.12 as relevant to intratrack events, whereas the dose-squared dependence relates to intertrack events.

The oxygen effect was accounted for on the assumption that the production of sub-lesions is reduced by a factor ρ when oxygen is absent. This is equivalent to supposing that $(\rho - 1)/\rho$ energy deposits then fail to result in sub-lesions: in fact, the symbol ρ used by Kellerer & Rossi is equivalent to m, the symbol for o.e.r. used throughout this text. They took the view that ρ should become smaller, as LET_∞ increases, until it approaches one. While the postulated decrease in ρ, i.e. in $(\rho - 1)/\rho$, was not explicitly attributed to physico-chemical interactions between the loci of energy deposits, the implication is no different from my own proposal (Alper, 1956). On the other hand, the sub-lesions inflicted by the passage of independent electrons were presumed to be capable of interaction post-metionically, since the phenomenon known as 'repair of sub-lethal damage' (Elkind & Sutton, 1960; Chapter 13) was supposed by Kellerer & Rossi to demonstrate that the sub-lesions they postulated could be separately repaired at times up to hours after they had been inflicted. Thus the theory embodies the concept that intertrack and intra-track interactions are qualitatively different.

Because the Neary and Kellerer–Rossi theories both involve the notion of interaction between two sub-lesions, and their estimates of maximum interaction distances do not grossly disagree, it is not uncommon to find the two theories being regarded as substantially not very different. That view ignores an essential part of Neary's theory, which is that intra-track, like intertrack, events interact only post-metionically. That was indeed a new concept in models for LET effects, but it does not seem to have made much impact, despite the lapse of time since Neary put it forward, and the continuing, indeed increasing, interest in the effects of change in radiation quality.

Neary recognized that his conclusion would invalidate the older explanation for decreasing o.e.r. with LET, since each radical formed within the track would have given rise to a fixed lesion, if oxygen were present, before interaction occurred. He therefore invoked an idea that had previously been mooted by several authors: that oxygen is formed by radiation chemical action in the tracks of densely ionizing particles through water. As a result, measured values of o.e.r. at high LET could not be based on the response of truly anoxic cells (see fig. 6.2): and the higher the LET, the greater the effective concentration of oxygen engendered in the neighbourhood of the site of energy deposition within the target, therefore the lower the measured o.e.r. would be.

The importance of the 'oxygen-in-the-track' hypothesis

This hypothesis for reduction in o.e.r. with increasing LET is of special importance because evidence of its validity would support Neary's unorthodox interpretation of RBE effects; whereas if it should prove to be untenable, his proposal that intra- and intertrack events interact in the same way (Neary, 1965) would be much less plausible. Thus a test of the oxygen-in-the-track hypothesis can be regarded to some extent as an in-direct test of Neary's main hypothesis.

How may the subsidiary hypothesis be tested? A test suggested by Neary himself has received the most attention, namely that the value of K (equation 6.1) should be greater at 'high' than at 'low' LET by just the concentration of oxygen formed in the tracks of the high LET particles.

Let the measured values of m and K at 'high' and 'low' LET be respectively m_y and m_x, K_y and K_x. It follows from equations 6.5 and 6.6 that

$$K_y = K_x + [O_2]' \tag{11.1}$$

$$m_y = \frac{m_x([O_2]' + K_x)}{m_x[O_2]' + K_x} \tag{11.2}$$

Rearranging equation 11.2 we have

$$[O_2]' = K_x \cdot \frac{m_x - m_y}{m_x(m_y - 1)} = K_y - K_x \tag{11.3}$$

Thus the test depends on establishing equality between $(K_y - K_x)/K_x$ and $(m_x - m_y)/m_x (m_y - 1)$. The error of the difference between K_y and K_x will depend on the errors in their measurement, which in their turn will depend on the magnitude of m_y and m_x. A large difference between m_y and m_x should give a large difference in K values, which is desirable; but if m_y is near to 1, the error in $(m_y - 1)$ will be large and the error in measuring K_y will also be large.

On the face of it, the difference usually observed between values of m for X-rays and neutrons would seem suitable for the test; and beams of fast neutrons are more readily available, and on the whole easier to use, than other forms of radiation of 'high LET'. Perhaps for these reasons the few measurements of K for radiations other than those of 'low LET' have been with neutrons, except for two by Alper, Moore & Smith (1967) for helium ions. Unfortunately the use of neutrons imposes such uncertainties as to invalidate the test. As described in Chapter 10, oxygen enhancement ratios with fast neutrons will be a weighted average of the ratios proper to the various ionizing particles they produce, the weight being determined by their effectiveness. For simplicity we may regard the ionizing particles as falling into the two categories, knock-on protons and α-particles plus heavy ion recoils. In determinations with mammalian cells, the o.e.r. pertaining to the latter was found to be about one (Broerse, Barendsen & van Kersen, 1968; Bewley, McNally & Page, 1974). With such cells, therefore, any change in overall sensitivity with P_{O_2}, that is to say, any determination of K_y, must pertain to the effect of oxygen on that part of the damage contributed by the knock-on protons. Thus the m_y and $[O_2]'$ of equation 11.3 should refer to the knock-on protons, while the overall value of m for neutrons will depend on contributions to the total dose of the various ionizing particles and on their RBEs. Equation 11.3 can therefore not be used to test the oxygen-in-the-track hypothesis if m_y is known only for the combined effect of the two classes of ionizing particle: i.e. it cannot be applied when neutrons are the source of radiation of 'high LET'.

The only determinations of K_y made to date with monoenergetic particles have been for *Shigella flexneri* irradiated by helium ions, values of
$Z^{*2}\beta^{-2}$ about 260 and 610. The values of K, with 95% confidence intervals,
were respectively equivalent to values of P_{O_2} (in the gas phase) of 0.105
(0.085–0.138), 0.165 (0.115–0.282) and 0.234 (0.145 to 0.554)% for fast
electrons and the helium ions. Only the confidence intervals for the lowest
and highest K values failed to overlap, the difference between the means
being 0.13% O_2. According to Alper & Bryant (1974), the P_{O_2} formed in
the tracks of particles of $Z^{*2}\beta^{-2} = 610$ should be about 0.07% (see below).
Allowing for the inevitably wide confidence intervals, the results could be
regarded as agreeing with the predictions of the oxygen-in-the-track
hypothesis; but they serve mainly to demonstrate that any test based on
measurements of K values with radiations of different quality is unlikely
to be definitive. Equation 11.3 suggests alternative tests of the hypothesis,
one of which is usable with a neutron beam as a source of 'high LET'
radiation. Whatever form of radiation is used, in a series of experiments
$[O_2]'$ will be constant. This yields a linear relationship between $1/m_x$ and

Fig. 11.1. Test of the relationship between oxygen enhancement
ratios with neutrons and photons, as predicted from the 'oxygen-in-
the-track' hypothesis. (After Alper & Moore, 1967).

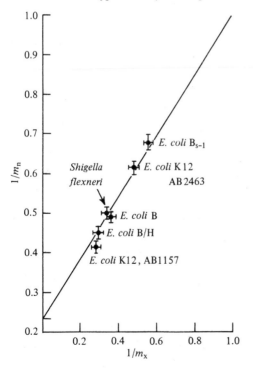

$1/m_y$, if those parameters can be varied. Alper & Moore (1967) used a genetically related group of *Escherichia coli* that differed in respect of their radiosensitivity and in the values of o.e.r. at 'low LET', but which they assumed would yield the same values of K_x (see Johansen, Gulbrandsen & Pettersen, 1974). The predicted relationship between $1/m_x$ and $1/m_y$ was observed (fig. 11.1).

Another test based on equation 11.3 is applicable if K_x is known for two (or more) types of cell and values of m_y are known for several different qualities of radiation. The calculated value of $[O_2]'$ for each quality should depend on the physical characteristics of the radiation and should therefore be the same for all types of cell; however, the cells might vary in respect of the parameters K_x, m_x and m_y. That test does involve measuring K_x, but only at 'low LET', with which the greatest accuracy is attainable. Alper & Bryant (1974) calculated values of $[O_2]'$ for monoenergetic particles of several values of LET_∞, using data on the radiobiologically dissimilar organisms *Chlamydomonas reinhardii* and *Shigella flexneri*, for both of which determinations had been made of K_x and of the values of m_y for the different ionizing particles. The calculated values of $[O_2]'$ with the various particles were consistent as between the two types of cell.

At the time those inferential tests were made there was evidence from previous radiation chemical studies that oxygen was produced when anoxic water was irradiated by densely ionizing particles; but none that the oxygen was produced specifically within the tracks of the particles, rather than in the bulk liquid. Baverstock & Burns (1976) made measurements of the oxygen produced in anoxic solutions of ferrous ammonium sulphate when it was irradiated by γ-rays and by heavy ions of various LET. They used increasing concentrations of Fe^{2+} ions, until the lowest limiting yield was reached: the Fe^{2+} ions served to scavenge all the radicals that would react to produce oxygen in the bulk of the liquid, so the limiting yields of O_2 were regarded as having been formed as primary products within the tracks. The yield measured by this means varied from zero, with ^{60}Co γ-irradiation, to 0.031 molecules per 100 eV absorbed with the most densely ionizing radiation used. They concluded that the yields they had measured were in reasonable agreement with the effective oxygen concentrations calculated by Alper & Bryant (1974), if they took into account the probable rates of diffusion of oxygen out of the tracks and also the times for which incipient lesions in cells remained available for reaction with oxygen, as measured by Michael *et al.* (1973).

The results of Baverstock & Burns provide encouragement for the oxygen-in-the-track hypothesis, but it must be borne in mind that their measurements were made in dilute aqueous solution. They may therefore not be quantitatively applicable for the effects of ionizing radiation on cells, since it is not even known whether the sensitizing action of oxygen in killing cells stems from its being dissolved in an aequeous phase.

If the oxygen-in-the-track hypothesis holds, reduction in o.e.r. with increasing LET should be seen as an artefact attributable to radiation chemical action and there should be no reason why sensitizing agents other than oxygen, and protective agents, should decrease in effectiveness as LET increases, provided they exert their effects by metionic reactions. The limited evidence available, discussed in Chapter 10, suggests that they do not, in fact, decrease in effectiveness until values of LET are so high that the RBE values decrease because of the more-than-necessary number of energy deposits per target (referred to often as 'overkill', or 'saturation').

In summary: inferential evidence supports the oxygen-in-the-track rather than the interacting-radicals hypothesis as an explanation of reduction in o.e.r. with increasing LET and to that extent it favours also Neary's concept of post-metionic intratrack interactions.

Inter- and intratrack interactions and the death of mammalian cells

Whereas Neary *et al.* (1964) had investigated the induction of chromosomal aberrations by monochromatic X-rays of very low energies, Cox, Thacker & Goodhead (1977a) used them to get information on how the killing of mamalian cells, and the induction of mutations in these cells, depends on radiation quality. They were led to question the magnitude of 'interaction distances' or 'site radii' as estimated in terms of the Kellerer & Rossi model; to that extent, their criticisms apply also to Neary's calculations, at least insofar as those might be considered relevant to the killing of eukaryotic cells.

The basis for the criticism was the observation of Cox *et al.* that aluminium K characteristic X-rays, of photon energy 1.5 keV, were more effective in killing Chinese hamster cells and also freshly explanted human fibroblasts *in vitro* than ^{60}Co γ-rays, and about as effective as helium ions of $LET_\infty = 20$ keV/μm. It was calculated by Goodhead & Thacker (1977) that the electrons produced by these photons had a combined maximum track length of 0.07 μm in biological tissue. Parameters for the survival of Chinese hamster cells irradiated by γ-rays were analysed in accordance with the model of Kellerer & Rossi, yielding a site diameter or interaction distance of about 0.4 μm, in good agreement with analyses in terms of that model of other results with the same cell line. According to the model, the effectiveness of 20-MeV helium ions is greater than that of γ-rays because of intratrack interactions as the ions traverse the site; but the tracks of the electrons produced by the low-energy X-rays could traverse only a fraction of a site of that diameter, so their effectiveness should not be expected to be very different from that of X-rays, and certainly not as great as that of the 20-MeV helium ions.

An additional consideration was the exponential nature of the survival curves for diploid human fibroblasts *in vitro*, with X-rays as with all radiations. According to the model that relates the D^2 term in equation 5.12 to intertrack interaction, the site radius for cells that are killed exponentially must approach zero; that is to say, the killing of the cells does not require interaction of the loci of energy deposition events, and effectiveness should therefore not increase with increasing LET. Cox *et al.* (1977a) and Goodhead (1977) posed the specific problem of reconciling the Kellerer & Rossi theory with their results on the human fibroblasts and in particular with their observations with soft X-rays; but although exponential survival curves for mammalian cells exposed to sparsely ionizing radiation are much less common than shouldered ones, they have been observed with several cell lines (table 5.1); furthermore, as in the results reported by Cox *et al.*, the RBE has been shown to increase with increasing LET (e.g. Silini & Maruyama, 1965; Caldwell, Lamerton & Bewley, 1965). Shouldered survival curves at 'low LET' can therefore not be regarded as a basic requirement for increasing RBE. These facts are difficult to reconcile, not only with the Kellerer & Rossi model, but with any generalization about shouldered survival curves that invokes solely physical aspects of track structure to account for them.

Conclusions

As far as the evidence goes, it supports Neary's view that inter- and intratrack interactions are not qualitatively different, if they occur. It may be necessary to invoke such interactions to account for 'two-hit' chromosomal aberrations, but the evidence provided by exponential survival curves observed with some mammalian cell lines calls into question the necessity for invoking inter- or intratrack interactions at all, in the physical sense, when the killing of mammalian cells is under consideration.

A decrease in extrapolation number with increasing LET_∞ has been a focal point in some model building, but models devised to interpret that phenomenon are somewhat exclusive, in terms of basic radiobiological mechanisms. With bacteria, survival curves are often exponential at all values of LET_∞, or, if they are shouldered, extrapolation numbers remain constant over a wide range of LET_∞. Models focussing on changes in survival curve shape would have to exclude also a whole class of eukaryotic cells, namely yeasts. Survival curves for haploid yeast were exponential at all values of LET_∞ tested, up to and beyond that at which the RBE had its maximum value (Manney, Brusted & Tobias, 1963); and those pertaining to diploid yeast were shouldered, with the value of n

unchanged up to values of LET_∞ of 4900 MeV cm g^{-1}, the most densely ionizing radiation used by Lyman & Haynes (1967).

Interaction between loci of separate energy-deposition events, at least with sparsely ionizing radiations, is implied in the derivation of most of the equations set out in Chapter 5; but the repair model (equation 5.13) accounts for shoulders to survival curves on quite different assumptions. However, models in that category have not been well explored in relation to the phenomena of increasing effectiveness and, usually, reduction in extrapolation number, as the density of ionization increases. This topic is discussed more fully in Chapter 14.

12

Variation in radiation response with growth phase or cell cycle stage

The phase of growth

Effects of radiation on cells are subject to modification not only by agents present during irradiation, but also by biochemical processes occurring over a much longer time span. There is a considerable body of evidence that some metionically fixed lesions may be repaired or by-passed, and several mechanisms may operate.

Profound changes in radiation response may attend the progress of cells through different phases and stages of growth. These may be associated with changes in the intracellular environment, i.e. with an increase or decrease in the probability of certain metionic reactions, and/or with changes in the nature or concentration of cellular constituents that contribute to repair or fixation of damage by a biochemical pathway. Variations in radiation response with phase and stage of growth have been observed with prokaryotic as well as eukaryotic cells; but there is no direct information as to whether the changes are in the progress of events at the physico-chemical or biochemical levels. In a few cases inferences may be drawn from experiments in which additional modifying treatments have been applied.

Free-living cells cultured *in vitro* go through well-defined phases of growth. When first introduced into a suitable nutritional environment, cells that are capable of proliferating will start to synthesize the components required for division: this period, the length of which depends on many factors, is commonly called the lag phase. Once the first division has occurred, most of the daughter cells will proceed to the next division, the average time between successive divisions usually being fairly constant. The number of cells in the population therefore increases geometrically or exponentially, so this is known as the logarithmic or exponential phase. When the population has reached a certain size, the environment will impose restrictions on further proliferation. For example, the required metabolites may be exhausted, or there may be an ever-increasing concentration of inhibitory substances secreted by the

146

cells, or, if they require to be attached to a surface, as is true of most mammalian cells grown *in vitro*, there may be no more available space. The cells will now be in the stationary phase, sometimes called the plateau phase.

Variation in radiation response with growth phase: bacteria

When survival curves have been examined for cells in relation to their progress from initial seeding to cessation of growth, populations selected for comparison have usually been from exponential or stationary phases; more exceptionally, also from lag phase. In experiments on their radiation response cells are usually categorized as being in one or other phase of growth. This suggests an underlying assumption that response is characteristic of the phase in question, but such an assumption is not justified by evidence from experiments with either bacteria or mammalian cells. In a detailed investigation with *Escherichia coli* B/r, Stapleton (1955) irradiated samples of a culture every hour from the time of seeding up to 12 hours, by which time the bacteria had been in the stationary phase for 3 hours. Examples of variation in response within the growth phases are given in fig. 12.1, selected from Stapleton's data. The pair of survival curves for cultures rather early and rather late in the exponential phase differ widely, and it is particularly noteworthy that one is shouldered

Fig. 12.1. (*a*) Growth of *E. coli* B/r. (*b*) Survival curves for *E. coli* B/r during different phases of growth. The curves with the greatest and least slopes were taken in respectively early and late exponential phase. (Adapted from Stapleton, 1955.)

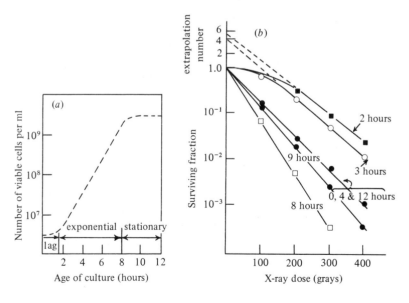

while the other is not. The pair from cultures early and late in the stationary phase are not very different, but they fall between the pair taken on cells in exponential growth.

The literature on the radiation responses of bacteria in different growth phases reveals some contradictory results. Sometimes survival curves are steeper for cells in exponential than in stationary phase, sometimes the reverse is true. Shoulders to survival curves may be seen when cells are in stationary phase, but not when they are growing exponentially, or vice versa. Differences may be attributable to individual biological characteristics of the strain used, but in the light of Stapleton's results it may be that differences in the responses of cells within one of the growth phases may be even greater than any difference assumed to be associated with the phases from which the cells were harvested.

Variation in radiation response with growth phase: mammalian cells

When the first few survival curves for mammalian cells were delineated, mostly for cells *in vitro*, but including mouse leukaemia cells *in vivo* (Hewitt & Wilson, 1959), there was a close similarity in parameters for different lines. This gave the impression that there was a standard, at that time often referred to as 'the' mammalian cell survival curve, with D_0 between 100 and 150 rad (1 to 1.5 Gy) and 'hit number' about 2. But it soon proved illusory that survival curves for all mammalian cells were roughly the same, and indeed the response of cells within a given population was shown by Terasima & Tolmach in 1961 to be very heterogeneous. Still later, major changes in survival curve parameters were shown by Madoc-Jones (1964) to attend the progress of mammalian cells through their phases of growth, as had been shown long previously to be the case with micro-organisms. With a line of cells originating in an induced rat tumour Madoc-Jones found variations in the value of D_0 and n throughout the period of growth, with the maximum value of n approaching 100 for cells in the stationary phase: although this was difficult to judge, because, as he pointed out, the survival curve for these cells was continuously bending within the range of survival levels at which observations were made.

Those great changes in survival curve parameters, particularly in the value of n, may be contrasted with the rather small differences found by Berry, Hall & Cavanagh (1970) between the values of D_0 and n for exponential and plateau phase cultures of cells of two lines: Chinese hamster lung and HeLa, originally derived from a human tumour. With the former, values of n were of the order of 8, for exponential as well as 'fed' and 'unfed' plateau phase cells; with the latter, values of n were of the order of 2. There were small differences relating to the nature of the growth media used.

It was suggested by Hahn & Little (1972) that mammalian cells *in vitro* serve better as a model for the radiobiology of cancerous tumour cells when they are treated in the plateau phase of growth than in the exponential phase, since studies of tumour cell populations have shown only small fractions to be proceeding through the mitotic cycle at any one time. The review of Hahn & Little includes information on the physical and biochemical characteristics of such cells as well as on their radiation response, including repair of damage. They summarized values of survival curve parameters for several cell lines that had been irradiated both in the plateau and exponential phases of growth. But it must again be emphasized that such parameters are not necessarily single-valued for the phases with which Hahn & Little associated them. A more recent example is afforded by measurements of Taylor & Bleehen (1977a) with cells of EMT6 tumour, cultured *in vitro*. For cells in exponential phase and early and late plateau phase they quoted values of *n* respectively in the ranges 18 to 147 (mean 51), 20 to 97 (mean 44) and 6 to 28 (mean 13). This shows that differences in survival cell parameters may be greatest for cells in early and late plateau phases; and also provides confirmation for the observation of Madoc-Jones (1964) that extrapolation numbers are not necessarily at their highest when cells are in the exponential phase, although this has come to be regarded as generally true for mammalian cells *in vitro*.

Modification of radiation response through the growth phase

If it is illusory to regard the radiation response of cells as being characteristic of one or other growth phase, it is hardly to be expected that there should be phase-specific patterns of modification by physical or chemical factors. With some strains of bacteria the magnitudes of oxygen enhancement ratios have been found to be related to the phase of growth from which the cells were harvested, but the o.e.r.'s may also vary with pre- and post-irradiation culture conditions and there are no clear-cut data specifically associating values of o.e.r. with, and only with, growth phase. As regards mammalian cells, Berry *et al.* (1970) found no differences in values of o.e.r. for the Chinese hamster cells they used, whether the cells were irradiated in exponential or plateau phases. With HeLa cells values of o.e.r. were about three, except for cells that had grown to stationary phase in a rather deficient medium. Those cells were rather easily killed by being made hypoxic, and the best-fitted survival curves for the hypoxic and aerobic cells failed to exhibit dose-multiplication. The cells grown to stationary phase in the comparatively deficient medium were less well protected by hypoxia, the estimate of o.e.r. being about 1.8.

Cell cycle stage

Variation in radiation response with stage in the cell cycle, or the 'age response' as it is sometimes called, has received a good deal of attention. Perhaps the earliest observation of a marked dependence on position in the cell cycle was that observed with haploid yeast of a strain which divides by budding. The survival curve for an actively growing, freshly harvested population is complex (fig. 12.2), but it was resolved by Beam *et al.* (1954), who identified the highly resistant component as the budding cells; those with even the smallest buds belong to the comparatively resistant population. This great difference in the response of the two populations is seen when ionizing radiation is used, but barely at all with germicidal ultraviolet light, which is specifically absorbed in nucleic acid. The o.e.r. for the resistant budding cells is considerably less than for the single ones: about 1.7 as against 3.4 (Alper, 1959; fig. 12.2). This means that the difference in sensitivity is less marked for anoxic than oxygenated cells, and still less marked when germicidal UV is used.

Fig. 12.2. The survival of freshly harvested haploid yeast exposed to ionizing radiation or germicidal UV. The difference between the responses of single and budding cells was most marked for electron irradiation when oxygen was present, and not detectable with UV irradiation. Superposition of points marked × on the oxygen curve shows that the o.e.r. for the single cells was about 3; but it was much less for the budding cells.

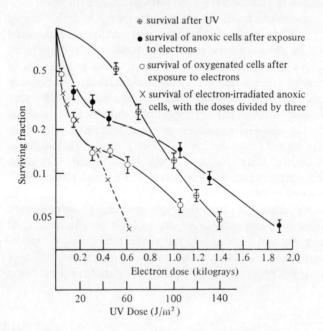

⊕ survival after UV

● survival of anoxic cells after exposure to electrons

○ survival of oxygenated cells after exposure to electrons

× survival of electron-irradiated anoxic cells, with the doses divided by three

The stages of the growth of cells deriving from plant and animal tissues have come to be defined exclusively in terms of the DNA synthetic process plus the period of mitosis, which itself has well defined stages that have long been recognized from microscopical observation. The details of the DNA synthetic cycle were first observed by Howard & Pelc (1953), who used meristematic cells of bean roots; their conclusions apply also to animal cells. The nomenclature of Howard & Pelc, which is in general use, is as follows:

M: the mitotic stage
G_1: the 'first gap' – a period before DNA starts to be synthesized
S: the period of DNA synthesis
G_2: the 'second gap' – period after the DNA content has been doubled, before the first stage of mitosis.

A population that is in exponential growth will, in general, comprise cells in all these stages, since the length of the lag phase as well as of the intermitotic stages will not be identical, but will be distributed about the mean for any phase or stage. If cells are growing exponentially, the proportion of cells in any one stage of the intermitotic cycle will be approximately the same as the fraction of the cycle time occupied by that stage. Such populations are described as asynchronous, and the majority of survival curves have been delineated for asynchronous populations. This is the case with many to which the equations formulated in Chapter 5 have been applied. To observe the radiation response of cells at specific stages of the cycle it is necessary to use a technique that will yield a great majority of the population in the same intermitotic stage. Synchronization of mammalian cells *in vitro* was achieved first by Terasima & Tolmach (1961), who made use of their observation that cells normally growing attached to a surface round up during mitosis and are easily detached by shaking. A population of cells collected in this way will proceed through the next division cycle as a cohort, so they can be treated by radiation or other agents at selected stages. Various other methods for attaining synchronous populations have been used as alternatives, or jointly with the shaking-off method. For example, some drugs will inhibit DNA synthesis; or the cells may be supplied with an injurious agent that will be taken up specifically during DNA synthesis, so that all cells in, or entering into, that stage will be killed. Synchrony obtained by these methods is seldom maintained for more than two division cycles.

Methods of attaining synchronized mammalian cell populations *in vitro*, and their radiation response as they proceed from mitosis to mitosis, were extensively reviewed by Elkind & Whitmore (1967) and subsequently by Sinclair (1968, 1970).

Results of experiments on synchronized cells are often presented in terms of the relationship between surviving fractions after a given dose and the time after synchronization at which the sample was exposed. However, such representations will in general depend on the size of the dose. If changes in surviving fractions are to be associated with the nature of the change in radiation response it is necessary also to have information on full survival curves taken on cells in various stages of the cycle.

The various cell lines used in such investigations have demonstrated different patterns in respect both of survival after a single dose and of

Fig. 12.3. Upper curve: survival of HeLa cells after they had been exposed to 3 Gy while in different stages of the cell cycle, the starting population being mitotic cells detached from the dishes in which they were growing. Lower curve: fraction of cells labelled by tritiated thymidine after treatment for 20 minutes at different times after the population was selected. (After Terasima & Tolmach, 1963.)

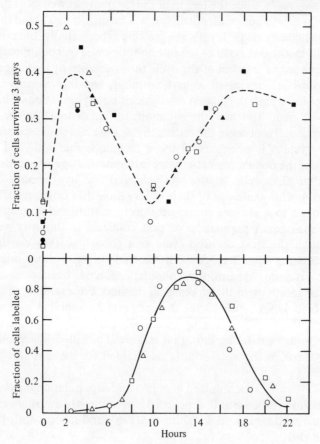

variability in survival curve shapes through the cycle. The first observations were with a sub-line of HeLa cells. Mitotic cells were selected by shaking them off the surface of petri dishes (Terasima & Tolmach, 1961). Figure 12.3 reproduces results from Terasima & Tolmach (1963) for survival after a single dose administered at different times. Whitmore, Gulyas & Botond (1965) synchronized mouse L cells by feeding them with tritiated thymidine of high specific activity, so that all cells synthesizing DNA at the time would be killed. No separate survival curves were reproduced in their report, but the parameters D_0 and n were shown to

Fig. 12.4. Changes in survival curve parameters for mouse L cells, partially synchronized by exposing them to tritiated thymidine in high enough concentration to kill all cells synthesizing DNA. 'Cold' thymidine was then added to stop uptake of radioactive material. (Adapted from Whitmore, Gulyas & Botond, 1965.)

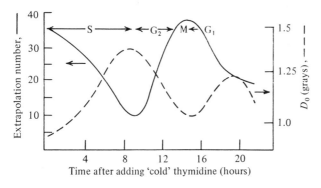

Fig. 12.5. Survival of a population of L cells after a dose of 8 Gy of X-rays or 45 J/m² of germicidal UV. Synchronization as described in the caption to fig. 12.4. (After Rauth & Whitmore, 1966.)

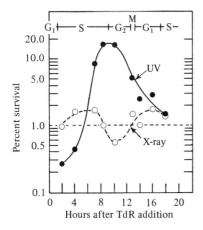

vary through the cycle in reciprocal fashion (fig. 12.4). The pattern of survival of the synchronized cells after doses of 8 Gy is reproduced in fig. 12.5.

In an extensive review of all the information available to him, Sinclair (1968) reproduced patterns of survival after a single dose for several sublines of L cells, and showed that these were all rather different; but there was a common tendency for survival to go to a minimum at some stage of the DNA synthetic period. In that respect, the pattern of survival is almost the opposite of the well-known pattern pertaining to Chinese hamster cells. In their detailed investigation, Sinclair & Morton (1966) used the shaking off technique to harvest mitotic cells of the V79 line, then removed any of the cells that were synthesizing DNA by feeding tritiated thymidine. Full survival curves pertaining to single cells are reproduced in fig. 12.6. 'Idealized' patterns of survival after two doses were reconstructed from these survival curves (fig. 12.7).

In all such experiments the timing of the DNA synthetic phase was the

Fig. 12.6. Survival curves for Chinese hamster V79 cells at different stages of the cell cycle. Synchronization by shaking off mitotic cells, then removing any cells synthesizing DNA by feeding tritiated thymidine. (After Sinclair & Morton, 1966.)

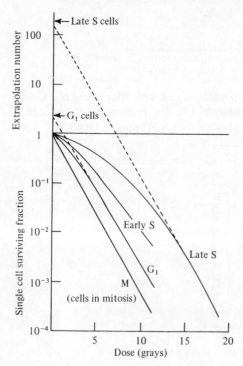

only marker for the progress of cells through the mitotic cycle, so there seems little reason for the assumption of a causal relationship between the stage of synthesis and radiation response, particularly in the light of the widely differing relationships seen with different cell lines. The assumption is all the more surprising when account is taken of variations through the cell cycle in response to germicidal UV. Such comparisons have been made, for example, for mouse L cells (Rauth & Whitmore, 1966), for HeLa (Djordjević & Tolmach, 1967) and for V79 hamster cells (Han & Elkind, 1977). As may be seen (figs. 12.5, 12.8 and 12.9), responses to UV and to X-rays varied in a different manner through the cell cycle, with both lines.

Fig. 12.7. 'Idealized' curves of response to X-rays of Chinese hamster cells as they proceed through the cycle. Curves A and B after doses respectively of 10 and 6.6. Gy. Curve C is for less sharply synchronized cells. (After Sinclair & Morton, 1966.)

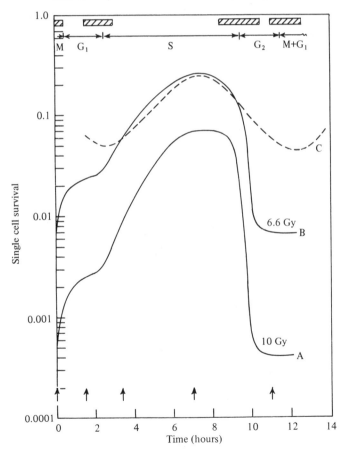

Sinclair (1972) expressed the opinion that there was some factor, which he named Q, 'varying during the S period, which also plays a part in controlling lethal radiation damage'. This proposal is discussed more fully in Chapter 14. At this point, it is relevant to note a summary by Mitchison

Fig. 12.8. Upper curves show survival of synchronous HeLa cells after exposure to 7 J/m^2 of germicidal UV (circles) or 5 Gy of 220-kVp X-rays (squares). Lower curve shows the comparative extent of DNA synthesis (i.e. of incorporation of labelled thymidine) after mitotic cells were collected. (After Djordjević & Tolmach, 1967.)

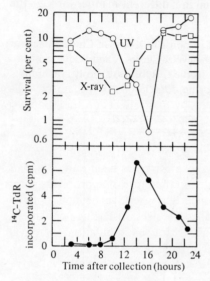

Fig. 12.9. Survival of synchronous Chinese hamster V79 cells after exposure to 4.5-Gy X-rays or 17.5 J/m^2 germicidal UV. (Adapted from Han & Elkind, 1977.)

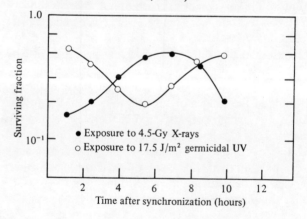

(1972) of research on an alternative set of markers for the progress of cells through the mitotic cycle, namely the synthesis of various enzymes. He remarked that the majority of enzymes are synthesized discontinuously and that each one has its own 'S period'. Furthermore, they do not appear to have a close connection with DNA replication. It seems reasonable to assume that analogous considerations will apply to the synthesis of other products involved in cell replication, and that the patterns of synthesis will be different in different cell lines, just as the DNA synthetic cycle itself varies as between cell lines. The repair of radiation damage may easily be envisaged as depending on enzymes or other products the synthesis of which is discontinuous.

Changes in survival curve shape through the cycle

As will become evident, certain modifying treatments have been found to reduce variability in the response of V79 hamster cells and of HeLa cells, through the cycle, by their effects on the 'shoulder widths' of the survival curves. It is clearly important to know whether the changes in radiation response, as cells progress, is, in fact, mainly in respect of that parameter, characterized by the extrapolation number, n, in the results of all the authors whose work has been quoted so far. The fitting of a survival curve model, and therefore of appropriate parameters, is bound to be particularly uncertain when synchronous populations of mammalian cells are used, since the sizes of starting populations are even more limited than with asynchronous populations. Survival cannot therefore be observed to levels much below 10^{-2} or 10^{-3}.

With hamster cells of the V79 line, at least, changes in survival curve with progression through the cycle appear to be characterized mainly by changes in 'shoulder width'. This feature is evident in results of Bird & Burki (1975) on effects of X-rays on synchronized V79 Chinese hamster cells of the sub-line maintained by Sinclair (fig. 12.10). It should be noted that those curves refer to the survival of microcolonies, whereas those of Sinclair & Morton (1966), reproduced in fig. 12.7 were corrected to reflect survival of single cells.

Synchronized Chinese hamster cells of the V79 line were used also by Gillespie *et al.* (1975), whose object was to test the applicability to survival curves of equation 5.12, $\ln f = -(\alpha D + \beta D^2)$, at different stages of the cell cycle. Within the limits of the ranges of survival they used, the fit was satisfactory. The question, 'which survival curve model?' is discussed in Chapter 14. At this point, it is relevant to remark that the full survival curves shown by Gillespie *et al.* might well have displayed approximately parallel terminal regions if survival had been reduced somewhat further, particularly in the case of cells in the most resistant stages.

With HeLa cells, the survival curves reproduced by Terasima &

Tolmach (1963) could be regarded also as showing changes mainly in respect of shoulder width; and that was the description applied to the mode of change in the response of those cells by Han, Sinclair & Kimler (1976).

According to Whitmore *et al.* (1965) there were substantial changes in both D_0 and n as the mouse L cells they used progressed through the cycle (fig. 12.4), and the changes in survival curves could accordingly not be characterized in the same way as those for V79 hamster and HeLa cells. A sub-population with a markedly higher value of D_0 might be expected to manifest itself as a 'resistant tail' to the survival curve for the asynchronous population. However, Whitmore *et al.* calculated that the survival curve to be expected from combining those for the components would be of the usual form, shouldered, with a terminal exponential region.

Sensitization by oxygen

The substantial change in o.e.r. for haploid yeast cells, as they proceeded from the single to the budding stage, was shown above (fig. 12.2). With mammalian cells, earlier investigations antedated the report of Chapman *et al.* (1970) that plastic petri dishes act as oxygen reservoirs. For that reason, such modest variations in o.e.r. through the cell cycle as

Fig. 12.10. Survival curves for microcolonies of synchronous Chinese hamster cells exposed to 145-kVp X-rays, at 1.9 Gy per minute, at various times after synchronization (after Bird & Burki, 1975).

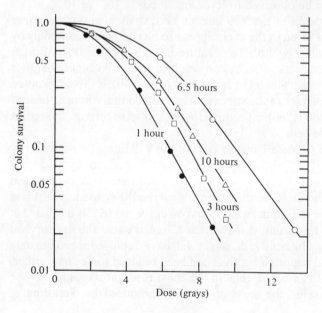

are discernible in the reports of Kruuv & Sinclair (1968) and of Legrys & Hall (1969) may be attributable to artefact, since those authors irradiated cells attached to plastic dishes. For example, the value of K (as used in equation 6.1) may change as cells progress through the cycle, by analogy with the differences observed for mouse ascites tumour cells cultured in different ways before irradiation (Cullen & Lansley, 1974).

There is evidence to support the suggestion that oxygen exerts its sensitizing role mainly through primary lesions in membranes (see e.g., Alper 1970), and in higher cells the configurational association between nuclear membrane and chromatin changes most markedly when cells go into mitosis. If the o.e.r. changes at all through the cell cycle, the greatest difference might be expected as between mitotic and interphase cells. But unfortunately it is difficult to prevent an initially homogeneous population of mitotic cells from progress out of that stage, and the o.e.r. for those cells was not measured by Kruuv & Sinclair (1968) nor, more recently, by Kimler, Sinclair & Elkind (1977), who found no changes in o.e.r. through the remainder of the cell cycle. Legrys & Hall (1969), however, paid particular attention to the o.e.r. for mitotic cells. In their experiments, populations of cells attached to petri dishes were treated with hydroxyurea, to kill the cells in S, and were then irradiated. The mitotic cells were harvested after irradiation by shaking off, and surviving fractions were measured. The o.e.r., about 2.7, was no different from that pertaining to the cells in S; but it may be remarked that, in contrast with the results of Sinclair & Morton (1966), for example, the survival curve parameters for the aerobic mitotic cells harvested in that way were also not very different from those for the cells in S.

A significantly reduced o.e.r. for mitotic cells was, however, observed by Sapozink (1977), who used HeLa cells. He took advantage of the longer G_1 stage in that line than in hamster cells, and of the delay in progression out of M that could be induced by holding the cells in calcium-free medium. This gave him the time required to render the cells hypoxic. Hypoxia was induced by using dense suspensions of non-viable but respiring cells (see Chapter 6); but Sapozink controlled possible artefacts induced by that method in two ways. Firstly, he re-admitted oxygen to the suspensions that had been rendered hypoxic by respiration. The survival after irradiation was the same as that for cells not previously rendered hypoxic. Secondly, some cultures were irradiated after nitrogen bubbling, to control the possibility that the reduced o.e.r. observed for mitotic cells was attributable to less adequate oxygen removal in the vessels containing those cells. Values of o.e.r. for mitotic and interphase cells were respectively about 2 and about 3, whether or not nitrogen bubbling had been used. Some of Sapozink's results are reproduced in fig. 12.11 and 12.12.

Fig. 12.11. Response of synchronous HeLa cells at different stages of the cell cycle (after Sapozink, 1977).

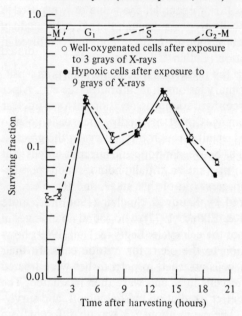

Fig. 12.12. Survival of oxygenated and hypoxic HeLa cells in M and G_1. The survival curves for the two populations of oxygenated cells (curve A) superposed by using different dose scales. Curve B anoxic mitotic cells (lower dose scale). Curve C anoxic G_1 cells (upper dose scale). (Adapted from Sapozink, 1977.)

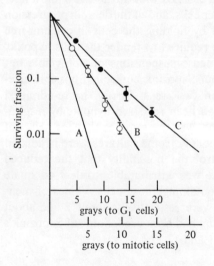

Sensitization by para-nitroacetophenone and misonidazole
Uniformity in the degree of sensitization of hypoxic V79 cells through the cycle were noted when these hypoxic cell sensitizers were used (Chapman, Webb & Borsa, 1971; Asquith *et al.*, 1974). Mitotic cells were, however, not specifically tested.

Effects of cysteamine and sensitization by n-ethylmaleimide (NEM) through the cell cycle
The experiments of Sinclair and his colleagues with these two compounds contributed to and supported his deduction of the existence of Q factor (Sinclair, 1972) accounting at least in part for variation in radiation response through the cell cycle. Both compounds acted in a sense to level this out, cysteamine by exerting a larger protective effect on cells in the more sensitive stages (Sinclair, 1969; fig. 12.13) and NEM by acting predominantly by reducing the shoulders to survival curves both with V79 Chinese hamster cells (Kimler *et al.*, 1977) and with HeLa cells (fig. 12.14; Han *et al.*, 1976). In general, the largest effects were on the survival curves that had the widest shoulders. The selective action of cysteamine first led Sinclair to postulate that changes in the concentration of intracellular –SH

Fig. 12.13. Cysteamine as a protective agent during the cycle of Chinese hamster V79 cells. The protective effect is greatest on those cells for which survival after a given dose is normally least. (After Sinclair, 1969.)

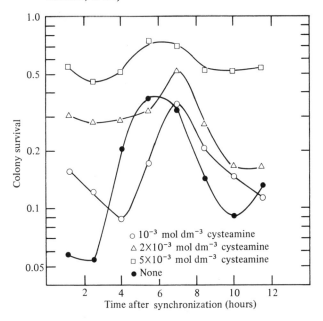

groups were associated with variation in radiation response through the cycle and that Q was to be identified with a fraction of intracellular sulphydryl which he presumed to be involved in repair processes. Subsequent results with NEM were thought to support that proposal, and in a review of these ideas Sinclair (1975) concluded that the –SH groups concerned were protein-associated.

Changes in radiation quality

With increasing density of ionization, survival curves for mammalian cells tend to approach the more nearly to the exponential (Chapter 9). As noted above, the changing radiation response of V79 Chinese hamster cells through the cycle is characterized mainly by changes in the shoulder width of the X-ray survival curves. It might therefore be expected that, as density of ionization increases, differences in response at different stages of the cell cycle should decrease; and the expectation is fulfilled by the results of Bird & Burki (1975), who exposed V79 cells to accelerated heavy ions. They found that the response of cells in all stages was uniform when they used ions of LET_{∞} 179 keV μm^{-1} and above. But Skarsgard (1974), using a different line of Chinese hamster cells (designated 'CH2B$_2$'), found different responses at different stages of the cell cycle, even when the LET_{∞} of the bombarding ions (neon) was greater than 500 keV μm^{-1}.

Fig. 12.14. The effect of n-ethylmaleimide on the radiation response of a synchronous population of HeLa cells. NEM at 0.75 μmol dm^{-3} was added to an unirradiated population (curve N$_0$); or 5.2 grays was delivered to cells while the NEM was present (curve XN); or cells were exposed to NEM in that concentration for 30 minutes at the time shown, after which fresh medium was put on, and the cells irradiated at 18 hr (curve NX). The curve marked X shows the response of untreated cells to 5.2 grays. (After Han, Sinclair & Kimler, 1976).

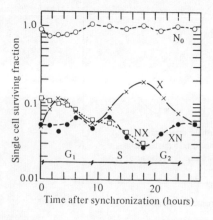

Time after synchronization (hours)

With X-rays, the response of cells in G_1, G_2 and early S was much the same, the survival curve for cells in late S having a much wider shoulder and a somewhat lesser slope. As the LET_∞ of the bombarding ions was increased, the difference in survival of G_2 and late S cells decreased and was then reversed. With oxygen ions ($LET_\infty \sim 350$ keV μm^{-1}) as well as neon ions the G_2 cells were the most resistant. Survival curves in early and late S were shouldered, even with the radiation of highest LET_∞.

13

Dose fractionation and dose rate

Elkind recovery

Very soon after X-rays began to be used clinically it was recognized that their biological effectiveness was usually reduced if lower dose rates were used, or if a total dose was delivered in fractions rather than as a single shot. Chronologically, it was the radiobiological investigation of dose fractionation that helped to elucidate the sparing effect of a reduction in dose rate, and to show that it is linked with the manifestation of shoulders in survival curves.

The relevant observations were first made with an alga, *Chlamydomonas reinhardii* (Jacobson, 1957) and with Chinese hamster ovary cells *in vitro* (Elkind & Sutton, 1959). It was observed that of those cells surviving a first or conditioning dose, D_c, then allowed a radiation-free interval, more would survive subsequent irradiation than if it had been administered immediately after D_c. Thus the cells had recovered the property that had made them more refractory to radiation at the outset.

The manifestation of shoulders to survival curves was widely assumed to be evidence that radiation-induced damage must accumulate within the cells before a final event could prove lethal. The damage that was presumed to accumulate, while still leaving the cell viable, came to be designated as 'sublethal damage', and the recovery that restored the cells' capacity to accumulate such damage was therefore called 're-covery from (or repair of) sublethal damage'. However, as shown in Chapter 5, 'repair models' for shouldered survival curves embody an alternative interpretation. It is preferable, therefore, to refer to the re-covery phenomenon under discussion by a term that carries no implication for the mechanisms of cell killing, and I shall call it Elkind recovery. Its description is greatly faciliated if it is assumed that survival curve parameters can be used that are proper to equations 5.8, 5.9 or 5.13; namely, the extrapolation number, n, the quasi-threshold dose, D_Q, and the initial and final slopes, i.e. the reciprocals of $_1D_0$ and D_0.

Much of the research into the phenomenon, in particular the detailed work of Elkind and his colleagues, was done with mammalian cells *in vitro*. In addition, a good deal has been learned from experiments with some strains of algae and of diploid yeasts, which are very suitable because survival curves manifest high extrapolation numbers. Observations that apply to all three classes of cell may plausibly be regarded as having general validity.

Split-dose timing curves

A method frequently used for investigating Elkind recovery is to measure survival as a function of time between two doses. The conditioning dose, D_c, is followed by a second delivered at different intervals thereafter, with survival measured after the single and the total doses. This will yield a split-dose timing curve. In their first and subsequent reports on the phenomenon of recovery processes in Chinese hamster cells surviving a first dose of radiation, Elkind & Stutton (1959, 1960) showed that as the delay before a second dose was increased so survival

Fig. 13.1. Split-dose timing curves for mouse L cells (upper panel: Whitmore, Gulyas & Botond, 1965) and for Chinese hamster cells (Elkind & Sutton, 1959 & 1960).

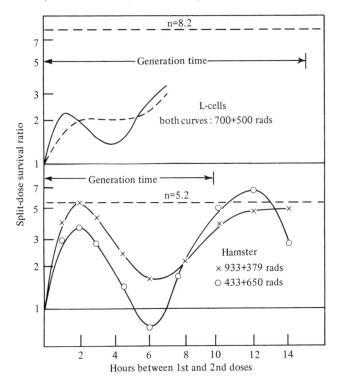

increased, until a maximum was attained after about two hours' recovery. With further increases in the length of the recovery interval survival decreased to a minimum then rose to the plateau value (fig. 13.1). With the chinese hamster cells they used, the plateau survival level after each additional dose was the same as if the cells surviving D_c had not before been exposed to radiation (fig. 13.2). Split-dose timing curves of that nature were explained (Sinclair & Morton, 1964; Elkind & Whitmore, 1967) by taking into account the partial synchrony imposed on a hetero-geneous population of cells by the first dose of radiation. Those cells that are in sensitive stages of the cell cycle will be preferentially killed, so the survivors, relatively resistant at the time of the first dose, will progress as a cohort through the mitotic cycle. A second dose delivered when they are in a more sensitive stage of the cycle will therefore leave fewer survivors than if it is delivered later.

Fig. 13.2. Survival curves for Chinese hamster cells synchronized by treatment with hydroxyurea (after Elkind *et al.*, 1967b).

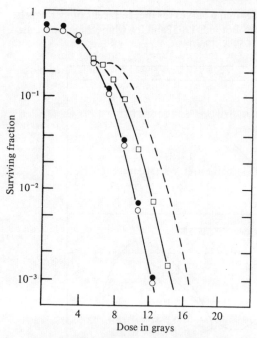

o, ● Single dose at 3.75 and 5.25 hours after the removal of hydroxyurea. Microcolonies averaging 3.3 cells each were irradiated.

□ Cells exposed to a conditioning dose of 5.42 Gy, then a recovery interval of 1.5 hours.

– – Survival curve for cells given a conditioning dose of 5.42 Gy followed by a recovery interval of 12 hours ('full recovery')

Split-dose timing curves of the pattern illustrated by fig. 13.1 have been observed with other cell lines, and also with cells irradiated *in vivo*. The exact pattern is clearly determined by several factors: for example, the conditioning dose will impose a lag on the progression of cells through the cycle, and cells in different stages at the time of irradiation will be differently affected in that respect. In some instances the peak-and-trough pattern is not observed, as shown in one of the curves in fig. 13.1. Where it is observed, the times of appearance of the first peak may be different for different cell lines, even when there is no great difference in the length of the mitotic cycle.

Given the complicated relationships governing the overall response of an asynchronous population of higher cells to a single dose of radiation, it is striking that in some cases survival curves are accurately reproduced when survivors of a conditioning dose have been allowed sufficient time for full recovery. Presumably any synchrony imposed by the conditioning dose disappears fairly soon.

Fig. 13.3. Split-dose recovery in mouse ascites tumour cells exposed to 200-kVp X-rays. Single doses of 5.58 and 11.16 grays reduced fractional survival to about 10^{-1} and 9×10^{-4} respectively ('S_1' and 'S_2' in the figure). Surviving fractions after a conditioning dose of 5.58 grays, followed by a recovery interval and a second, equal, dose are shown by the points. Assay by counting numbers of cells required to originate tumours in baby mice; cells injected into mouse peritoneum for period of holding between doses. (After Hornsey & Silini, 1962.)

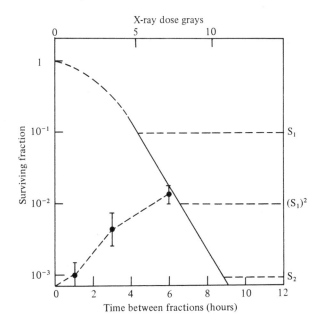

Observation of the ratios of survivors of a total dose given with and without a recovery interval have been used as a means of verifying survival curve parameters. For example, an unusual finding was equality in the extrapolation numbers of neutron and X-ray survival curves for mouse ascites tumour cells irradiated *in vitro* and assayed for viability *in vivo*. As shown by figs. 13.3 and 13.4, the 'recovery ratio' was about the same, whichever form of radiation had been used (Hornsey & Silini, 1962). Conversely failure to find Elkind recovery of any significance served as a check on the validity of describing the survival curve for freshly cultured human fibroblasts as exponential (Cox & Masson, 1974). Those authors used the technique demonstrated by fig. 13.2, namely the delineation of complete survival curves taken on survivors of a conditioning dose after different periods for recovery.

Fig. 13.4. Split-dose recovery in mouse ascites tumour cells exposed to neutrons, mean energy 6 MeV. Single doses of 3 and 6 grays reduced fractional survival to 3.5×10^{-2} and 2.25×10^{-4} respectively ('S_1' and 'S_2' in figure). Surviving fractions after a conditioning dose of 3.0 grays, followed by a recovery interval and a second, equal, dose are shown by the points. Details as in X-ray experiment, fig. 13.3. (After Hornsey & Silini, 1962.)

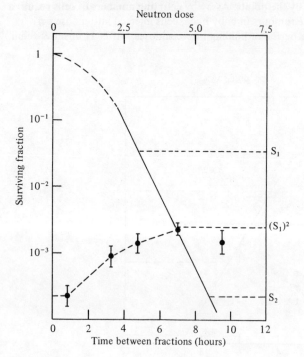

Survival curves after recovery intervals

If the conditioning dose is followed by a predetermined recovery interval, a series of doses may then be given so that a new survival curve can be constructed with the survivors of D_c now regarded as the starting population. The parameters of the new survival curve will depend on the time allowed for recovery. In some instances, exemplified by fig. 13.2, the single-shot survival curve will be exactly repeated within a time that is long enough for recovery to be complete, but not so long that the starting population (the cells surviving the conditioning dose D_c) will have proliferated.

Assuming that full recovery has occurred, the result of the fractionation may be expressed by comparing the number of cells surviving doses $D_c + D$ given immediately with the survivors of D_c + recovery interval + D. As shown by fig. 13.5, the ratio of surviving fractions will depend on the magnitudes both of D_c and D unless both are large enough to reduce survival to the exponential region of the single-shot curve. In that case, the ratio defined above would be given by n, for all values sufficiently large of D_c and D. Alternatively, the results may be expressed in terms of doses to give the same survival with single dose shots and with a split dose. If it is accepted that shouldered survival curves do, in general, have terminal exponential regions, then, with the same provisos, it may be seen that the total split dose, say D_2, is more than the single dose to

Fig. 13.5. Dependence of split-dose survival ratio on magnitude of first and second doses.

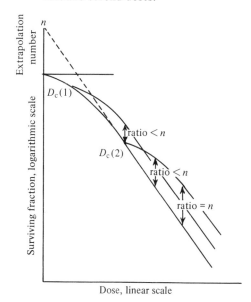

give the same effect, say D_1, by the amount D_Q: i.e. $D_2 - D_1 = D_Q$. Thus a split dose experiment can be a useful method for the direct measurement of n or D_Q (if those are real parameters!), perhaps with more accuracy than can be attained by calculation based on the single-shot survival data points. However, those values could not be relied upon unless it were known that the necessary conditions had been met.

The conditions can, of course, not be met if a cell survival curve is more properly described by equation 5.12; nor if the usable range of doses is insufficient, as might well be the case when a survival curve fails to approach a terminal exponential region within the range of measurable levels of survival. In practice, the observed changes in the value of $D_2 - D_1$ will be similar for the two cases, as illustrated by fig. 13.6.

There are also instances of $D_2 - D_1$ failing to reach a constant value because the surviving cells that have been exposed to a conditioning dose have become more refractory to radiation; in such cases, $D_2 - D_1$ increases with increasing D_c. This effect of dose fractionation was observed in experiments on a sub-line of HeLa cells by Lockhart, Elkind & Moses (1961; fig. 13.7), and has been observed also with *Chlamydomonas reinhardii* (Hillová & Drášil, 1967). Investigations into the mechanism of that phenomenon in *Chlamydomonas* were made by Bryant (1974, 1975, 1976) but its manifestation with mammalian cells has hardly been remarked. This is surprising, because its occurrence in other mammalian

Fig. 13.6. Curves A and B (left panel) show how the value of $D_2 - D_1$ varies with conditioning dose D_c (doses a, b, c etc., right panel). 'End-point' A is at a level of cell survival low enough for $D_2 - D_1$ to be constant, and equal to D_Q, over a range of doses; but 'end-point' B is not.

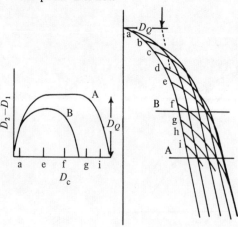

cell lines could have serious implications for radiotherapy (Chapter 17). The occurrence of 'over-repair' could have a bearing also on the deduction of survival curve parameters from fractionation experiments *in vivo*. The phenomenon is more fully discussed in Chapter 16.

Much of the basic work involving the construction of full survival curves on previously irradiated mammalian cells was done by Elkind and his colleagues, and the detailed and exhaustive studies have been extensively reviewed, for example by Elkind & Sinclair (1965); Elkind & Whitmore, Chapter 6 (1967); Elkind (1967, 1970); Elkind & Redpath (1977). Only a few selected aspects of the recovery phenomenon will therefore be dealt with in the ensuing discussion.

Do surviving cells recover fully after each of many dose fractions?

An answer to this question was among the first to be sought by Elkind & Sutton (1960) in their investigations with Chinese hamster cells. Consider the single-shot shouldered survival curve A (fig. 13.8). If cells are subjected to repeated dose fractions followed by enough time for the survivors to recover fully, the predicted overall survival

Fig. 13.7. Increasing value of $D_2 - D_1$, with increasing dose, for two sub-lines of HeLa cells. (*a*) shows how survival after a second dose increased with the length of the recovery interval, above the level of full recovery expected if parameters for secondary and single-shot survival curves were the same. (*b*) shows divergence between the secondary and single-shot curves. Control experiments showed that the effects could not be attributed to cell division during the recovery interval. (After Lockhart, Elkind & Moses, 1961.)

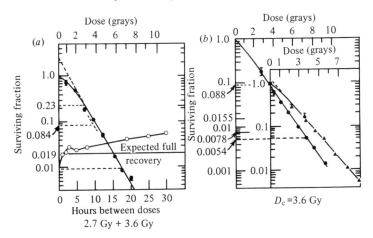

Fig. 13.8. Idealized curve (F) for survival after fractions D_1, D_2, ...,
each followed by an interval for full recovery (after Elkind & Sutton,
1960).

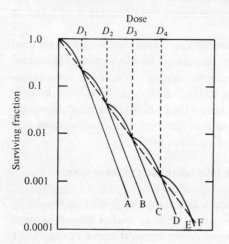

Fig. 13.9. The rationale for deducing initial slopes of cell survival
curves by multi-fraction experiments.

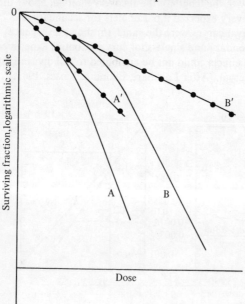

curve will be treated by the exponential curve F. The prediction illustrated by fig. 13.8 was fulfilled in experiments on Chinese hamster cells in which the total dose was delivered in five equal fractions, with a recovery period of 22 hours following each fraction (Elkind & Sutton, 1960).

However, much larger numbers of dose fractions are commonly used in radiotherapy treatments, and they have been used also in certain experiments aimed at delineating initial regions of survival curves. This is technically difficult to do by direct measurement on cells *in vitro*, and, as explained in Chapter 3 the low dose range of survival curves for most clonogenic cells *in vivo* cannot be directly observed at all.

The rationale for deducing initial regions of survival curves from multi-fraction experiments is illustrated by fig. 13.9. If each small fraction is followed by a sufficiently long recovery interval, and if all survivors of each fraction respond as if they had not previously been irradiated, the initial slopes of the single-shot survival curves A and B will be given respectively by the exponentials OA' and OB' where A' and B' are levels of survival that are low enough so that the relevant observations can be made with confidence. But those exponentials will in fact be a guide to the initial slopes of the survival curves only if the basic assumption underlying the experimental method is valid.

An experiment was designed by McNally & de Ronde (1976) partly for the purpose outlined above, and the results suggest that, at least in the

Fig. 13.10. The overall survival of Chinese hamster V79 cells exposed to ten small fractions followed by recovery intervals. For the first five fractions survival follows the prediction shown in fig. 13.8. (After McNally & de Ronde, 1976.)

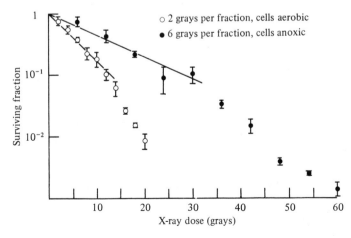

o 2 grays per fraction, cells aerobic

● 6 grays per fraction, cells anoxic

conditions of their experiment, full recovery between dose fractions did not occur in surviving cells after they had been exposed to many fractions. They used Chinese hamster V79 cells in the plateau phase, and measured survival after each of 10 fractions of 1.5 or 2 Gy to aerobic cells, and of 4.5 or 6 Gy to anoxic ones. (fig. 13.10). The first five or six points for survival do define straight lines, as they should, on the assumption that all surviving cells are restored to their pre-irradiation state during a recovery interval after each dose fraction. But evidently this process was not maintained in the conditions of the experiment of McNally & de Ronde when still more dose fractions were delivered. These results may be indicative of a general aspect of Elkind recovery, in which case conclusions drawn from some multi-fraction experiments will have to be reconsidered.

The role of cellular metabolic activity in Elkind recovery

Whether or not surviving cells can completely and indefinitely recover from previous, non-lethal, effects of exposure to radiation must clearly depend firstly, on the nature of the trauma from which they recover, and secondly, on the recovery process itself, which requires time of the order of hours with eukaryotic cells. Evidently some aspects of cellular biochemistry are involved.

Early experiments of Elkind and his colleagues (e.g. Elkind *et al.*, 1965) suggested to them that recovery in surviving cells was only weakly related, if at all, to cellular metabolic processes. Recovery of Chinese hamster V79 cells during the first 2 hours after irradiation followed nearly the same course, and was almost of the same extent, when cells were held at room temperature instead of at 37 °C. Some recovery occurred even in surviving cells held at 0 °C. Recovery was only partly inhibited when the partial pressure of oxygen was as low as 0.08 mm Hg during both irradiation and the recovery interval. These results led Elkind, Moses & Sutton-Gilbert (1967a) to the tentative conclusion that recovery was a 'passive', i.e. non-enzymic process.

However, the question as to whether Elkind recovery was indeed passive, or whether some metabolic activity was required, was first answered in a clear-cut way by results of two series of experiments with algae (Bryant, 1968, 1970; Howard, 1968). In Bryant's experiments, survival curves for oxygenated and anoxic cells could be fitted with the same extrapolation number, and the pattern of recovery was the same provided that air was available to the cells during the interval between doses. When the cells were irradiated on membrane filters, on nutrient agar, recovery was considerably reduced in cells irradiated in an atmosphere containing oxygen in partial pressure about 8×10^{-3} mm Hg and kept in that atmosphere (fig. 13.11). An interesting observation was that admission of air after the first two hours of the recovery interval resulted in an immediate increase in the fraction of cells surviving the second dose. Bryant also traced split-dose timing curves for cells kept in buffer during and between doses, with various gas mixtures bubbled through the suspension during the recovery interval (fig. 13.12). Those results led

Fig. 13.11. Survival ratio as a function of time between two doses to *Chlamydomonas reinhardii*, with oxygen present or absent throughout the recovery interval, or admitted after two hours' hypoxia. Cells on membrane filters resting on nutrient agar. (After Bryant, 1970.)

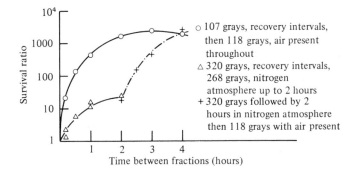

Time between fractions (hours)

Fig 13.12. Survival ratio for *Chlamydomonas reinhardii* exposed to 240 grays while anoxic (nitrogen bubbling through the suspension), then allowed a recovery interval while in equilibrium with oxygen at various partial pressures, followed by 150 grays delivered when the cells were oxygenated. (After Bryant, 1970.)

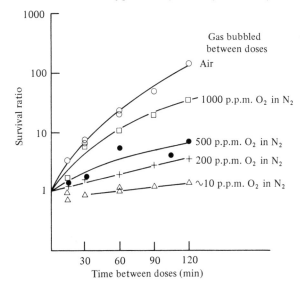

Time between doses (min)

Bryant to conclude that the speed of recovery depended on the partial pressure of oxygen to the extent that this affected the respiration rate of the *Chlamydomonas* cells: and therefore that Elkind recovery was an energy-dependent process.

The initially somewhat controversial issue as to whether or not the presence of at least some oxygen was required for Elkind recovery was confused with another one, namely whether or not the shoulders of survival curves taken in rigorously anoxic conditions were as wide as those for oxygenated cells. If the presence of oxygen were necessary for Elkind recovery, and if fractionation were used as a method for verifying the magnitude of the extrapolation number, clearly this would appear to be less in cells that were irradiated while hypoxic, and held hypoxic afterwards, or irradiated at a low enough dose rate so that Elkind recovery could occur during irradiation. It would then occur to a greater extent in oxygenated than hypoxic cells. The underlying question, whether or not oxygen does usually act in dose-multiplying mode, is referred to in Chapter 14.

If the role of oxygen in Elkind recovery is metabolic, it is to be expected that the partial pressures of oxygen at which recovery is inhibited will vary as between classes of cell and as between cell lines, since these differ greatly in respect of the values of P_{O_2} at which respiration rates are affected. With rigorous conditions for achieving anoxia between doses, Koch & Kruuv (1971) observed no Elkind recovery in synchronized Chinese hamster cells, at whichever stage in the cycle they were irradiated, but Koch, Kruuv & Frey (1973) found that, with oxygen present at a P_{O_2} of only 0.15 mm Hg during a radiation-free interval, recovery proceeded at about the same rate as in cells exposed to air. It seems reasonable to conclude that, in mammalian as well as algal cells, Elkind recovery does require the presence of oxygen, but a very low P_{O_2} may suffice. At present, information on this point regarding mammalian cells is restricted to Chinese hamster V79 cells.

With diploid yeast it would appear that energy metabolism is required for Elkind recovery, rather than the presence of oxygen *per se*. Kiefer (1971) made use of the ability of those cells to respire either aerobically or by anaerobic glycolysis. Irradiated cells that were held in medium containing glucose recovered to the same extent, whether or not suspensions were anoxic or oxygenated during the recovery interval; but if nonfermentable substrates such as lactate or acetate were used instead of glucose, cells exhibited Elkind recovery only if oxygen was present (fig. 13.13). In contrast, the supply of glucose to 'extremely hypoxic' mammalian cells did not restore any capacity for Elkind recovery (Koch, Meneses & Harris, 1977). This different result may reflect differences in metabolic pathways in the two classes of cell; but it may be that the degree of hypoxia achieved by Koch *et al.* ($P_{O_2} < 0.008$ mm Hg, reduced further by cell respiration) was much more extreme than in Kiefer's experiments.

Kiefer (1971) examined the effects of several metabolic inhibitors, but came to no clear-cut conclusion as to the metabolic pathway on which Elkind recovery depends. Nevertheless, an essential requirement seemed to be the production during the recovery period of energy-rich metabolites, for example adenosine triphosphate (ATP).

Results of experiments with diploid yeast have led to similar conclusions by Jain & Pohlit (1973), Jain, Pohlit & Purohit (1975) and by Reinhard & Pohlit (1976). In particular, the latter authors noted the effect of culturing diploid yeast cells in the presence of the glucose antimetabolite 2-deoxy-D-glucose. However, the effects they noted were on the parameter D_Q of the single-shot survival curves. Such effects cannot strictly be regarded as giving clues to the mechanism for Elkind recovery if the phenomenon is considered to be 'recovery from sublethal damage':

Fig. 13.13. Diploid yeast, *Saccharomyces cerevisiae*: dependence of Elkind recovery on presence of oxygen during interval between doses, or on fermentable substrate. Ordinate is based on D_Q for the single-shot curve. It gives the D'_Q (for the secondary curve) as a fraction of D_Q. (Adapted from Kiefer, 1971.)

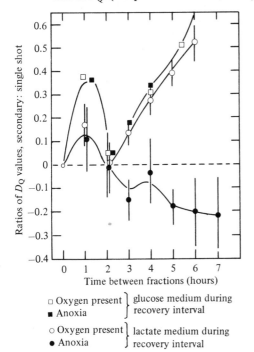

□ Oxygen present ⎫ glucose medium during
■ Anoxia   ⎭ recovery interval

○ Oxygen present ⎫ lactate medium during
● Anoxia   ⎭ recovery interval

on that interpretation, a treatment which results in a reduction in D_Q must act by 'reducing the capacity for sublethal damage'. This important distinction is related to the choice between 'repair models' and 'multi-sublethal lesion models' for shouldered survival curves, and is discussed in Chapter 14.

Investigations on the role of protein and nucleic acid synthesis

The compound 5-fluorodeoxyuridine is considered to be a specific inhibitor of DNA synthesis, and Kim, Eidinoff & Laughlin (1964) used it in concentration 4×10^{-7} mol dm^{-3} to treat HeLa cells for 14 hours before the first of two doses of radiation separated by varying intervals of time. There was no difference between treated and untreated cells in the time course or extent of change in survival, although the treatment by the drug resulted in reduced survival after the first of the combined doses. Similarly, Berry (1966), using HeLa cells, found no effect of 5-fluorouracil on recovery, nor of two inhibitors of protein synthesis, puromycin and cycloheximide, over the first 2 hours' treatment of the cells between dose fractions; although incubation of the cells with those drugs for more extended periods before irradiation affected the parameters of the single-shot survival curves.

In general agreement, Elkind *et al.* (1967a) found split-dose recovery to be unaffected by treatment with excess thymidine, another way of inhibiting DNA synthesis. They also found puromycin to be ineffective. Their object in using puromycin was to inhibit protein synthesis, but, in the concentration they used, it was found to be as effective in inhibiting RNA synthesis as actinomycin D, which Elkind, Whitmore & Alescio (1964) had previously been found to inhibit recovery. This had tentatively been interpreted as showing a specific requirement for unimpaired RNA synthesis; but the later work suggested that the inhibitory effect of actinomycin D might result rather from its property of binding to DNA.

Does distortion of DNA structure inhibit Elkind recovery?

Reduction in the extent of recovery in diploid yeast was observed by Kiefer (1973) as a result of introducing various drugs into cell suspensions after conditioning doses both of X-rays and of germicidal UV. He too associated their action with ability to bind to DNA. However, two of the same compounds (acriflavine, actinomycin D) and also ethidium bromide were found by Arlett (1970) to reduce the shoulders of single-shot survival curves for a line of Chinese hamster cells when they were applied after X-irradiation, just as Elkind *et al.* (1967b) had found actinomycin D to do. It is clearly important to distinguish such effects from perturbations in the actual process of recovery.

If binding to DNA is indeed a cause of perturbation, there may be some analogy between that action and the effect of small doses of germicidal UV, which were found by Bryant & Parker (1977), working with *Chlamydomonas reinhardii*, to inhibit subsequent Elkind recovery from conditioning doses of ionizing radiation. UV is known to kill cells primarily by inducing pyrimidine dimers, which are monomerized by photoreactivating enzyme in the presence of light. Bryant & Parker found that a small (barely lethal) dose of UV reduced the extent of Elkind recovery; but if cells were then illuminated for 4 hours, before they were exposed to the conditioning dose of electrons, Elkind recovery was unimpaired. Bryant & Parker concluded that 'the site of accumulation and repair of sublethal radiation damage is in the DNA'. In view of the evidence against requirements for either DNA or protein synthesis, it could hardly be supposed that induction of pyrimidine dimers, or the binding of certain molecules to DNA, could inhibit Elkind recovery through effects on the replicative or transcriptional functions of DNA. Elkind & Redpath (1977) suggest that local DNA strand distortion might account for the effects of agents (including UV) the effects of which are 'additive' with those of ionizing radiation.

Interpretation of the results cited in this section is complicated by several factors. Pyrimidine dimers cannot in general be regarded as agents for blocking Elkind recovery, since that phenomenon, or something so similar as to be operationally indistinguishable, occurs also between fractions of UV doses, in diploid yeast (Kiefer, 1968, 1973) and in bacteria (Alper & Hodgkins, unpublished). Germicidal UV induces many photoproducts other than those that are monomerized by photoreactivating enzyme; and 'reactivation' of UV effects by light of shorter wavelength is a complex process, sometimes involving mechanisms other than enzymic monomerization of dimers (see Swenson, 1976, for review). As regards agents presumed to act by binding to DNA, most of them have properties that cause them to bind firmly also to other cell components. Acriflavine, for example, has been shown profoundly to modify effects of UV as well as ionizing radiation by mechanisms unrelated to binding specifically to DNA (e.g. Cramp, 1970; Forage & Alper, 1973).

Evidently it will be necessary to take account of the inhibition of Elkind recovery by a variety of treatments, including removal of an energy source, if the process is to be understood.

Elkind recovery after irradiation at 'high LET'

Reduction in 'shoulder width' of survival curves, as density of ionization is increased, has sometimes been described as indicative of reduction in the extent of Elkind recovery. But several split-dose experiments have shown that recovery is generally related to the magnitude of the shoulder to the cell survival curve, whatever quality of radiation has been used. There were some exceptions in the results of Skarsgard *et al.* (1967), who exposed Chinese hamster cells to beams of heavy ions. Survival curves for asynchronous populations were shouldered for all qualities up to an LET_∞ of 1890 MeV cm^2g^{-1} (see table 9.2). While recovery was observed in cells that had been irradiated by helium and lithium ions (extrapolation numbers to survival curves respectively 2.6 and 1.9), it was not seen between fractions of irradiation by boron and carbon ions (extrapolation numbers 1.3 and 1.6). However, there were technical complications in their experiments, for which they used synchronized cells. The experimental design called for the delivery of the first of two dose fractions at the time the cells were released from the S stage in which they had been blocked, but this timing was not always achieved. Todd (1968) did similar experiments with T1 cells, using beams of ions with values of LET_∞ up to 5800 MeV cm^2g^{-1}. Survival curves were shouldered for values of LET_∞ up to 1650 MeV cm^2g^{-1}, and Elkind recovery occurred between split doses of radiation in all those cases, but not when the single-shot survival curves were exponential.

Not many observations have been made on Elkind recovery in mammalian cells irradiated by fast neutrons *in vitro*, probably because survival curves have often approached an exponential or near-exponential form. The observations of Hornsey & Silini (1962) quoted previously (p. 168) are exceptional in that respect. Sometimes results have been ambiguously described as showing 'little if any repair of sublethal damage' when there may be little, if any 'sublethal damage' to repair. However, results of Durand & Olive (1977) suggest differences at least in the time course of Elkind recovery after neutron and γ-irradiation. They tested Chinese hamster V79 cells irradiated in monolayers (single cells) or in suspension as 'spheroids', small aggregates of cells that remained in contact until they were separated by treatment with trypsin. Durand & Sutherland (1972) had found that survival curves for spheroid cells plated singly after irradiation had considerably higher extrapolation numbers than if the cells were single when irradiated, and Durand & Olive made analogous observations on spheroids and single cells irradiated by neutrons. Survival curve parameters for high dose approximations to exponential regions are given in table 13.1, and their split-dose timing curves are reproduced in fig. 13.14. In contrast, Ngo, Han & Elkind (1977) concluded that comparative degrees of recovery observed after neutron and X-irradiation of V79 Chinese hamster cells were qualitatively consistent with the differences in the extrapolation numbers to the single-shot survival curves.

Values of D_Q for X-ray survival curves taken *in vivo* on some clonogenic tissue cells tend to be considerably larger than on the majority of mammalian cell lines *in vitro*, and the same seems to be true for survival curves after neutron irradiation. Fractionation experiments *in vivo* are sometimes designed to measure the differences between single and total

Table 13.1. *Parameters fitted to terminal regions of survival curves for Chinese hamster V79 cells irradiated singly, or as small spheroids, dispersed into single cells after irradiation*

	250-kVp X-rays		MRC cyclotron neutrons (mean energy 7.5 MeV)	
	n	D_0	n	D_0
Single cells	9.4	176	3.8	113
Spheroids	63	177	14.9	116

After Durand & Olive (1977).

Fig. 13.14. Surviving fractions for Chinese hamster V79 cells exposed as single cells in monolayers or as one-day spheroids to γ-rays or fast neutrons given in two doses separated by intervals of time. No recovery was detected for 6 hours after a first dose of neutrons, but there was recovery after 24 hours. Doses: γ-rays: 2×4.5 and 2×7.5 grays to single cells; 2×6.0 and 2×9.5 grays to spheroids. Neutrons: 2×1.41, 2×2.80 and 2×4.21 grays to single cells; 2×2.35, 2×3.74 and 2×5.21 grays to spheroids. (After Durand & Olive, 1977.)

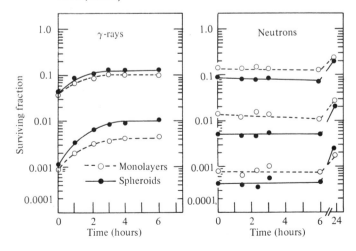

fractionated doses required to cause the same level of damage, i.e. $D_2 - D_1$ (p. 170). In several such experiments, when many hours have been allowed between two dose fractions, values of $D_2 - D_1$ for neutron irradiation have been substantial, although information may not be available to compare these with the value of D_Q for a corresponding single-shot survival curve. Examples are given in table 13.2.

Effects of changes in dose rate

It is implicit in the analysis of single-hit action that the average dose per lethal event should not be affected by the rate at which the dose

Table 13.2. *Differences between total doses of fast neutrons in two fractions, and single shots, i.e. $D_2 - D_1$ for effects* in vivo. *Interval between fractions was 24 hours in experiments cited*

1. *Values of $D_2 - D_1$ from fitted survival curves for clonogenic cells*

	$D_2 - D_1$ (grays)	Reference
Mouse strain and tissue		
SAS/TO, epidermal cells	1.6	Denekamp, Emery & Field (1971)
CDF1, stem cells, jejunum crypts	1.5	Withers, Brennan & Elkind (1970)
C_3H, stem cells, jejunum crypts	0.5 to 1.8	Hornsey (1973b)
C_3Hf/Bu, stem cells, jejunum crypts	0.7 (a) 1.0 (b)	Withers *et al.* (1974b)
Rat, growing cartilage	1.0	Kember (1969)

2. *Values of $D_2 - D_1$ from dose needed to realize a given end-point*

End-point	$D_2 - D_1$ (grays)	Reference
Skin reactions, mouse legs, 8 to 30 days after irradiation	1.4 to 1.9	Denekamp *et al.* (1966)
Skin reactions, rat feet		
7–30 days after irradiation	1.8	Field, Jones &
5–23 weeks after irradiation	1.2	Thomlinson (1968)
Stunting, rat tails	not significant	Dixon (1969)
Death of 50% of C_3H mice by 5 days after whole body irradiation (intestinal death) (c)	0.5 to 1.5	Hornsey (1973b)
Death of 50% of TO mice between 40 and 180 days after irradiation of the thorax (death from lung damage)	1	Hornsey, Kutsutani & Field (1975)

(a) Neutrons from 16 MeV deuterons on beryllium target.
(b) Neutrons from 50 MeV deuterons on beryllium target.
(c) Mice breathed oxygen during irradiation.

is delivered (Lea, 1946). However, there are circumstances in which the response by extracellularly irradiated one-hit detectors may be altered by changes in dose rate, as mentioned in Chapter 4. When macromolecules with biochemical or biological function are irradiated in solution or suspension, long-lived toxic products may be formed. Inactivation by toxic agents increases with increasing time, so the macromolecules under test may appear the more sensitive to radiation, the lower the rate at which it is delivered. In some instances irradiation may sensitize the test objects to these products without actually inactivating them (e.g. fig. 13.15), so that the greater effectiveness of low dose rates is further increased. Another artefact may be caused by radiation-chemical depletion of oxygen in the irradiated volume, through its combination with radiation-induced radicals. If oxygen is consumed at a rate which is greater than that at which it is replaced, and if radiation effects on the system

Fig. 13.15. Left-hand panel shows that bacteriophage S13 is sensitized to H_2O_2 by X-irradiation. Curve a, exposure to H_2O_2. Curve b, exposure to X-rays, 1.4 grays/min. Curve c, exposure to X-rays with H_2O_2 present. Dashed curve shows expected survival if inactivation by X-rays and H_2O_2 acted independently. (After Alper, 1955.) Right-hand panel shows that the toxic effect of formed H_2O_2, which requires time for its action, results in increased effectiveness with decrease in dose rate.

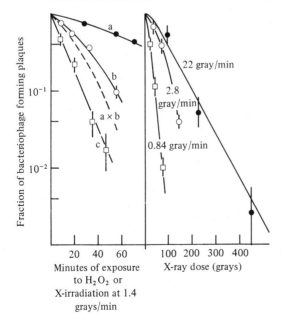

under test are influenced by the presence of oxygen, a dose-rate effect may be observed.

At the cellular level, however, effects of changes in dose rate are mostly linked with the operation of Elkind recovery. This is manifested as a reduction in the effectiveness of radiation when the dose rate is reduced. When it is low enough so that all recovery possible occurs while irradiation is in progress, further reduction should not decrease effectiveness any more unless cell division can proceed concomitantly. Conversely, the effectiveness of radiation is likely to increase with increasing dose rate, until it is high enough for only a negligible amount of recovery during exposure. A further increase in dose rate may in some instances result in a decrease in effectiveness if the rate at which oxygen is removed by chemical action is greater than that at which it is replaced. Both these mechanisms were illustrated in experiments on the dependence on dose rate of intestinal death in mice, at 4 to 5 days after irradiation. The $LD_{50}/4$ decreased with increasing dose rate, up to 60 Gy/minute; correspondingly, recovery was observed within times of the order of minutes

Fig. 13.16. Survival of colony-forming units in mouse bone-marrow after whole-body exposure to γ-radiation at various dose rates (after Fu *et al.* 1975a.)

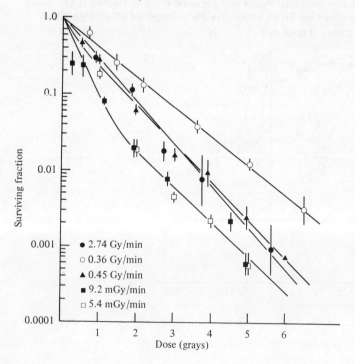

(Hornsey & Alper, 1966). With still higher dose rates the cells became less sensitive because of oxygen depletion: the $LD_{50}/4$ increased for the mice breathing air during irradiation, but not for mice that were made anoxic beforehand (Hornsey & Bewley, 1971). Dose rate-associated effects attributable both to oxygen depletion and to the operation of Elkind recovery were comprehensively reviewed by Hall (1972). He included also some data on the relatively much reduced sparing effects of low-dose rates with radiation of 'high LET'. This is to be expected, if reduction in effectiveness results from recovery during irradiation, and in some classes of cell, including most mammalian cell lines, radiation at 'high LET' is characterized to a lesser extent by the type of damage operationally manifested through the recovery phenomenon. If Elkind recovery occurs in oxygenated cells, but is inhibited in hypoxic ones, the protective action of hypoxia will be reduced when irradiation is at a low enough dose rate for recovery to occur concomitantly (p. 176). Several observations in accord with that expectation were cited by Hall (1972).

A result of Fu *et al.* (1975a) showed an unusual effect of a change in dose rate: the opposite of what is to be expected from the operation of intracellular recovery. The colony-forming ability of cells from the bone

Fig. 13.17. The effect of dose rate on diene conjugation for a fixed dose of 10 Gy delivered at different rates. Linoleic acid concentration 10^{-2} mol dm^{-3}, pH 10.3. (After Raleigh, Kremers & Galboury, 1977.)

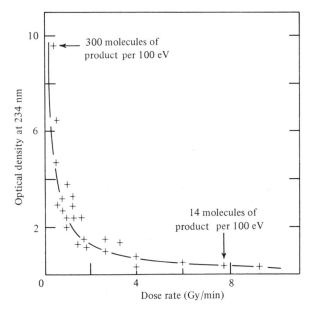

marrow of mice irradiated *in vivo* was more effectively destroyed after irradiation at dose rates below 0.36 Gy/minute than at higher dose rates (fig. 13.16). Such a result might be the consequence of a chemical or biochemical effect of radiation that becomes more pronounced, the lower the dose rate. One possibility is suggested by observations of Raleigh, Kremers & Gaboury (1977) on oxidative damage to lipids. This might contribute to damage to cell membranes and hence to cell death. At dose rates below 1 Gy/minute Raleigh *et al.* found a sharp and inversely dose-rate dependent increase in the products of oxidation of linoleic acid, as reflected by the extent of diene conjugation (fig. 13.17). If such an effect were to contribute to the killing of cells by radiation, very carefully controlled experiments would be required to avoid its being obscured by the normally sparing action of a reduction in dose rate.

Another unusual effect of dose rate was observed by Harrop *et al.*, (1978) who found the o.e.r. of a repair-deficient variant of *Escherichia coli* K12, AB2480, to be less than one when they exposed it to electrons at the very high dose rate used in the gas-explosion technique (p. 65). The total dose from the field emission source used in those experiments was delivered within five nanoseconds. Dose rates were varied by using beams of electrons from a linear accelerator, with which dose per pulse and pulse frequency could be varied. Sensitivity of anoxic, but not of oxygenated, cells increased with increasing dose per pulse, with a consequent fall in o.e.r. (fig. 13.18).

Fig. 13.18. Dependence of o.e.r. on dose rate for the repair-deficient variant *E. coli* K12 AB2480 (after W. D. Rupp, B. D. Michael, H. A. Harrop & R. L. Maughan, unpublished data.)

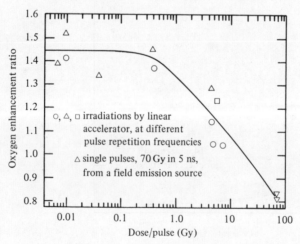

Protection by oxygen was observed if the bacteria were in contact with oxygen for times up to 4 μsec before a pulse of radiation was delivered, but with later times of contact the greater anoxic sensitivity was observed. A similar phenomenon occurred only with one other repair-deficient strain, and then to a lesser extent. AB2480 is deficient in two pathways of repair of damage to DNA by UV, so those results suggest that oxygen acts as scavenger of a species which can damage DNA by a reaction that has been very fast. The implications are discussed in Chapter 15.

14

Survival curve shoulders and survival curve models

The subject matter of preceding chapters will have demonstrated how difficult it is to discuss radiobiological phenomena relating to cell killing without referring to survival curve parameters. Most of the authors quoted have used parameters associated with terminal constant slopes to survival curves (for survival plotted logarithmically against dose plotted linearly). A challenge to the assumption that this is a valid description at high dose of shouldered survival curves, particularly for mammalian cells, has come in particular from the proposal that survival should rather be expressed by equation 5.12, $\ln f = -(\alpha D + \beta D^2)$, which cannot yield a constant slope at high dose. A brief account was given in Chapter 11 of the derivation of that equation by Kellerer & Rossi (1972), who based it on consideration of changes in the RBE of neutrons as a function of dose. The same equation was proposed also on other grounds, namely that it could be derived theoretically on the assumption that a cell is killed if radiation causes a double-strand break in DNA (Chadwick & Leenhouts, 1973).

Inadequacy of curve-fitting as a means of validating a survival curve model

It has been shown that equation 5.12 is an admirable fit at least to the lower dose regions of some shouldered survival curves and this has been regarded as evidence in favour of one or other method of theoretical derivation (e.g. Kellerer, 1975; Gillespie et al., 1975). Gillespie et al. as well as Kellerer made the point that the survival of a heterogeneous population could not be expected to conform with the equation if the survival of the sub-populations did so conform. Nevertheless, the results of Gillespie et al. on an asynchronous population of Chinese hamster V79 could be fitted by the equation, a result described by the authors as fortuitous.

But such a result is to be expected, whatever model for survival curves is adopted, provided it allows of a non-zero slope at zero dose. Any curve passing through the origin of co-ordinates may be fitted by the polynomial

$$y = Ax + Bx^2 + Cx^3$$

provided the slope is not zero at the origin and provided also that the slope does not become demonstrably constant within the range of observations. If the degree of curvature near the origin is not very pronounced, a good enough fit can easily be obtained to a series of observations by the use of two terms of the polynomial, since unavoidable experimental errors will mask small departures from the theoretical curve. Thus the description of survival curves on mammalian cells *in vitro* by the equation $\ln f = -(\alpha D + \beta D^2)$, when survival is followed to 10^{-2} or 10^{-3}, does not necessarily support a hypothesis underlying a theoretical derivation of that equation. In fact, the 'multi-target' and 'multi-hit' equations set out in Chapter 5 can all be expanded to give an equation of the form $\ln f = -(\alpha D + \beta D^2 + \ldots)$ provided they are used in the modified forms that yield non-zero slope at zero dose.

Fig. 14.1. Curve A: survival curve for *Chlamydomonas reinhardii* according to the parameters α and β computed by Bryant & Lansley (1975) in fitting data for *Chlamydomonas reinhardii* by the equation $\ln f = -(\alpha D + \beta D^2)$. Curve B: exponential terminal region drawn on the assumption that the measured survivals after 0.2 and 0.25 kGy were in that region. Survivals would have been given by the crosses. Curve C: exponential terminal region extrapolated back on the assumption that the last two hypothetical data points on curve A were in such a region.

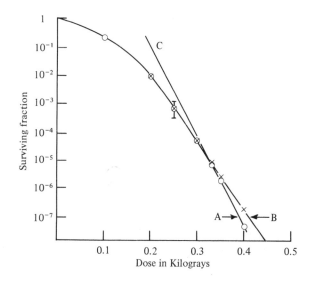

It might be thought that the measurement of survival to lower levels than is usual, with mammalian cells *in vitro*, could provide an answer to the question whether survival curves do or do not approximate to terminal exponential regions. Figures 14.1 and 14.2 have been drawn to show that this approach will not serve, at any rate when shoulders to the survival curves are 'wide', which is to say, *n* is large, in terms of models predicting a terminal slope. The form of equation 5.12 precludes such a description, but the ratio α/β will be the greater, the less pronounced the curvature near the origin. Figure 14.1 shows the theoretical curve of the form 5.12 fitted to their experimental points for the survival of *Chlamydomonas reinhardii* by Bryant & Lansley (1975). The best fitting parameters were $\alpha = 6.3$ kGy^{-1}, $\beta = 0.89$ kGy^{-2}. The lowest measured survival point shown was about 5×10^{-4}, the dose being 250 Gy. If it is assumed that points at 200 and 250 Gy were, in fact, on a terminal exponential region, survival would have followed curve *B*. Curves *A* and *B* would be unlikely to be experimentally distinguishable for any level of survival above 10^{-6}; and even then, slight adjustments to the parameters α and β would give a reasonably good fit of equation 5.12 even if the data to such low levels

Fig. 14.2. Curve A: survival curve for synchronous late S Chinese hamster V79 cells, constructed from parameters α and β fitted by Hall (1975). Curve B: exponential terminal region on the assumption that measured survivals at 14 and 16 Gy were in that region. Survivals would have been given by crosses. Curve C: exponential terminal region extrapolated back on the assumption that the last hypothetical data points, curve A, were in such a region.

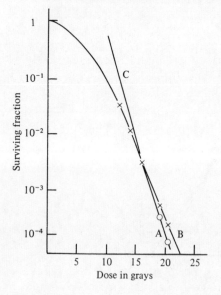

of survival had been available. Conversely, if they had been, it would not have been possible to determine whether or not survival was exponential in the range 10^{-4} to 10^{-6}.

The basis for fig. 14.2 is the survival curve taken by Hall (1975) on 'tightly synchronized' V79 hamster cells in late S phase. Hall took particular pains to avoid the experimental artefacts known to give 'continuously bending' survival curves (p. 16). The curve fitted by equation 5.12 gave $\alpha = 0.053$ Gy^{-1} and $\beta = 0.019$ Gy^{-2}. The last two experimental points were for surviving fractions at 12 and 14 Gy. As in fig. 14.1, curve B has been drawn on the assumption that the last two experimental points were on a terminal exponential region. It would appear that equation 5.12 could be regarded as a reasonable description of Hall's survival curve to a much lower level of survival than was shown in his diagram; and it says much for the accuracy of his work that he was able to say: 'the continued curvature of [the model $\ln f = -(\alpha D + \beta D^2)$] does strain the fit [to the survival curves reproduced] at higher doses'.

Figures 14.1 and 14.2 demonstrate the difficulty in deciding by curve-fitting whether or not a shouldered survival curve does have a terminal exponential region; but they illustrate also how parameters fitted to 'straight' regions may give poor estimates of the true values of D_0 and extrapolation number. The curves marked B in figs. 14.1 and 14.2 were drawn on the assumption that the last two experimental points were in each case on a terminal exponential region; whereas such a region might have existed, but might not have been reached until survival had been reduced still further, as shown by the curves marked C.

Evidently curve-fitting, whether by computer or by eye, cannot be relied upon to provide a decision as to the validity of adopting a model for a shouldered survival curve that approximates to an exponential region at high dose, nor are the parameters derived on the assumption of such a model necessarily reliable. If enough experimental points are available, the constancy of a terminal slope can be tested by measuring it over two reasonably well-separated sections (Bryant, 1973); but there are few instances in which a sufficiently wide range of observations is available, particularly when survival curves are taken on mammalian cells *in vitro*.

However, some results of experiments on dose fractionation provide inferential evidence of the validity of models that give exponential survival at high dose approximations.

Evidence from fractionation experiments

It was shown in Chapter 13 (p. 170) that comparison between a single-shot survival curve and one taken on survivors of a conditioning dose would give a constant value of $D_2 - D_1$, i.e. of the difference between two fractionated doses and the single dose to give the same survival, only if there were terminal constant slopes to the survival curves and if these were equal. The pair of curves shown in fig. 13.2, from Elkind *et al.* (1967b) shows constant values of $D_2 - D_1$ over a 20-fold range in survival: about 4 Gy, after full recovery, and about 2 Gy, for irradiations 1.5 hours after the conditioning dose. Other examples are to be found, including some from experiments on cell survival *in vivo*. For instance, the results of Bacchetti & Mauro (1965) on Elkind recovery in diploid yeast show $D_2 - D_1$ to have been constant, about 44 Gy, over a range of ten in fractional survival, when the cells had a 75 minutes' recovery interval after a dose of 150 Gy. Withers (1967c) measured surviving clones in mouse skin for doses given 24 hours after a conditioning dose. $D_2 - D_1$ was constant, about 3.5 Gy, over a range of 10^4 in the number of surviving clones per unit area.

In some instances the relationship between $D_2 - D_1$ and D_c has been of the form of curve B in fig. 13.6 and one such result has been cited in support of the validity of equation 5.12. However, as shown in Chapter 13, a result of this kind is to be expected if the range of survival within which measurements can be made does not permit sufficiently large dose fractions to be used. This difficulty is manifest particularly when survival curves are taken on cells in some tissues *in vivo*. Thus the failure of $D_2 - D_1$ to reach and maintain a constant value, in fractionation experiments, can be accounted for even if a model with constant terminal slope is adopted; whereas equation 5.12 cannot accommodate experimental evidence of constancy in $D_2 - D_1$.

Theoretical derivations of equation 5.12 fail also to allow for the occurrence of exponential survival curves in asynchronous populations of mammalian cells. It might be postulated that the coefficient β approaches zero for cells in a particular stage, like mitosis, but the premises underlying both methods of derivation (Kellerer & Rossi, 1972; Chadwick & Leenhouts, 1973) preclude the possibility that β should be zero throughout the cell cycle, for normally growing eukaryotes, or indeed specifically for mammalian diploid cells. In summary, therefore, there is evidence against the validity of equation 5.12 as a model; but it is a useful empirical description of survival curves in the low dose region. On the other hand, fractionation experiments support the view that many shouldered survival curves, plotted semi-logarithmically, do approximate to straight lines in the high dose regions. But it may be difficult or even impossible in some cases to be sure that a 'final slope' fitted to such a curve yields meaningful values of n and of D_0.

The use of equation 5.12 to determine initial slopes

The radiobiological significance attached to the parameters of equation 5.12 by some of its proponents may be questioned, but its success in describing at least the low dose regions of shouldered survival curves has been amply demonstrated. When it does fit the experimental observations the value of the coefficient α gives the initial slope of the curve. This is important in applications of cellular radiobiology to calculations of radiation hazards, and also to radiotherapy, since patients are commonly treated by rather small fractions of dose. But in researches into effects of changes in quality of radiation, or of sensitizing and protective agents, changes wrought in survival specifically in the low dose regions have only rarely been measured.

Reported values of α and β pertaining to survival curves have mostly been derived by the use of appropriate computer programs. It is useful for those who do not have ready access to computers and, indeed, more fun for the experimenter, if estimates of survival curve parameters can readily be made by visual inspection of a suitable plot of data points. It also makes for easier communication if results are presented in a manner allowing of easy visual comprehension. An ingenious method of plotting

Fig. 14.3. Plot of $\ln f/D$ ($f = S/S_0$) against D for synchronous Chinese hamster V79 cells in G_1 stage. (All irradiations of aerobic cells.) This method of plotting shows that the BUdR does not change the initial slope of the survival curve; but the cells are sensitized in the higher dose region. The protective agent changes both the initial slope and the response of the cells in the higher dose regions. (Figure provided by Dr J. D. Chapman.)

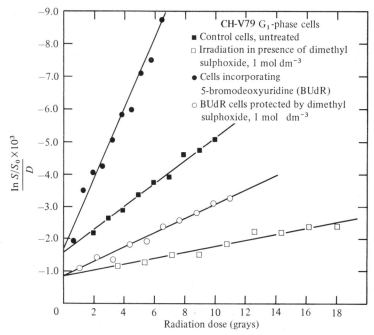

results on cell survival was devised by Dr J. D. Chapman (personal communication). Equation 5.12 may be arranged to give $(\ln f)/D = -\alpha - \beta D$ where f is surviving fraction and D the dose. A plot of $-(\ln f)/D$ against dose should therefore yield a straight line, within the dose region in which the equation applies, with α given by the intercept on the axis $D = 0$. An example of this method of plotting, kindly provided by Dr Chapman, is shown in fig. 14.3.

Douglas & Fowler (1976) used an analogous plot in analysing results of an experiment involving many dose fractions. On the assumption that cell survival could be defined by the equation $\ln f = F_e \times D$, where F_e is a function of dose, they plotted $\ln f/D$ against dose per fraction, and from the straight line so derived they deduced that F_e was equal to $-(\alpha + \beta D)$.

It is important to note that the method of plotting used in fig. 14.3 is no better for determining the dose region within which equation 5.12 applies than a conventional plot, such as those of figs. 14.1 and 14.2. Figure 14.4 is a hypothetical plot of $-\ln f/D$ for a survival curve such that equation 5.12 describes the results in the range 0 to 4 dose units; whereas it approximates to an exponential form (equation 5.4) at higher doses. A plot of $\ln f/D$ would yield the curve ABCDE, with a point of inflection at B. The accuracy with which fractional survival can be determined is unlikely to allow that curve to be distinguished from the straight line ABD, and this method of plotting might therefore give the impression that equation 5.12

Fig. 14.4. Hypothetical plot of $\ln f/D$ against dose to show the difficulty of detecting departure from the equation $\ln f = -\alpha D - \beta D^2$, used to describe the initial region of a shouldered survival curve.

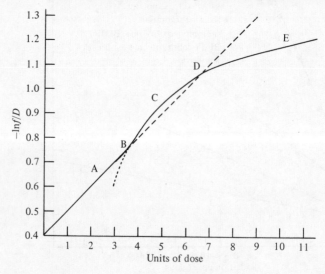

was a good description over a dose range almost twice as great as was actually the case.

Interpretation of extrapolation numbers according to multi-sublethal lesion models

As shown above, experimental evidence tends to support models for shouldered survival curves that predict a terminal exponential region. In all such models the slope of that region is given by the inverse of the average dose required to inflict a final lethal event on the cell; and all except repair models (equation 5.13) embody the premise that sublethal injury, in one or other form, must accumulate within the cell before that final event can be lethal. Since energy is deposited at random, there is no reason to suppose that a final lethal event is of a different nature from the sublethal ones, or that there are differing average energy requirements for sublethal and the final lethal events. Indeed, the theoretical derivations of all multi-sublethal lesion models are based on the assumption of equal probabilities of occurrences of sublethal and lethal events.

The association of Elkind recovery with 'repair of sublethal damage' has been widely accepted, but not much consideration has been given to analysis of the recovery phenomenon in the light of any radiobiological meaning that might be attached to the value n of the extrapolation number.

In single-hit, multi-target models (equations 5.3 or 5.8) n is a measure of the number of targets all of which have to be inactivated if the cell is to be killed. It may, of course, be assumed to have other meanings, as in equation 5.5, in which n is a combined measure of a number of targets all of which are essential for survival and the number of hits required to inactivate each target. The value of n derived from the survival curve for a heterogeneous population may be an average of the values for all the sub-populations. But whichever multi-sublethal lesion model is adopted, n is in some way associated with a number of sublethal lesions that have to be inflicted before cells are killed by final events. If Elkind recovery is identified with the repair of such lesions, there is some conceptual difficulty in the consequent assumption that reparability should be lost when, and only when, the last lesion has been inflicted.

Changes in extrapolation numbers

Association of shoulders to survival curves with sublethal lesions involves some problems also in the interpretation of the mode of action of agents or procedures which bring about a change in the value of n pertaining to a population of cells. Even if attention is confined only to observations with mammalian cells *in vitro* many instances of such action have been recorded.

It is particularly difficult to envisage a plausible mechanism for changes in extrapolation numbers of single-shot survival curves by post-irradiation treatments, while the value of D_0 remains constant. Examples were quoted in Chapter 13 of the reduction in n caused by exposure of irradiated cells to various drugs (Elkind *et al.*, 1967b; Arlett, 1970). An example of an increase in n resulting from a post-irradiation treatment is afforded by results of Miletić *et al.* (1964). When irradiated mouse L cells were supplied with highly polymerized DNA the survival curve showed a slight but significant increase in extrapolation number, compared with the control. It is hardly plausible that post-irradiation treatments should have a retrospective effect on cells' capacity to accumulate sublethal damage!

Modifying treatments applied before irradiation have often been observed to change only the value of n of single-shot survival curves, and in several instances they have been described as changing the cells' capacity to accumulate damage in sublethal form. It was argued above that, in terms of multi-sublethal lesion models, the extrapolation number must be related, if not equal, to a real number of lesions, all but one of which can be repaired in the course of Elkind recovery. If that is accepted, a pre-irradiation treatment that reduces n must be sublethally damaging, in that its action substitutes for one or more of the sublethal lesions that have to accumulate if the cell is to be killed. In that case, it might be expected that n should be reduced by the same treatment applied after irradiation.

The radiosensitizing compound diamide, if incubated with V79 Chinese hamster cells before irradiation, causes a reduction in n of survival curves (Chapter 7, p. 76). When cells in ice-cold suspension were treated with diamide at a concentration of only $4 \times 10^{-5}\,\mathrm{mol\,dm^{-3}}$, the survival curve had a reduced extrapolation number, as compared with the control curve, but there was no difference in the value of D_0 (Harris, Power & Koch, 1975). But there was no effect of diamide added to a suspension of V79 cells immediately after irradiation (Watts, Whillans & Adams, 1975).

Analogous effects were observed by Taylor & Bleehen (1977a, b) in their research into the mode of action of a compound known as ICRF159 (see Taylor & Bleehen, 1977a, for references). They found that its effect was to reduce the value of n in the survival curve for EMT6 tumour cells irradiated *in vitro*. When the drug was used at 200 μg/ml, prolonged pre-irradiation treatment was required for the manifestation of that effect, none being observed until after contact times longer than 10 hours. After 24 hours' contact the value of n for survival curves pertaining to logarithmic phase cells was reduced from about 51 to about 3, with no significant change in D_0 (Taylor & Bleehen, 1977a). Figure 14.5 shows that result and also that the drug killed some of the cells before they were irradiated; so the reduction in n might indeed have indicated the infliction of sublethal as well as lethal lesions by the drug. But Taylor & Bleehen (1977b) found no effect on the extrapolation number to the survival curve if the cells were exposed to the drug for as long as 24 hours after irradiation, and the extent of Elkind recovery matched the magnitude of *n* to the same extent, whether that had been high or low for irradiations respectively in the absence or presence of ICRF159. These results therefore fail to support an identity between 'reduction of capacity for sublethal injury' and the infliction of supernumerary sublethal lesions by the drug.

Another phenomenon that is difficult to reconcile with the logical implications of multi-sublethal lesion models is the one reported by Durand

Fig. 14.5. Survival of exponential-phase EMT 6 cells with and without a previous 24 hours exposure to 200 μg/ml of ICRF 159. The drug killed 40 to 80% of the cells before irradiation. The dashed line shows the survival curve based on the number of viable cells just before irradiation. (After Taylor & Bleehen, 1977a.)

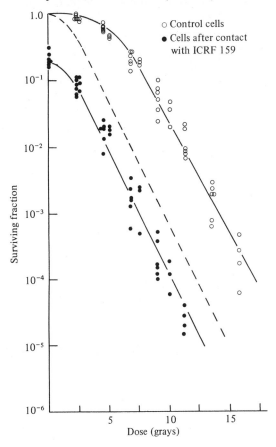

& Sutherland (1972), referred to in Chapter 13: survival curves for cells irradiated as spheroids, then plated as single cells, often have higher values of *n* than if the cells are irradiated singly; but the terminal regions to the survival curves are parallel. A particularly interesting aspect of their observations was the retention by the spheroid cells, after they had separated, of some part of the property that had conferred the higher extrapolation number on the survival curve. This property gradually disappeared with time after irradiation (fig. 14.6). It would seem that a very elaborate hypothesis would have to be contrived to account for those observations in the context of any of the multi-sublethal lesion models.

Survival curve shoulders according to repair models: Q factor

As shown in Chapter 5, shoulders to survival curves are quite differently interpreted in terms of 'repair models' (equation 5.13). They are based on the assumption that the shoulder is evidence both that repair of potentially lethal lesions occurs, presumably at some time before the cell goes into its first division after irradiation, and that repair capacity

Fig. 14.6. The reduction in surviving fraction of cells separated from spheroids, as a function of the time after separation. This was attributable to a gradual decrease in extrapolation number of the survival curves with increase in time. The upper and lower dashed lines show the levels of survival after 11.2 Gy, respectively for cells irradiated as spheroids and as single cells. (After Durand & Sutherland, 1972.)

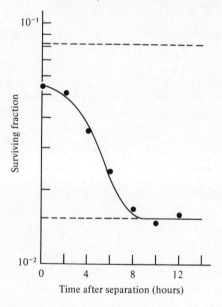

declines with increasing dose. Pre- and post-irradiation procedures that change the values of n in survival curves might then be accounted for on the assumption that the capacity of the cell to perform such repair has been decreased or increased as the case may be. Indeed, that 'capacity' might be identified as representing concentrations or efficacy of a specific factor, like the Q factor postulated by Sinclair (1972) (see p. 161).

The assumptions associated with Sinclair's Q factor bear some similarity to those made by Orr, Wakerley & Stark (1966) and by Laurie, Orr & Foster (1972) in their development of what they called the 'pool' model, after the proposal of Powers (1962) that Elkind recovery is the replenishment of a 'pool or intracellular constituents or states' that are required for repair. But it seems likely that the recovery phenomenon as accounted for in repair models must depend on a factor, or factors, of greater specificity than is associated with the word 'pool' and it is a simplification to adopt the term Q factor.

The suggestion that extrapolation numbers to survival curves are governed by concentration or efficacy of Q factor gains plausibility from several observations, especially if it is assumed to be diffusible across cell membranes. This could provide an explanation of the results of Durand & Sutherland (1972), since Q factor would be more likely to remain in equilibrium, in higher concentration, in cells in contact with each other than in single cells surrounded by medium. Analogous considerations could account for the general experience that cells in organized tissues respond as if the extrapolation numbers to these survival curves were usually larger than for most mammalian cells irradiated *in vitro* (Withers, 1970; Alper, 1974; Chapter 17).

If Q factor were diffusible, changes in the nature of the nuclear and/or plasma membrane could result in changes in the values of n for survival curves taken at different stages in the mitotic cycle. Survival curves for some cell lines have been shown to be closest to an exponential form for cells in mitosis, when the nucleus is not defined by a membrane (see figs. 12.6, 12.12). Alternatively, or in addition, Q factor might be discontinuously synthesized through the mitotic cycle, as has been shown with some enzymes (Mitchison, 1972).

Exponential survival curves for mammalian cells

Insofar as it is correct to assume that shouldered survival curves terminate in exponential regions, it follows from a repair model that radiation-induced killing, even of eukaryotic cells, must proceed basically by the exponential mode, even with sparsely ionizing radiation (see fig. 5.3). Thus an exponential survival curve could be interpreted as showing that Q factor was wanting, or ineffective, in conditions of the experiments concerned. With mammalian cells, that would seem the most likely explanation of exponential survival curves for which the value of D_0 is in the range applicable to most exponential regions of shouldered curves. However, an exponential curve with a much higher value of D_0 might be observed in special cases, if, for example, the efficacy of Q repair remained constant within the dose range examined. It might then be expected that the value of D_0 would fall in the range of values of $_1D_0$, i.e. of the inverse of initial slopes of most survival curves for mammalian cells.

Dose-multiplying agents

With all multi-sublethal lesion models, the shape of the survival curve is determined by constants applied to the term D, so the effect of a dose-multiplying agent is to yield a new curve which is superposable on the unmodified one, if the dose scales are in the appropriate ratio. If the curves have exponential terminal regions, these will extrapolate to the same number (n, in an n-target model) on the zero-dose axis or, if not, tangents on the two curves at the same survival level will extrapolate to the same number. The point was discussed on p. 47, but the special aspects of survival curve modification relevant to repair models were not elaborated.

Like multi-sublethal lesion models a repair model will require n to be constant when a dose-multiplying agent is used, provided that the treatment does not change n in some way that is independent of its dose-multiplying effect. In repair models, the extrapolation number, n, derives its meaning from the assumption that, in a given set of conditions, there will be a number of potentially lethal lesions per cell that can be repaired, but no assumption is involved as to whether that number may, or may not, be affected by a modifying agent. If it is, there is no *a priori* reason to suppose that the change is related to the dose-multiplying action of the agent in respect of the expression of reparable or irreparable lesions.

As mentioned previously (p. 176) there has been a controversy as to whether or not oxygen is truly dose-multiplying, when cells are irradiated under 'severe anoxia' or after a period of prolonged hypoxia. In those conditions, some experimenters have found survival curves to be exponential, although those for the cells irradiated with oxygen present were shouldered (Littbrand & Révész, 1969; Nias *et al.*, 1973). In the light of the oxygen-fixation hypothesis, which has been supported by experimental evidence, such results have defied interpretation in terms of multi-sublethal lesion models, since these embody the association of the parameter n with the variable quantity, dose. But those results would have a different significance if they were interpreted in accordance with a repair model plus the Q factor hypothesis. The reduction in n might signify that the pre-irradiation treatment of the cells had wrought a change in the concentration of Q factor, perhaps by its effect on cell membranes, or a change in its ability to effect repair. If that were the case, it might be expected that cells exposed to severe anoxia, or prolonged hypoxia, would still demonstrate a lower value of n, even if irradiated aerobically, than if they had not been so treated before irradiation.

Support for that suggestion comes from results of Foster *et al.* (1971). They kept HeLa cells hypoxic, at 37 °C, for 22 hours, after which time the cells were exposed to air. Survival curves were taken after 10 minutes and 5 hours. As shown by fig. 14.7, the extrapolation number was considerably higher when the cells had been aerobic for some hours, a result carrying the implication that the previous period of prolonged hypoxia had caused it to be reduced.

Fig. 14.7. Survival curves for HeLa cells kept hypoxic for 22 hours before irradiation. Curves A and B refer respectively to irradiations 10 minutes and 5 hours after admitting oxygen to previously hypoxic cultures. (After Foster *et al.*, 1971.)

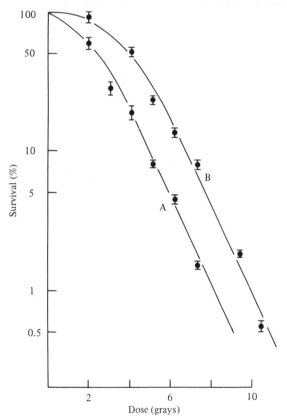

'Repair of potentially lethal damage'

According to repair models, lesions repaired by Q factor are potentially lethal, so it might be argued that Q repair should come under this heading, a phrase that has come into general use, after Phillips & Tolmach (1966). But 'repair of potentially lethal damage' has come to be associated with observations of changes in terminal slopes to survival curves, which means that the presumptive repair is not subject to depletion. To avoid confusion, it is convenient to use terms that distinguish between the two classes of postulated repair, say, P and Q repair. In case the lesions dealt with differ in nature, they can be referred to respectively as P and Q lesions.

It is useful to rewrite equation 5.13 so that the consequences of the different modes of repair can be dealt with analytically. It could be rewritten

$$\ln f = -(_1\lambda_p + {}_1\lambda_q)D + F(D) \tag{14.1}$$

$_1\lambda_p$ and $_1\lambda_q$ are the inactivation constants for P and Q lesions, with the magnitude of $_1\lambda_p$ depending on the extent of P repair in a given set of conditions. $F(D)$ refers to repair of Q lesions, and at very low dose it will approximate to $_2\lambda_q$, giving the low dose approximation

$$\ln f = -{}_1\lambda_p D - (_1\lambda_q - {}_2\lambda_q)D \tag{14.2}$$

The high dose approximation will be

$$\ln f = -(_1\lambda_p + {}_1\lambda_q)D + \ln n \tag{14.3}$$

It follows that the operation of a process currently categorized as 'repair of potentially lethal damage' could affect the final slope, by virtue of the modification $_1\lambda_p \to {}_2\lambda_p$; but n would not be affected unless the modifying process had an independent effect on the efficacy of Q repair.

It is easily seen that a modification causing $_2\lambda_p$ to be less than $_1\lambda_p$ will result in a smaller change in the initial than the final slope of the survival curve. If a dose-multiplying agent acts, the initial and final slopes of the survival curve could be changed to a different extent if P and Q lesions are indeed different in nature and are differently affected by the agent. If the effect on the P lesions is the greater, there will be a greater dose-multiplying effect on the initial than on the final slope, and vice versa. That result is derived analytically in the appendix.

Repair models and change in radiation quality

Unless overt 'multiplicity' exists in an irradiated population, as it would if clumps of cells or microcolonies were irradiated, repair models embody the concept that the basic mode of cell killing is exponential, and therefore attributable to single-hit action. In that case effectiveness should decrease with increasing density of ionization (Chapter 4), i.e. terminal slopes to survival curves should decrease, which is contrary to experience. Thus the adoption of a repair model to account for shoulders to survival curves involves an apparent paradox which requires to be explained. But, as mentioned in Chapter 11, there are in any case examples from several classes of cell, including mammalian cells, of exponential killing with sparsely ionizing radiation; nevertheless the effectiveness of radiation increases with increasing LET. Thus the paradox is not exclusively to be associated with repair models.

It is concluded that such models are potentially much better able to accommodate a variety of experimental observations than any of the multi-sublethal lesion models, and that they are no worse, in respect of effects of change in radiation quality. The validity of a model of the repair type will therefore be assumed in the following chapters.

Appendix: effect of a dose-multiplying agent which acts differently on P and Q lesions

Let the dose-multiplying constants be m_p and m_q for P and Q lesions respectively.

The low dose approximations, with and without the operation of the agent, will be respectively:

$$\ln f = -(m_p \lambda_p + m_q \cdot {}_1\lambda_q - m_q \cdot {}_2\lambda_q)D$$

and

$$\ln f = -(\lambda_p + {}_1\lambda_q - {}_2\lambda_q)D,$$

so the dose-multiplying factor for the agent will be

$$\frac{m_p \lambda_p + m_q \cdot {}_1\lambda_q - m_q \cdot {}_2\lambda_q}{\lambda_p + {}_1\lambda_q - {}_2\lambda_q} \; (= \mathrm{DMF_i}) \text{ in the initial region.}$$

The corresponding high dose approximation will be:

$$\ln f = -(m_p \lambda_p + m_q \cdot {}_1\lambda_q)D + \ln n$$

$$\ln f = -(\lambda_p + {}_1\lambda_q)D + \ln n,$$

the extrapolation number remaining constant, provided the agent has no (independent) effect upon the repair mechanism that determines n.

The dose-multiplying factor at high dose will be

$$\frac{m_p \lambda_p + m_q \cdot {}_1\lambda_q}{\lambda_p + {}_1\lambda_q} \; (= \mathrm{DMF_f}) \text{ in the terminal exponential region.}$$

If m_p and m_q are both > 1, and if $m_p \neq m_q$, the dose-multiplying factors will not be the same in the initial and terminal regions of the pair of survival curves.

Let m_p be greater than m_q, i.e. $m_p = m_q + \Delta$.

$$\text{DMF}_i = m_q + \frac{\Delta\lambda_p}{\lambda_p + {}_1\lambda_q - {}_2\lambda_q}$$

and

$$\text{DMF}_f = m_q + \frac{\Delta\lambda_p}{\lambda_p + {}_1\lambda_q}$$

$\therefore$ The DMF applying to the initial regions will be greater. Conversely, if $m_q > m_p$, the DMF applying to the initial regions will be less than that applying to the final ones. The extrapolation number would be unchanged if the agent did not change the effectiveness with which Q lesions were repaired.

15

Targets and mechanisms for cell killing

The genome as target

If a cell is to divide, and its daughters are to be viable, the whole of biological experience is evidence enough that the genome must be intact, at least in respect of the information needed for the biochemical processes preceding division. It is axiomatic, therefore, that the genome is a target, or indeed 'the' target, for the killing of cells by radiation. That assumption does not depend on knowledge of the chemical nature of genes: it was an obvious one to make, within the framework of the target concept, also at the time when genes were assumed to be proteins (cf. Lea, 1946).

Whether or not the genome should be regarded as 'the' target is a matter of semantics. Should it be identified strictly with the components that carry the genetic code, i.e. with DNA alone; or should unique cellular components that are essential to its functions, particularly its replicative function, be included? It could hardly be argued that a cell with a perfectly intact set of chromosomes would be viable if reproduction could not be supported! It must follow that any structural component of the cell essential to the successful replication of DNA will, like the DNA itself, have the quality of uniqueness that categorizes it as a target, in which a primary energy deposition event may lead to proliferative failure. To understand the whole chain of events leading to that end-point, including mechanisms of sensitization and protection, the chemical constitution of primary targets clearly needs to be known.

The association between DNA and cell membranes

For many years evidence has accumulated on the essential role of membranes not only with respect to the topology of DNA within cells, but also for its replication. Attachment to the bacterial membrane is necessary also for the replication of foreign DNA, like bacteriophage DNA which, after its entry into the host, thereafter dictates transcription and replication (e.g. Puga & Tessman, 1973). To effect a heritable change

205

in a recipient bacterium, transforming DNA similarly requires to be attached to the cell membrane before it is replicated and incorporated into the genome (Dooley & Nester, 1973).

In eukaryotic cells there is evidence of very close adherence of DNA, in the form of condensed chromatin, to the nuclear membrane (fig. 4.1; plates 1 and 2, p. 20). There has been contradictory evidence about the role of the nuclear membrane in the synthesis of DNA in eukaryotic cells. In some plant cells replication at the nuclear membrane occurs in late S, and the DNA synthesized at the location remains there until mitosis (Sparvoli *et al.*, 1976; plates 3 and 4, p. 207). Those authors established an almost one-to-one correlation between the fraction of cells incorporating radioactive label in the DNA, in late S, and the fraction with labelled chromosomes at the subsequent mitosis.

The importance of membranes as sites of attachment and replication of DNA in prokaryotes has been recognized for many years (e.g. Ryter & Jacob, 1966; Tremblay, Daniels & Schaechter, 1969; Smith, Schaller & Bonhoeffer, 1970). Analogous observations and conclusions with respect to higher eukaryotic cells came somewhat later, but it is nevertheless surprising to find that in the mid-1970s it should still have been regarded by some radiobiologists as controversial, and even implausible, that primary energy depositions in membranes could result ultimately in cell death.

Inferential evidence for a primary target other than DNA

With DNA established as the cell component embodying the genetic code (Avery, MacLeod & McCarty, 1944; Hershey & Chase, 1952) it hardly seemed necessary to adduce evidence that DNA was 'the' target for the killing of cells by radiation. Consequently cell death came to be almost universally attributed to chemical or physical changes resulting either from direct energy deposition in the DNA, or from indirect attack upon it by external products, in particular products of the radiolysis of water. But there were already some troublesome experimental results that required to be explained away (and still do!) if DNA was to be regarded as the only primary radiation target of any importance. These were associated in particular with the radiosensitizing action of oxygen for the killing of cells of all classes, as well as for the induction of chromosomal aberrations in higher eukaryotes, i.e. plant and animal cells.

In contrast, there was no sensitization by oxygen of the inactivation of DNA irradiated extracellularly. This was true of bacteriophage (Hewitt & Read, 1950) and also of transforming DNA (Ephrussi-Taylor & Latarjet, 1955). Furthermore, oxygen was found actually to protect bacteriophage against inactivation when the amount of radical-scavenging material in the suspension was low, and this was later found to be true also of DNA

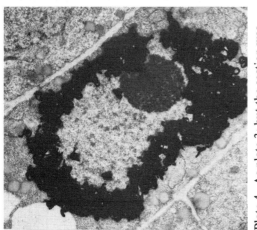

Plate 4. As plate 3; but the section was left in contact with photographic emulsion for a longer period.

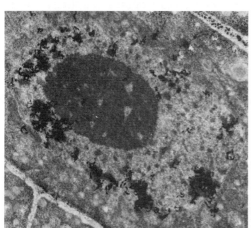

Plate 3. Electron microscope autoradiograph of a cell in the meristem of a *Haplopappus gracilis* embryo. Tritiated thymidine applied for 2 minutes was incorporated into DNA being synthesized 8 hours after the seed coat was broken, at which time the rate of DNA synthesis had passed its maximum. Labelling (from disintegrations of tritium) is mostly over condensed chromatin near the nuclear membrane (from Sparvoli *et al.*, 1976).

released from bacteriophage, as well as of RNA, when damage was assessed by a functional test (table 4.1).

The assumption that primary events only in DNA were responsible for cell killing was difficult to reconcile also with results showing that the o.e.r. for the killing of some micro-organisms depended on physiological factors: for example, the difference in the o.e.r. for single and budding yeast cells, and the fact there was effectively no difference in response of the two populations to UV (fig. 12.2). After all, it is a part of the 'central dogma' that the genome is constant in composition during the life of the cell and its progeny; it varies only to the extent that it is duplicated during the interval leading up to cell division.

Results with *Escherichia coli* strain B, given in table 6.2, were even more difficult to interpret on the assumption of metionic reactions between oxygen and radicals formed solely in DNA, because they suggested some degree of qualitative difference in the nature of the lethal events inflicted in the presence and absence of oxygen. Procedures enabling the cell to bypass the consequences of those events were always more effective when the cells had been anoxic during irradiation, and they were more effective still when lesions had been inflicted by germicidal UV, and therefore specifically in DNA (figs. 15.1, 15.2).

It was those observations, among others, that seemed best accounted for by the hypothesis that ionizing radiation exerted its overall lethal effects by virtue of energy deposits in two loci that differed in their chemical composition. Damage which was modified in a similar way to that inflicted by UV, and which contributed more to the killing of anoxic cells, was postulated to result from primary events in DNA (type N damage). The other type, which was not modified in the same way as UV damage, was mainly responsible for overall sensitization by oxygen and this was named type O damage (Alper, 1963b). At the time I formulated the hypothesis I could make no plausible suggestion as to the site of primary events leading to type O damage. It had been known for many years that lipids were particularly subject to oxidative changes when irradiated in the presence of oxygen, so cell membranes might have seemed obvious as sites of damage that would be responsible for most of the oxygen effect. But it was difficult to envisage that a bacterial or nuclear membrane could respond to radiation as a one-hit detector, as it would have to do, if it were the site of type O damage envisaged in the simple model. By hindsight, it should, of course, have been obvious that these membranes had more important functions than to act simply as containers! However, accumulating evidence on the role of membranes in DNA attachment and replication has suggested the identification of membrane as type O sites that could well respond as single-hit detectors.

Two systems have been used as models to test directly the hypothesis that damage of membrane function (as distinct from purely chemical change) by radiation is greatly enhanced by the presence of oxygen. Watkins (1970) assessed damage to lysosomes in terms of the relationships between radiation dose and amounts of bound lysosomal enzymes released. Values of o.e.r. varied between five and about fifteen, depending on the enzyme assayed. Smith *et al.* (1970) and Knippers & Stratling (1970) had reported that new DNA continued to be synthesized by preparations of bacterial membrane to which some DNA remained attached. Cramp, Watkins & Collins (1972) used such preparations to investigate the effect of ionizing radiation in inhibiting that synthetic process. They found that

Fig. 15.1. The increase in survival of *E. coli* B grown anaerobically after irradiation by X-rays or germicidal UV. Dose scales chosen to give approximate superposition to survival curves taken on anaerobically grown organisms. To do this, scales for irradiation of anoxic and aerobic bacteria had to be in the ratio 1 to 3.4. Effective dose-reductions through imposition of anaerobic growth: 32, after UV; 2.3, after X-irradiation of anoxic bacteria; 1.2, after X-irradiation of aerobic bacteria.

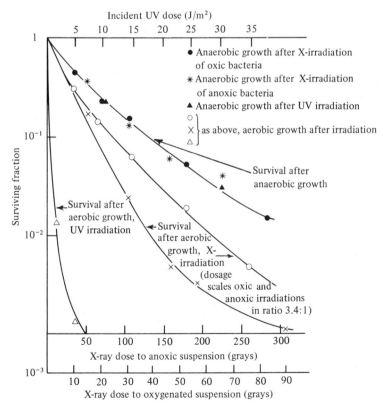

the o.e.r. was about eight; that the doses required to induce inhibition were of the same order of magnitude as those used in constructing survival curves; and that effects on DNA–membrane preparations were no different when they originated in a repair-proficient or repair-deficient strain.

It is an important aspect of both sets of observations that the effects could be observed only when time elapsed between irradiation and measurement. Maximum yields of lysosomal enzymes were recorded 24 hours after irradiation. Cramp *et al.* observed that, if bacteria were lysed immediately after irradiation, there was no inhibition of synthesis by DNA–membrane complexes, in fact it was increased. The maximum inhibitory effect was seen only when the bacteria were incubated for about 40 minutes after irradiation, and if they were kept in the cold the development of that mode of damage was delayed.

The justification for attributing the effects observed by Cramp *et al.* to primary events in the membrane fraction, not the DNA, comes from work of Cramp & Walker (1974), who found no difference in the distri-

Fig. 15.2. Increase of survival of *E. coli* B through holding it in contact with chloraphenicol before final incubation or nutrient medium. Survival of UV-irradiated bacteria was lower than for those that had been X-irradiated; survival after chloramphenicol treatment was higher (After Gillies & Alper, 1959.)

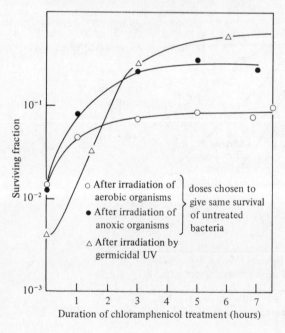

bution of sizes of the pieces of DNA synthesized by irradiated and unirradiated complexes; and of Cramp & Petrusek (1974), who made
measurements of DNA synthesis by DNA–membrane complexes after
UV irradiation. Even very large doses of UV failed to depress synthesis
by more than 50%, and the observed depression was unrelated to the
interval between irradiation and lysis of the bacteria.

A hierarchy of o.e.r.'s

With values of o.e.r. ranging from more than ten to less than
one, i.e. to effects against which oxygen protects, a rank order may be
established for o.e.r.'s as measured by functional tests of damage to cells
or cell components by radiation of 'low LET'. This is shown diagram-
atically in fig. 15.3, which gives ranges of values of o.e.r. for various

Fig. 15.3. Ranges of factors for sensitization or protection by oxygen
in respect of biological or biochemical functions: A, release of
lysosomal enzymes; B, inhibition of DNA synthesis by DNA–membrane complexes; C, inhibition of the induction of bacterial enzymes
requiring a permease to transport the inducer across the membrane;
D, killing of repair proficient bacteria; also of polymerase-deficient
strains; E, killing, mammalian cells; F, killing of repair-deficient
bacteria, yeast, an alga and a slime-mould; G, bacteriophage irradiated in suspension in the presence of radical-scavenging material
and a high concentration of –SH compound; H, bacteriophage irradiated within the host cell as soon as possible after injection of the
nucleic acid; I, J, bacteriophage and transforming DNA irradiated in
broth; K, loss of biological activities of RNA; L, bacteriophage
irradiated in buffer, with or without various solutes.

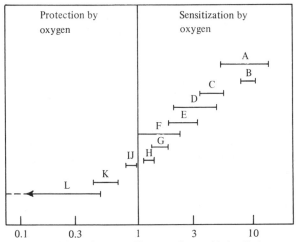

Ratio of doses for same effect, anoxic: aerobic irradiations

classes of observation. In respect of o.e.r.'s, cell killing falls between the high values for damage to membranes and values of less than one for nucleic acid irradiated in conditions in which inactivation could come from attack by radiolysis products of water.

Mutant strains that are particularly sensitive to UV have been shown, or may be assumed, to be deficient in capacity to repair damage to DNA, and the range of o.e.r.'s for such mutants extends down to one, and in one instance to even less than one (table 6.1).

As mentioned on p. 123 (see fig. 10.1) the o.e.r. for induction of some bacterial enzymes is higher than for cell killing. It was shown by Moore (1966) that the comparatively high o.e.r.'s for induction were related to a requirement for a membrane transport system (a permease) for entry of the inducing chemical; whereas if no permease was involved in induction, values of o.e.r. were less than for cell killing. Results of Fried & Novick (1973) afford evidence that an intact membrane is needed for successful induction of β-galactosidase. They showed that 'perturbation' of the membrane in *E. coli* K12, by exposure of the bacteria to ethanol, affected the permease required for the transport of lactose.

The so called 'pol' mutants of *E. coli* K12 are deficient in DNA polymerase activity, and are much more sensitive to both X- and UV-irradiation than their wild-type parents. As noted in Chapter 6 (table 6.1) and in fig. 15.3 values of o.e.r. may be higher for these strains than for their wild-type parents. The pattern of their increased sensitivity is also very different from those mutants which are deficient in ability to excise damaged regions of DNA. In the latter, increased sensitivity is much more marked with UV- than X-irradiation, and more marked for X-irradiation of anoxic than oxygenated cells. Paterson, Boyle & Setlow (1971), in experiments with the mutant of K12 designated W3100 thy⁻ polA1⁻, compared sensitivities to UV and X-rays with those of the pol⁺ parent. The increase in sensitivity was slightly greater when X-rays were used. The high o.e.r.'s observed with pol⁻ strains could be accommodated by the hypothesis that sensitization by oxygen is associated with damage to membranes.

Polymerase-deficient strains grow normally and must therefore be able to synthesize the minimum of DNA polymerase requisite for replication. However, the quantity of polymerase produced is too low to effect some kinds of repair of DNA. Since those enzymes are membrane bound, a membrane lesion that would be bypassed in a pol⁺ strain might prove lethal to the pol⁻ mutant. Furthermore, effective repair of DNA by polymerase might enable membrane lesions to be bypassed that would be lethal to deficient strains.

Speculations on the role of radiolysis products of water

When the model for type N, type O damage was first put forward it was suggested that the intrinsic o.e.r. for type N damage was about 1.4. But if there are mutant strains which are not sensitized at all by oxygen, or even protected (table 6.1), that estimate will not serve. Furthermore, oxygen protects some UV-sensitive strains of *E. coli* irradiated in the presence of nitroxyl compounds (p. 73); and one UV-sensitive strain, deficient in two repair pathways, was found by Rupp *et al.* (unpublished; see Harrop *et al.*, 1978) to be protected when it was irradiated at a very high dose rate (p. 186). Since type N damage presumably plays the major part in the killing of such bacteria, those results show that in some circumstances metionic reactions in the presence of oxygen can protect against damage specifically to cellular nucleic acid, as they do, in appropriate conditions, when damage to the biological functions of extracellular DNA and RNA is under test (table 4.1). Oxygen is the most effective of all scavengers of the species e_{aq}^- and H, so it will protect against damage by those radicals. Such damage might be inflicted through direct attack on DNA, by reducing species, or they might interact with radicals formed directly in the DNA so as to give a lethal product. This seems likely in the light of the results of Rupp *et al.* (unpublished). The lifetimes of e_{aq}^- and H are very short, so indirect action of that kind is likely to come only from radicals formed within the sheath of pure water thought always to be intimately in contact with nucleic acid (Luzzatti, Nicolaieff & Masson, 1961).

The hypothesis that reducing radicals can damage intracellular DNA and that oxygen is therefore protective against damage specifically to DNA is just the opposite of that put forward to account for the protective action of chemical agents by their ability to scavenge OH radicals. As mentioned in Chapter 8, the attribution of a major fraction of damage to interaction of OH with DNA necessitates a subsidiary hypothesis to account for the sensitizing action of oxygen, and Chapman *et al.* (1973) postulated the reactions

$$DNA + OH \rightarrow DNA \text{ radical}$$

$$DNA \text{ radical} + O_2 \rightarrow \text{lethal product.}$$

On that hypothesis, it is difficult to account for the low o.e.r.'s manifested not only by UV-sensitive bacteria, but also by certain sensitive mutant eukaryotes.

The hydroxyl radical hypothesis was proposed by Johansen & Howard-Flanders (1965) and by Sanner & Pihl (1969) on the basis of the correlation between the efficacy of protective agents and their rates of reaction with OH; whereas there was no correlation with rates of reaction with other radiolysis products of water. The same point was made by Reuvers *et al.* (1973) in respect of rates of reaction of dimethyl sulphoxide with OH and e_{aq}^-. However, if sensitization by oxygen is attributed to its interaction at the loci of energy deposits in membranes, protective chemical agents may be assumed to protect specifically against membrane damage, because they act more effectively when oxygen is present. Low rates of reaction of chemical protective agents with e_{aq}^- or H are therefore not in conflict with the suggestion that, if there is 'indirect action' on nucleic acid, it comes from reducing species.

Reactions of radiolysis products of water with DNA in suspension have been extensively studied for many years, but damage to DNA function, extra- or intracellularly, has not been identified with any reaction product as specifically as cell killing by germicidal UV, and the inactivation of transforming DNA, have been shown to be associated with the induction of pyrimidine dimers. However, R. B. Setlow and his coworkers found that γ-irradiation of DNA, extra- or intracellularly, creates lesions that are subject to attack by an endonuclease which also nicks single strands of DNA near pyrimidine dimers. These 'endonuclease-sensitive sites' are much more frequent after irradiation under anoxia than in the presence of oxygen (Paterson & Setlow, 1972; Setlow & Carrier, 1972), which suggests attack by e_{aq}^- or H. The consequence of the formation of the endonuclease-sensitive sites is the excision of 'long regions' of DNA, if the appropriate endonuclease is present, and repair can be detected by special techniques which allow this 'unscheduled' DNA synthesis to be distinguished from normal DNA replication (Setlow, Faulcon & Regan, 1976). Evidence as to the association of cell survival with that form of repair is discussed in Chapter 16.

As mentioned previously, any hypothesis attributing lethal action in cells to radiolysis products of water has to be reconciled with increasing effectiveness of radiation as LET increases, since the yield of products decreases (fig. 9.5). This is not a serious problem, however, if such reactions are postulated to contribute only to a minor extent to the overall lethal effect of radiation, and if, as suggested by the work of Setlow *et al.*, the damage is readily repaired in normally proficient cells. It may be relevant that the RBE for radiosensitive mutants increases much less than for wild-type strains (p. 122; table 15.1).

All in all, it would seem that if there is indirect action involved in the killing of cells by radiation, it is easier to fit experimental facts by postulating reaction of DNA with reducing species than with OH. Nevertheless, the observations that led to the formulation of the hydroxyl radical hypothesis may provide a clue to the mechanism of chemical protection. The correlation of protective efficacy with rates of reaction with OH suggest that effective protective agents might react readily also with radical species arising as a result of the initiation of peroxidation in membranes and subsequent chain reactions.

However, it cannot be assumed that the whole contribution of membrane damage to radiation-induced cell killing is due to lipid peroxidation. This point is discussed later.

Table 15.1 *Comparative RBEs of radiation of 'high LET', repair-proficient and deficient variants of ssame strain/*

Radiations compared	Repair proficient strain	RBE	Deficient variant	RBE	Reference
MRC cyclotron neutrons, and X-rays	E. coli B/r	1.4	B(a)	1.2	Alper & Moore (1967) (b)
	E. coli B-H	1.4	B_{s-1}	1.1	
			B_{s-12}	1.0	
	E.coli K12AB115Y	1.5	AB2463	1.06	
MRC cyclotron neutrons, and X-rays	Mouse lymphoma L5178Y/R	2.2	L5178Y/S	1.3	Hesslewood (1978)
MRC cyclotron neutrons, and 7-MeV electrons	E. coli B/r	1.8	B_{s-1}	1.3	Cramp, Watkins & Collins (1972) (b)
20-keV/μ protons and ^{60}Co-γ	E. coli B/r WP_2 HCr$^+$	1.4	WP_2 HCr$^-$	1.3	Munson et al. (1967) (c)
			B_{s-1}	1.1	
			B(a)	1.0	
	E. coli K12 AB1157	1.7	AB2463	1.0	
			AB2500	0.87	Yatagai &
29-keV/μ α-particles, and ^{60}Co-γ	E. coli K12 AB1157	1.04	AB2470	0.78	Matsuyama (1977)
			JC1553	0.48	(d)

(a) *E. coli* B grown at low temperature after irradiation, giving high sensitivity: see table 6.2.
(b) Values of RBE relative: neutron dosimetry subsequently revised.
(c) Values of RBE calculated from fig. 3 of Munson *et al.*, 1967.
(d) Values of RBE calculated from table 1, Yatagai & Matsuyama, 1977.

Fig. 15.4. Schematic representation of how type O and type N damage contribute to sensitivities and o.e.r.'s in bacteria. Vertical bars represent sensitivities. For the purpose of illustration the presence of oxygen is presumed to enhance type O damage by a factor of about 8 and to protect against type N damage by a factor of about 2. Type N damage which is expressed is shown by shaded bars. For each case a and b represent sensitivities respectively of aerobic and anoxic bacteria.

Case 1: normally proficient in repair of both types of damage, o.e.r. ~ 3.

Case 2: abnormally deficient in repair of type N damage, o.e.r. ~ 1.3, e.g. a sensitive mutant of *E. coli* K12. Bar c represents the sensitivity in the presence of a nitroxyl-free radical sensitizer. This enhances type O damage, but does not scavenge e_{aq}^- (or H) and so does not protect against type N damage. Thus the enhancement ratio exceeds the o.e.r. By reducing type N damage, oxygen would be 'protective' in the presence of the sensitizer.

Case 3: unusually proficient repair of type O damage, with resulting reduction in o.e.r. Examples: *Micrococcus radiodurans* (o.e.r. ~ 1.8); *E. coli* B/r held at 18 °C for several hours after irradiation, a treatment that is more effective on aerobically irradiated organisms.

Case 4: *E. coli* B, cultured in conditions that inhibit protein synthesis so that type N damage is bypassed (see p. 246).

Case 5: *E. coli* B, cultured in conditions that bring type N damage to light; o.e.r. reduced, as in case 2.

Case 6: deficiency in repair of type O as well as type N damage, o.e.r. 4 to 5. Polymerase-deficient strains?

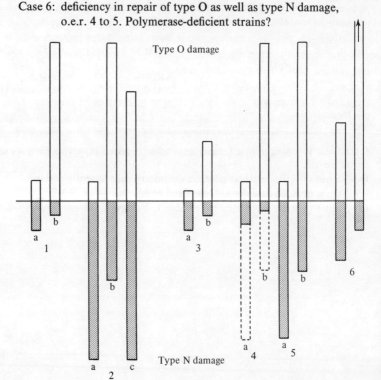

Application of the simple type O, type N damage model to results with micro-organisms

Because of the evidence that values of o.e.r. for cell killing can be less than one, it will be assumed that besides greatly enhancing the contribution of type O damage, oxygen affords some protection against damage to nucleic acid (type N damage). Figure 15.4 shows in an oversimplified way how contributions of damaging events in membrane (type O) and in DNA (type N) could combine to give overall sensitivities of micro-organisms to radiation in the absence and presence of oxygen. The efficacy with which each type of damage is repaired or bypassed depends in some cases on genetic composition and in others on special factors relating to conditions of culture. Some examples of variability in the oxygen effect, and of exceptionally low or high o.e.r.'s, are illustrated by diagrams according with the assumptions of the model.

Does the type N, type O model hold for higher eukaryotic cells?

Metionic reactions consequent on primary energy deposits in membranes and DNA should be the same for all classes of cell, although the subsequent steps leading to cell death must depend on spatial relationships as well as on biochemical pathways.

There is ample evidence that the killing of eukaryotic cells results mainly from energy deposited within the nucleus, so the applicability of the model depends on the extent to which initial damage to the nuclear membrane can ultimately be lethal. But the observations on the killing of bacteria which first of all prompted the model, and those which were then made to test its predictions, have no analogies in respect of higher eukaryotic cells. Experiments in which P repair has been observed with mammalian cells *in vitro* have seldom been done both on aerobic and anoxic cells; nor have comparative results with germicidal UV often been reported. Radiosensitive mutants appear to have been selected from only one line established *in vitro*, namely the L5178 Y strain of mouse lymphoma cells (Alexander, 1961). This strain, and others subsequently derived from it (Courtenay, 1969) have been used by experimenters to test certain

Table 15.2 *Sensitivities to X-rays and germicidal UV of the mouse lymphoma strains L5178 Y/R and S*

	X-rays			Germicidal UV		
	D_0 (Gy)	D_Q (Gy)	n	D_0 (J m^{-2})	D_Q (J m^{-2})	n
L5178 Y/R	0.96	0.46	1.6	20	0	1
L5178 Y/S	0.56	0.43	2.1	200	130	1.9

After Beer, Szumiel & Walicka (1973).

hypotheses about radiation-induced cell death, but the only comparative experiments with UV seem to be those reported by Beer, Szumiel & Walicka (1973; table 15.2). Since the strain more sensitive to X-rays was much the less sensitive to UV, the genetic defect in L5178Y/S cannot be identified as reduced ability to repair or bypass damage to DNA, if the model is applicable. The strain more sensitive to X-rays (which were delivered with oxygen present) might rather be postulated to be deficient in capacity to repair type O damage; but of course there may be more than one genetic defect.

Skin fibroblasts cultured from human patients with certain heritable diseases have been found to be unusually sensitive to irradiation either by germicidal UV or ionizing radiation. Cultures of 'XP' cells, from patients suffering from xeroderma pigmentosum, are in general deficient in ability to repair DNA after irradiation by UV. This defect appears to account for the extraordinary sensitivity of the patients' skin to sunlight. However, XP cells are not particularly sensitive to ionizing radiation. The converse is true of AT cells, from patients with the heritable disease ataxia telangiectasia, who are themselves unusually sensitive to ionizing radiation. Values of D_0 for AT cells were found by Taylor *et al.* (1975) to be about one-third to one-quarter of those for cells cultured in the same way from normal skin fibroblasts, from a paper of Weichselbaum, Nove & Little (1978), are given in table 15.3. Information on values of o.e.r. for these cells, and also of the RBEs of densely ionizing radiation, would help in evaluating the applicability of the type N, type O damage model to mammalian cells; but such information is not to hand at the time of writing.

One or two treatments that have changed the sensitivity of mammalian cells have at the same time resulted in a reduced o.e.r. Berry (1965) found that growing HeLa cells in the presence of methotrexate increased their sensitivity, with a concomitant reduction in o.e.r. from 2.8 to 1.8. Hornsey (1963) measured the survival of irradiated mouse ascites tumour cells by counting the numbers required to originate tumours in the brains of adult mice, and found that survival curves were steeper than when assay was by subcutaneous injection into baby mice, and the o.e.r. was reduced

Table 15.3. *Survival curve parameters for skin fibroblasts cultured from normal humans and patients suffering from ataxia telangiectasia*

	D_0, X-ray (Gy)	D_0, UV (J m^{-2})
Normal	1.49 ± 0.07	3.10 ± 0.06
AT	0.46 ± 0.03	2.9 ± 0.3

After Weichselbaum, Nove & Little (1978).

from 3.1 to 2.4. Such results are suggestive of changes in the overall contributions of type N and type O damage; but without more data this can be a suggestion only.

Inferences from observations on chromosomal aberrations

Damage to higher eukaryotic cells can be measured by one means not available with lower cells, namely the induction of chromosomal aberrations. The scoring of such aberrations was in fact used as a quantitative method for studying radiation damage to higher cells before methods were devised for counting colony-formers. The issue as to whether or not cell killing is a result of the induction of chromosomal abnormalities need not be discussed at length. Some forms of aberration must obviously leave the cell inviable, and good correlations have been observed between the induction of selected types of aberration and cell killing, when conditions have been varied, like quality of radiation (e.g. Skargard *et al.*, 1967) and dose fractionation (Evans, 1967). But there are instances also of poor correlation in the induction of the two forms of damage (e.g. Davies, 1963, plant cells; review by Elkind & Whitmore, 1967, mammalian cells). Many aberrations, induced by X-rays as well as other agents, leave the cells viable, and some induced chromosome abnormalities have been used as markers to identify whole populations (p. 3). It is hardly plausible, either, that every lethal event should be associated with a detectable change in chromosome morphology; so a one-to-one correlation is not to be expected.

What is relevant as a test of the model is the role of oxygen in the induction of chromosomal abnormalities. The first quantitative assessment of the sensitizing action of oxygen in higher eukaryotic cells was made by Thoday & Read (1947) on the basis of chromosomal aberrations induced in bean root cells. If, as has usually been assumed, chromosomal aberrations result solely from energy deposits within DNA, the model would have to be abandoned, at least in respect of higher eukaryotic cells; and to that extent it would be less plausible as an interpretation of the oxygen effect in micro-organisms. But the attribution of chromosomal aberrations to primary events solely in DNA is open to question, again on the grounds of inconsistencies with respect to the oxygen effect. The classical experiments on the effect of oxygen on induction of aberrations were all done by scoring cells at anaphase, that is, when one member of each pair of duplicated chromosomes, led by the centromere, has moved to each of the two poles of the cell that is about to divide into two daughters. Such observations are made by scoring 'abnormal anaphase', mostly 'bridges' and fragments. The bridges are formed by 'dicentrics', i.e. single structures with two centromeres formed by an 'asymmetrical' exchange between two chromosomes: one arm of each joins together, leaving fragments with no centromeres.

A 'bridge' is seen at anaphase if the two centromeres move to opposite poles. Thoday & Read (1957, 1959) scored abnormal anaphases in bean root cells to observe the effect of oxygen during irradiation by X-rays and α-particles. Abnormal anaphases in mammalian cells were scored, also in investigations on the oxygen effect, by Conger (1956) and Deschner & Gray (1959). However, when induction of chromosome abnormalities has been studied for the purpose of gaining insight into the mechanism of their production, many types of aberration have been scored, these being detectable only when all the chromosomes are visible, at metaphase. A good deal of such research has been done on dividing cells in plants, which have suitably large chromosomes; and it is an advantage if they are few in number. Oxygen enhancement ratios have been reported as varying with the type of abnormality; in some instances o.e.r.'s of less than one have been reported, which means protection by oxygen (e.g. Swanson & Schwartz, 1953; Swanson, 1955). A problem inherent in deducing effects of modifying agents on induction of chromosomal defects, with that kind of test system, is the virtual impossibility of constructing meaningful dose–effect curves. A dose of radiation delivered at any one time will delay the cells' progress towards mitosis, and the extent of the delay will depend on the stage in the cell cycle at the time of irradiation, as will their susceptibility to induction of different kinds of aberration (cf. Wolff, 1968; Savage & Miller, 1972).

Such problems are minimized when cells are not cycling at the time they are irradiated. Damage to chromosomes in the sperm of irradiated *Drosophila* (fruit fly) may be detected operationally through the appearance of inherited characteristics in the offspring, identifiable with known forms of chromosomal aberrations. By such means Leigh (1968) could measure values of o.e.r. for the induction of different kinds of aberration. For loss from rod-shaped chromosomes and translocations values were respectively just over one and about two.

Lymphocytes in circulating blood are normally not in cycle, so they can be irradiated *in vitro* before being stimulated to synthesize DNA and proceed to mitosis. All the aberrations induced are therefore chromosomal, and they are customarily scored for cells in their first metaphase after irradiation. The shapes of dose–effect curves are known to depend on the methods used for incubation after irradiation. Watson & Gillies (1970) found also that they were somewhat different for cells irradiated in the presence and absence of oxygen. The two conditions of irradiation also affected the dependence of aberration yield on the time at which the cells were fixed. For cells irradiated in oxygen, the yield of 'unstable' aberrations was the same, whether they were incubated for 50 or 60 hours; whereas anoxically irradiated cells manifested the maximum number only after the longer time. Thus there was no single value of o.e.r. The lowest ratio was about three, as measured for cells exposed to the small-est doses used, and incubated for 60 hours. The highest was about nine, for cells exposed to doses of 2.2 and 19.5 Gy respectively in oxygen and under anoxia, and incubated for 50 hours. All ratios, except the lowest, were higher than any yet recorded for the induction of chromo-somal aberrations, and higher than any recorded for the killing of mam-malian cells. Thus the range of o.e.r.'s for induction of chromosomal aberrations overall is very wide, from one (or perhaps less than one) to nine.

Interchanges between chromosomes have been regarded as chance events consequent on the random juxtaposition of two arms at the points of breakage of both by the passage of ionizing particles, though Revell (1959) postulated that the joining at the point of juxtaposition pre-ceded the breakage. On either view, an interchange would seem more likely if the chromosomes involved were close together not by chance, but by virtue of their attachment to the nuclear membrane. According to Evans (1967), dicentrics are the most frequent abnormalities induced by radiation in chromosomes, and the attraction of attributing a role to membrane attachment, in the process of chromosomal interchange, is increased by evidence he presented to show that Elkind recovery occurs with respect to numbers of dicentrics or anaphase bridges scored in irradiated plant cells of different species. Since times between dose fractions were of the order of hours, in such experiments, the process involved in the completion of interchanges must take a long time, and the metionically fixed lesions resulting in interchange must be subject to repair by Q factor. All in all, therefore, differences in values of o.e.r. for the induction of chromosomal aberrations of different types may well reflect the sites of the principal initiating primary events: nuclear membrane, for events leading to interchanges, but DNA itself, or even the associated proteinaceous material, for events resulting in aberra-tions the yield of which is much less enhanced by the presence of oxygen.

Evidence relating to the role of lipid peroxidation

From their results on the protection of a mycoplasma by super-oxide dismutase, Petkau & Chelack (1974) deduced that lethal effects on that organism were attributable in part to the initiation in membrane lipids of slow autoxidation, a process long known to occur after fats are irradiated (e.g. Hannan & Shepherd, 1954). Since time of the order of hours, at least, must elapse before any cell can be judged to have lost proliferative ability, a lethal effect of slow progressive chemical change can certainly not be ruled out. As mentioned above, the tests used for assessing damage to two model membrane systems required that time should elapse after the cessation of radiation. Lipid peroxidation may therefore represent one mechanism by which cell membranes, and therefore the associated DNA, are damaged. This was suggested also by Pietronigro *et al.* (1977). They observed a gradual decline in the biological activity of bacterial transforming DNA held in suspension along with liposomes containing poly-unsaturated fats. In complementary fashion, the process of autoxidation was greatly retarded by the presence of DNA, which, they suggested, acted as a trap for some of the reactive species formed during the chain reactions involved. Those observations offer some support for the suggestion of 'mutual repair' by membrane and attached DNA (Alper, 1976b) as discussed below.

However, some observations supporting the role of membranes as the site of type O damage show at the same time that there must be a

Fig. 15.5. Survival of *E. coli* B/r after exposure to X-rays, followed immediately by incubation in broth containing penicillin, 30 units/ml (after Gilles, Obioha & Ratnajothi, 1979).

——— After irradiation of oxygenated bacteria

− − − After irradiation of anoxic bacteria

● O_2 (150 Gy)   ▼ O_2 (400 Gy)

■ O_2 (600 Gy)   ▲ O_2 (800 Gy)

○ Anoxic (450 Gy)   ▽ Anoxic (1200 Gy)

□ Anoxic (1800 Gy) △ Anoxic (2400 Gy)

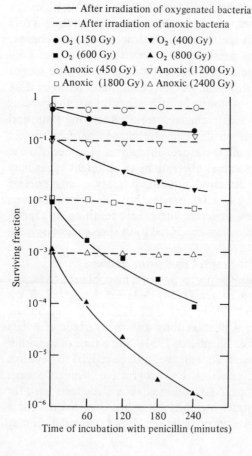

mechanism other than slowly progressing autoxidation. Gillies, Obioha & Ratnajothi (1979) have identified a lesion induced by ionizing radiation in the outer envelope (cell wall + membrane) of *E. coli* B/r, but only in the presence of oxygen. The lesion was detected by exposing irradiated cells in growth medium to penicillin in concentration that was just not lethal (fig. 15.5). Penicillin inhibits the growth of bacteria by preventing the synthesis of new cell wall material, and Gillies *et al.* observed that it caused the irradiated bacteria to grow into long filaments. There was no effect of penicillin treatment after UV irradiation. But the lesion could not be attributed to slow peroxidation because it was completely reparable and repair started soon after irradiation. The greater the interval between irradiation and the introduction of penicillin, the less the lethal effect (fig. 15.6).

Yatvin (1976) and Redpath & Patterson (1978) have used a mutant of *E. coli*, designated K–1060, which cannot synthesize or degrade unsaturated fatty acids. It can therefore be grown in media chosen so that membrane fatty acid content may be varied both as to the type and the total content of unsaturated fatty acid. According to Yatvin, survival curve parameters differed according to the fluidity of the membrane. Redpath & Patterson found that slopes of survival curves increased as the number of double bonds in the incorporated fatty acid increased, but they were not correlated with the content of unsaturated fatty acid

Fig. 15.6. Recovery of *E. coli* B/r from the penicillin sensitive-lesion induced by X-rays in oxygenated cells. (Adapted from Gillies, Obioha & Ratnajothi, 1979.)

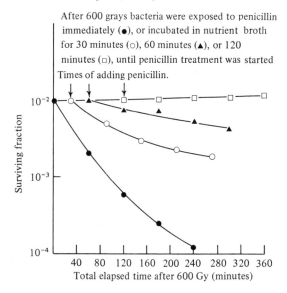

After 600 grays bacteria were exposed to penicillin immediately (●), or incubated in nutrient broth for 30 minutes (○), 60 minutes (▲), or 120 minutes (□), until penicillin treatment was started Times of adding penicillin.

in the membrane. However, there was no differential effect on the o.e.r. If anything, the results of Redpath & Patterson demonstrated proportionately more change in the value of D_0 for the anoxically irradiated bacteria.

Since Pietronigro *et al.* (1977) confirmed that DNA in contact with a poly-unsaturated fatty acid could act as an antoxidant, it should perhaps not be expected that the value of o.e.r. should be related to the degree of unsaturation, although that variable seems to affect susceptibility to irradiation even in the absence of oxygen. It should be noted that the type N, type O model embodies the concept that an incipient type O lethal lesion has a much higher probability of metionic fixation if oxygen is present; but some fixation will occur even in its absence.

If the enhancing action of oxygen is attributable wholly to its interaction at the sites of primary energy deposition events in membranes, the two 'components' of oxygen sensitization detected in mammalian cells (Shenoy *et al.*, 1975) and in bacteria (Michael *et al.*, 1978) may perhaps be associated with different modes of damage to membranes, one of them being the initiation of lipid peroxidation.

Increased effectiveness of densely ionizing radiation

As mentioned in previous chapters, target theory requires, and experiment confirms, that one-hit detectors are less effectively inactivated, per unit of dose absorbed, as the density of ionization increases. But the reverse is true for the killing of cells. Radiation is at its most effective for values of LET_∞ ranging from about 200 MeV cm^{2}g^{-1} for bacteria to 1000 to 2000 MeV cm^{2}g^{-1} for eukaryotic cells. This presents a problem, if the repair model for shouldered survival curves is valid, since the basic mode of cell killing is then by single-hit action. However, the problem exists independently of the model, since survival curves are often exponential for bacteria and haploid yeast, and sometimes also for eukaryotic cells. As was shown in Chapter 11, phenomena associated with radiation of high LET have not been satisfactorily accounted for purely in terms of the physics of track structure. It would seem necessary to take into account the operation of intracellular repair processes, which would be irrelevant in the application of target theory to the inactivation of test objects like bacteriophages. A general hypothesis of that nature seems reasonable, in the light of results given in table 15.1 which show that the RBE of more densely ionizing radiation is often less for repair-deficient mutants than for their wild-type parents. This supports the proposal that some repair processes became less effective as the average event size increases. It must be borne in mind, however, that this cannot be regarded as universally true, since there are some procedures resulting in repair or bypass of damage that operate equally well after exposure to radiations of high and low LET (p. 238).

It has been suggested (Alper, 1976b) that DNA and membrane in close contact can effect mutual repair or bypass of incipiently lethal lesions formed in either component of the complex; whereas simultaneous events in both components would inhibit such repair or bypass. The observations of Pietronigro *et al.* (1977) provide some encouragement for that suggestion. The 'simultaneous events' would have to result from the passage of a single particle, otherwise a 'two-hit' mechanism would be involved.

One form of repair of DNA has indeed been shown to become less effective as the LET of the radiation increases, namely the annealing or repair of main chain breaks in the DNA. This has been demonstrated by experiments involving different techniques. Ritter, Cleaver & Tobias (1977) irradiated mammalian cells by ions of varying LET_{∞}, and used velocity sedimentation in alkaline sucrose gradients to compare numbers of breaks induced in the DNA of cells lysed immediately and in those left for $8\frac{1}{2}$ hours in nutrient medium. They used doses between 30 and 450 Gy, which are large compared with the range in which survival can be measured. Ahnstφm & Edvardsson (1974) and Hesslewood (1978) measured strand breaks by separating single-and double-stranded DNA on hydroxylapatite columns after treatment that caused unwinding of the sections containing breaks (p. 241). That method permits the detection of breaks after doses in the cell survival curve range, and the annealing of breaks induced by those lower doses was effectively complete after an hour or less. In all three sets of experiments there were greater numbers of residual breaks when the inducing radiation was of higher LET. Residual breaks in DNA after completion of repair seemed to be relevant to cell killing according to the results of Ritter *et al.*, but not of Hesslewood. This aspect of work on DNA strand breakage and repair is discussed in Chapter 16. In the present context, the observations serve to support the general concept that some forms of biochemical repair may become less effective as the average energy deposited per event increases. If this is true also of Q repair, it provides a clue to the mechanisms by which shoulder widths of survival curves frequently decrease with increasing LET.

It should be emphasized, however, that the observations quoted above need not necessarily imply that it is the lesions themselves that become less reparable as the average amount of energy per event increases. Some repair processes are known to be mediated by enzyme action, and, as suggested many years ago by Bacq & Alexander (1955), radiation might affect enzyme function not so much by inactivation as by causing their untimely release from membranes to which they are normally bound. Attention to questions of track structure has dominated enquiry into cellular 'RBE effects' up to the present; clearly consideration must be given also to biochemical pathways and repair processes, if phenomena associated with changes in radiation quality are to be understood.

16

Repair or bypass of metionically fixed lesions

Inferences of 'repair of potentially lethal damage'

Departures from normal methods of culturing cells before or after irradiation are likely to increase or decrease the radiation response, and changes in both senses have come to be interpreted as evidence of repair processes. If cells are less effectively killed as the result of a modification in procedure, the change is often considered to have increased the efficacy of a repair system. Conversely, if they are more effectively killed, the modification is assumed to have suppressed or inhibited repair that would normally have occurred.

Reports of 'recovery', 'restoration' or 'repair' of irradiated cells, in the sense now commonly conveyed by the phrase 'repair of potentially lethal damage' (Phillips & Tolmach, 1966) have appeared in radiobiological literature for more than fifty years. Several reviews of earlier work may be found in the volume *Radiation Protection and Recovery*, edited by A. Hollaender (1960). There were no relevant investigations based on viable counting of mammalian cells before the 1960s; but some general conclusions could be drawn from results obtained with a wide range of other test systems. A majority of the conditions resulting in increased survival could be classed as sub-optimal for the growth of the cells; and changes in survival were seldom in the dose-multiplying mode, whether the departure from normal procedure had resulted in an increase or decrease in radiation response (Alper, 1961b).

Research during the intervening years has not shed much light on the biochemical processes governing such changes in the lethal effects of ionizing radiation as are currently regarded as affording evidence of positive repair. One reason may be that the word 'repair' is misleading in some instances. Cells may be able to originate colonies, in appropriate conditions, although a potentially lethal lesion has been inflicted, but has not been repaired: in other words, what was referred to in Chapter 14 under the generic term P repair may well encompass instances of bypass as well as of repair. This possibility is discussed below.

Considerable confusion stems from many and not well-defined uses of the words repair and recovery in radiobiological literature, and this has been aggravated to some extent by the implicit use made on occasion of the assumptions underlying repair models, although explicitly one or other form of multi-sublethal lesion model has been adopted. But the premises basic to the two groups of models are mutually exclusive: either a shoulder to a survival curve indicates that some potentially lethal lesions are repaired at the outset of radiation, and that the process of repair fails to be maintained; or it is symptomatic of an accumulation of sublethal lesions, but provides no evidence of repair, the sublethal lesions being repaired during a radiation-free interval.

Sources of confusion between P repair, Q repair and Elkind recovery

Observations on 'repair of potentially lethal damage' made only by measuring survival at different times after a single dose fail to allow a distinction to be drawn between the operation of P repair and a change in the efficacy of Q repair. In some instances both factors may contribute to changes in survival. But even if full survival curves are delineated with and without application of the modifying procedure under test, problems may arise if changes in survival depend on a lapse of time after irradiation, and the time required for the change depends on the magnitude of the dose. Figure 16.1, from the review of Korogodin, Meissel & Remesova (1967), on 'liquid holding recovery' in diploid yeast (see below), has been adapted to show also the dose–effect curve as it would have appeared if plating had been done after only one day's holding.

Insight into the mechanism of Elkind recovery has been sought in some experiments by the use of drugs or other perturbing treatments applied between dose fractions. But, as mentioned in Chapter 13, some drugs added after irradiation act to reduce the extrapolation numbers of the single-shot survival curves. Unless a drug or other treatment has been found to leave the single-shot survival curve unchanged, it cannot be assumed to have affected the recovery process *per se*.

Investigation of Elkind recovery involves a knowledge of how cells surviving a first or conditioning dose would respond to further irradiation given immediately, since it is the decreased response to further doses that marks the recovery. Sometimes observations are made only of survival after a second dose delivered at various times after the conditioning dose, and there are circumstances in which a 'split-dose timing curve' of that nature may be misleading. For example, the experiment may be on test cells maintained in an environment in which growth and division cannot occur, as when mammalian cells are in contact, in the plateau phase of growth. They must be separated and provided with nutrients before they can originate colonies. In such conditions, survival after a single dose often increases with the time after irradiation at which this is done (Hahn & Little, 1972); in other words 'repair of potentially lethal damage' may distort the changes attributed to Elkind recovery.

Holding or irradiating cells at temperatures sub-optimal for growth

Modification of radiobiological effects by such means have been observed since the 1920s. Reviewing the literature, Stapleton, Billen & Hollaender (1953) commented on the apparently contradictory reports on the effects of low post-irradiation temperatures. They therefore investigated in depth the effects of holding irradiated *Escherichia coli* B/r at various temperatures, before final incubation at 37 °C, and they studied also the relationship between times of holding and changes

Fig. 16.1. Survival of diploid yeast, immediate plating [dashed curve $S_0 (D)$] and after holding in water for 120 hours [dashed curve $S_\infty(D)$]. Solid curves show increase in survival as a function of holding time, scale marked on lower abscissa. Curve 1, 0.37 kGy; curve 2, 0.44 kGy; curve 3, 0.70 kGy; curve 4, 0.97 kGy; curve 5, 1.23 kGy; curve 6, 1.50 kGy; curve 7, 1.77 kGy. (Adapted from Korogodin, Meissel & Remesova, 1967.)

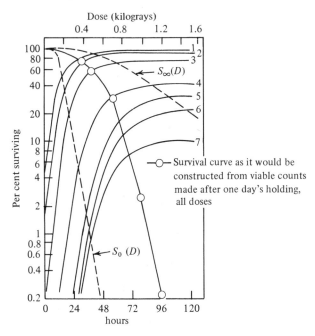

in survival. Some of their results are reproduced in figs. 16.2 and 16.3. The optimum holding temperature was 18 °C, and when the maximum time was allowed the effect was in the dose-multiplying mode, the modifying factor being 1.5. An interesting and important point was that the holding temperatures for maximum survival were different for two other strains of *E. coli* that they tested. The optimum holding temperature for one of them was about 12 °C, which, with B/r, gave almost the same survival as for irradiated organisms incubated at 37 °C (fig. 16.3). That series of experiments illustrates a common feature of effects on irradiated cells of changes in culture conditions: responses may be quite different in closely related lines.

There have been several reports of effects of holding irradiated mammalian cells for various periods at temperatures below 37 °C, and a few observations on changes in survival attributable to temperatures near

Fig. 16.2. Survival of *E. coli* B/r after 660 grays followed by periods of holding at temperatures lower than final incubation temperature (after Stapleton, Billen & Hollaender, 1953.)

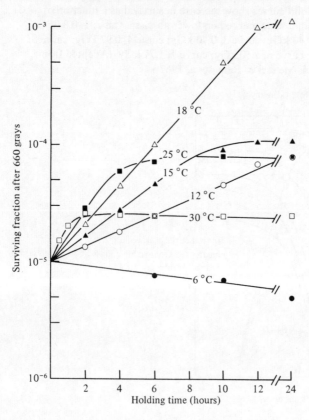

0 °C at the time of irradiation. However, if cells are attached to surfaces, covered with liquid medium or buffer solution, and irradiated at a temperature substantially lower than 37 °C, some time must necessarily elapse before the cells reach that temperature. Thus observations on effects of irradiating cells at low temperatures cannot be clearly separated from those made on irradiated cells held at defined temperatures before final incubation at 37 °C. Results of some investigations of both types are given in table 16.1; which serves only to show that no generalization is possible, even about observations on the same cell lines. In fact the data on mammalian cells are as confused and confusing as those reviewed by Stapleton *et al.* (1953) on lower organisms.

Increased survival from postponement of active growth

Some of the results given in table 16.1 could be interpreted as showing that survival may be increased by imposition of a delay before irradiated cells are in an environment conducive to vigorous growth. However, that is clearly not always the case when the delay is due to

Fig. 16.3. Survival after 800 grays as a function of post-irradiation holding temperature, three different strains of *E. coli* (after Stapleton, Billen and Hollaender, 1953.)

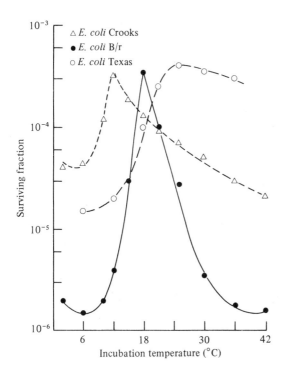

Table 16.1. *Some results of holding irradiated mammalian cells at temperatures below 37° C before final incubation*

Cell line	Temp. °C	Time held, or time of holding for maximum effect	Effect on survival or on survival curve parameters	Reference
Mouse lymphoma L5187 Y, sensitive	34	Maximum: 10 h	Increase in D_0, factor 1.6	Beer, Lett & Alexander (1963)
Mouse L cells exponential phase	5	Max. 30 min	D_0 increased, n decreased	Whitmore, Gulyas & Kotalik (1970)
stationary phase	5		No effect	
HeLa	29	4–12 h	Decrease	Phillips & Tolmach (1966)
HeLa	5	During irradiation	Increase	Belli & Bonte (1963)
Chinese hamster V79	3	Up to 6 h	Increase	Belli & Shelton (1969)
Chinese hamster V79	4	During irradiation	No effect	Koch & Burki (1977)
Chinese hamster V79	0	During irradiation	Slight decrease, dose-multiplying	Ben-Hur, Elkind & Bronk (1974)
Chinese hamster V79	25	Held 4 h	No effect	Arlett (1970)
Chinese hamster	20	Maximum: 1 h	Increase, 1.08, dose-multiplying	Winans, Dewey & Dettor (1977)
Chinese hamster ovary	4 (irradiation at 4)	Effect progressive with time held	Decrease	
Chinese hamster ovary	4 (irradiation at 37)		No effect	Koch & Burki (1977)

holding at a low temperature, with bacteria or mammalian cells. Another method of postponement that acts rather more uniformly is to maintain irradiated cells for a while in water, buffer solution, or medium of less than normal nutrient quality; or to irradiate cells in stationary phase, and impose a delay before conditions are provided for growth and division. Treatments of that kind, which for brevity will be called 'delay in plating', often result in an increase in survival, classifiable as demonstrating P repair.

The term 'liquid holding recovery' is used mainly to describe a phenomenon which, if not peculiar to yeasts with two or more sets of chromosomes, appears to operate to its greatest extent with those organisms. Korogodin & Malumina (1959) first reported that, if irradiated diploid *Saccharomyces vini* were maintained in tap-water for 24 to 48 hours before plating, survival was increased in the dose-multiplying mode, the modifying factor being 2.5 (fig. 16.1). The phenomenon does not occur with haploid yeast. The substantial dose-multiplying effect of that treatment of irradiated yeast cells makes it possible to investigate how the modifying procedure may itself be modified, and a considerable body of information has accumulated. Mammalian cells of some lines have likewise been found to survive irradiation more successfully if they are held for a while in balanced salt solution before they are supplied with nutrients (e.g. Belli & Shelton, 1969; Hahn *et al.*, 1973; Evans *et al.*, 1974), but no dose-reducing effect has been reported that is as large as that for liquid holding recovery in yeast. Similar mechanisms may perhaps operate in the 'repair of potentially lethal damage' sometimes observed when mammalian cells are irradiated while in contact, in the plateau phase of growth, but are not separated and plated until some time afterwards. This effect was found to be enhanced when irradiated mammalian cells were held in 'conditioned' medium, that is, medium in which cells had already grown (Little, 1971). Such medium inhibits the growth of cells in exponential phase, or may even have a slight lethal effect (Horowitz, Norwint & Hall, 1975).

Energy requirements for P repair associated with delay in plating

This question has been investigated by groups working with mammalian cells and with yeast. With the latter, there has been general agreement on a requirement for the presence of oxygen (Korogodin *et al.*, 1967; Mosin, 1969), though Jain & Pohlit (1972) concluded that the evident requirement for energy metabolism could be met through the fermentative as well as the respiratory pathway. Respiratory inhibitors reduced the extent of liquid holding recovery, but their effect was counteracted if glucose was supplied. Mutants of yeast that have defective respiratory function are easily selected; and these were shown to be defective also in their capacity for liquid holding recovery. This capacity was greatly increased if they were supplied with glucose, but the glucose anti-metabolite 2-deoxy-D-glucose counteracted the recovery potentiated by glucose (Jain, Pohlit & Purohit, 1973).

Results of Koch, Meneses & Harris (1977) suggest an analogous requirement for an energy supply to mammalian cells during the repair process that is postulated to occur while plating is delayed. 'Extremely hypoxic' cells (Chinese hamster ovary, HA1 line) failed to manifest that type of repair unless they were supplied with glucose. However, it would be facile to conclude from these results that any simple rule about oxidative metabolism applies to P repair. Various chemical compounds chosen for their property of blocking oxidative phosphorylation may affect repair differently, in the same cell line; the same compound may increase repair in one class of cell, but reduce it in another; and, in general, effects seem to be critically dependent on the precise experimental conditions. Some results are shown in table 16.2. It has been suggested that some of the apparent contradictions are attributable to qualitative differences in the effects of respiratory inhibitors depending on their concentration. For example, Durand & Biaglow (1974) showed that 2,4-dinitrophenol, in concentrations less than $5 \times 10^{-3}\,\text{mol dm}^{-3}$, stimulated oxygen consumption by Chinese hamster V79 cells, and depressed it only when the concentration exceeded $12 \times 10^{-3}\,\text{mol dm}^{-3}$ – higher than that used in any of the experiments in which treatment of mammalian cells by dinitrophenol was observed to increase survival (table 16.2).

P repair in clonogenic tumour cells

When tumours contain clonogenic cells that can be cultured *in vitro*, with a reasonably high fraction of colony-formers among the plated cells, experiments can be conducted that are analogous with those *in vitro* in which there is a delay in plating after irradiation. Irradiated tumours may be excised, and the cells separated and plated immediately after irradiation; or they may be left *in situ* and plated after some time has elapsed. With several different tumour lines the fractional survival has been found to be higher in tumours left *in situ* for a few hours (e.g. Little *et al.*, 1973). There has been a lack of uniformity as regards the relative capacities for P repair of tumour cells that are oxygenated or hypoxic during and after irradiation (Chapter 17).

P repair in relation to growth phase and cell cycle stage

These variables have not been investigated to any extent with diploid yeast. Wienhard & Kiefer (1977) found that liquid holding recovery was constant as the cells grew through exponential to stationary phase, although that was not true for Elkind recovery, which suggested to them that these were independent processes.

An ability to repair potentially lethal damage was regarded by Phillips & Tolmach (1966) as a property of normally growing HeLa cells. This was inferred from their observations of reduced survival resulting from post-irradiation treatments, mainly by inhibitors of specific biosyntheses. All their measurements were on synchronized cells, irradiated at various times through the mitotic cycle. Ratios of survival of normal and abnormal cells appear to have been much the same for all stages. However, since responses to only one dose were measured, it is not clear whether the differences were attributable to reduction of P repair, Q repair, or both, in the treated cells.

In view of the variability among mammalian cell lines in respect of changes in survival curve parameters with phase of growth, as well as stage in the cell cycle, it would be surprising if there were uniformity with regard to the influence of those variables on the operation of P repair. 'Repair of potentially lethal damage' has been inferred from results of a variety of treatments, and cells of different lines clearly respond differently. Hahn & Little (1972) took the view that mammalian cells in the plateau phase, induced by overcrowding and/or exhaustion of nutrients, accumulated mostly in G_1, or between G_1 and S, and they associated the occurrence of P repair mainly with the plateau phase and therefore with the G_1 or G_1/S stages. In apparent confirmation, Hahn *et al.* (1973) found that P repair resulting from holding Chinese hamster (HA1 line) cells in balanced salts solution was less when they had been irradiated in the exponential than in the plateau phase.

However, P repair as a result of delay before regrowth has been found to occur in cells irradiated while in exponential growth, with several lines (Belli & Shelton, 1969; Little & Hahn, 1973; Horowitz *et al.*, 1975; Hetzel, Kruuv & Frey, 1976). Furthermore, effects have sometimes been greatest for cells in S at the time of irradiation (Winans *et al.*, 1972; Hetzel *et al.*, 1976).

Several apparent contradictions in results of experiments on 'repair of potentially lethal damage' in mammalian cells were cited by Hetzel *et al.* From their own observations they concluded that many factors were involved in the phenomenon, including cell-to-cell contact during and after irradiation.

Influence of some modifying procedures on P repair

Other than in experiments with bacteria there have been few in which the operation of 'repair of potentially lethal damage' has been examined also for cells irradiated *in vitro* under anoxia. In their investigation of effects of delayed plating on wild-type diploid yeast and several

Table 16.2. *Some effects of respiratory inhibitors added to irradiated cells before final plating*

(1) *Survival* Cells	Inhibitor	Concentration (mol dm^{-3})	Effect on survival	Reference
HeLa	DNP(a)	5×10^{-4}	'Over-repair' during Elkind recovery	Berry (1966)
Mouse L	DNP	5×10^{-5}	Slight increase in D_0 and n	Dalrymple, Sanders & Baker (1967)
Mouse L	DNP	5×10^{-5}	'Over-repair' during Elkind recovery	Dalrymple *et al.* (1969)
Mouse L	DNP	10^{-4}	Increase	Baker *et al.* (1970)
	NaAsO$_3$	10^{-4}	Increase	
	NaN$_3$	10^{-2}	No effect	
	KCN	10^{-2}	Slight decrease	
Rat sarcoma	DNP	10^{-4}	D_0 increase, 2.3 to 3.3 Gy	Nash *et al.* (1974)
Diploid yeast	DNP	4×10^{-5}	Very slight reduction, l.h.r. (b)	Patrick & Haynes (1964)
Diploid yeast	KCN	5×10^{-4}	Slight reduction	
Diploid yeast	NaN$_3$	10^{-3}	Slight reduction	

Diploid yeast	DNP	4×10^{-4}	No effect	Mosin (1969)
Diploid yeast	NaAsO$_3$	2×10^{-2}	No effect	
Diploid yeast	DNP	10^{-3}	No l.h.r. within 60 hours (b)	Jain & Pohlit (1972)
Diploid yeast	KCN	2×10^{-3}		
E. coli K12 wild type	DNP	3×10^{-3}	D_0 reduced, 45 to 30 Gy	Van der Schueren, Smith & Kaplan (1973)
pol A1			D_0 reduced, 9 to 6 Gy	
recombination-deficient and X-ray sensitive strains			No effect	

(2) *Biochemical effects*

Annealing of single-strand breaks inhibited by DNP in irradiated mammalian cells and bacteria: Moss *et al.*, 1971; van der Schueren *et al.*, 1973.

Adenosinetriphosphate (ATP) content of irradiated yeast cells reduced by DNP and NaAsO$_3$: Mosin, 1969.

Synthesis of ATP, DNA, RNA in unirradiated mouse L cells considerably reduced by DNP, KCN and NaN$_3$ but only slightly by NaAsO$_3$: Baker *et al.*, 1970.

(a) DNP: 2,4-dinitrophenol.
(b) l.h.r.: liquid holding recovery.

radiosensitive mutants, Averbeck & Ebert (1973) found that the treatment was less effective when cells had been anoxic during irradiation. This was true of all the strains they tested. The increase in survival resulting from the delay in plating was less marked with the more sensitive yeast strains, for both conditions of irradiation. An analogous observation on 'repair of potentially lethal damage' in the radiosensitive skin fibroblasts from patients with ataxia telangiectasia was made by Weichselbaum, Nove & Little (1978); in their experiments cells were aerobic during irradiation.

Information is sparse also with respect to the influence of radiation quality on P repair. The radiation response of *E. coli* B is reduced by post-irradiation conditions that are sub-optimal for growth, and at least as much effect of such conditions was found after irradiation by neutrons as by photons. (Alper & Gillies, 1958; Alper, 1963c). With diploid yeast, results of several groups have demonstrated that 'liquid holding recovery' is equally effective after irradiation at 'high' and 'low' LET (Lyman & Haynes, 1967; Bertsche & Liesem, 1976; Schneider & Kiefer, 1976). Lyman & Haynes observed as large a dose-reducing factor after irradiation by neon ions, LET about 5000 MeV $cm^2 gm^{-1}$, as after 50-kV X-irradiation.

Rasey, Nelson & Carpenter (1977) observed repair of potentially lethal damage, in cells derived from EMT 6 tumour, after irradiation *in vitro* by both neutrons and X-rays. Thus there is no evidence of a dependence of P repair on radiation quality for cells irradiated *in vitro*. Cells from Lewis lung carcinoma, irradiated *in vivo*, showed 'repair of potentially lethal damage' after X-rays, but not after neutron irradiation (Shipley *et al.*, 1975). However, the main contribution was of Q repair (p. 270). Clearly no generalization is yet possible about effects of radiation quality.

Intracellular repair of lesions in DNA

This field of research is very wide, embracing several modes of intracellular repair of DNA in which lesions have been induced by a variety of chemical and physical agents (see, e.g., Hanawalt & Setlow, 1975; Lehmann & Bridges, 1977; Lohman, Bootsma & Bridges, 1977). A topic that has aroused special interest in radiobiology is the enzyme-mediated repair of breaks induced by ionizing radiation in one strand of DNA, or both together, i.e. single- and double-strand breaks. So much effort has been put into investigations in this field, and into consideration of the role of unrepaired DNA strand breaks in cell death, that this aspect of DNA repair is singled out for a brief description of the methods and of some results.

Breakage of the DNA main chain was noted as one of the effects of ionizing radiation in early experiments on DNA irradiated extracellularly. One way of quantifying results is to centrifuge the irradiated material, often labelled with a radioactive tracer, in a tube containing a solute such as sucrose in fairly high concentration. During centrifugation the pieces of DNA will reach different levels, according to their molecular weight. Fractions of the tube contents are collected in order and the amount of DNA in each fraction measured by one means or another, the most usual being to measure the numbers of disintegrations in unit time, or 'counts', from the radioactive material. Thus a sedimentation profile is obtained, with peaks corresponding with the positions of the average molecular weights of the major moieties of the DNA. When assays are made by counts of radioactivity, the count for each fraction may be related to the total activity of the sample before centrifugation; but it has been much commoner to relate it to the total activity recorded for the sum of all the fractions, and, as pointed out by Yatvin, Wood & Brown (1972) this may, in some circumstances, introduce a serious artefact. This is discussed below.

In early experiments aimed at investigating breaks induced in DNA irradiated intracellularly, complications were introduced because extraction procedures involved mechanical shearing forces which themselves introduced a great many strand breaks. This problem was overcome by the technique introduced by McGrath & Williams (1966). They made use of two properties of strong alkaline solutions. Firstly, cells are lysed by strong alkali, so no further handling is required for the release of DNA, and secondly, strong alkali breaks the hydrogen bonds holding the two strands of the DNA together. Thus, in theory, all single-strand breaks become detectable (but of course not double-strand breaks). The first step in the procedure for 'velocity sedimentation' is therefore to place test cells gently on the top of a strongly alkaline sedimentation gradient.

The report of McGrath & Williams aroused great interest not only because of the technique they had introduced but also because their results seemed to account for the radiosensitivity of a bacterial mutant of the *E. coli* B family, B_{s-1}, that had already been shown to be deficient in ability to repair lesions induced by UV. After a single dose of 200 Gy to suspensions of B_{s-1} and the repair-proficient strain B/r, the sedimentation profiles showed that the molecular weight of the DNA released immediately was uniformly less than that of unirradiated bacteria. However, when lysis of the cells was delayed for 20 minutes, the sedimentation profiles for the two strains differed. The DNA from B/r had apparently returned to its original molecular weight, whereas that from the sensitive mutant had not. Some of the results of Kapp & Smith (1970), of Sedgwick & Bridges (1972) and of Youngs & Smith (1973) appear to support the generalization, based on the report of McGrath & Williams (1966), that certain bacterial radiosensitive mutants are less efficient than their wild-type parents in repairing DNA single-strand breaks.

At the time all those experiments were done it was already well known that a large proportion of bacterial DNA may be extensively 'degraded', i.e. enzymically chopped into very small fragments, after the organisms have been exposed to ionizing radiation (Stuy, 1961; Miletić *et al.*, 1961). However, the effect of that process on observations such as those made by McGrath & Williams seems not to have been appreciated until it was pointed out by Yatvin *et al.* (1972). The extent of DNA degradation of bacterial DNA depends on the bacterial strain and on the dose; and, as shown by Cramp & Watkins (1970), dependence on dose is not monotonic. As dose is increased the amount of intact DNA decreases to a minimum, but then it increases until no further degradation is seen (fig. 16.4). Thus extensive degradation of DNA during a period of incubation, after bacteria have been irradiated, could be interpreted as an increase in molecular weight. Similarly, repair might be under-estimated because of DNA degradation. This artefact would be missed if sedimentation profiles before and after incubation were subjected only to internal comparison, i.e. if the radioactivity in each fraction were measured in each case as a percentage of the counts in all fractions.

Figure 16.4 shows that the single dose of 200 Gy used in the experiments

Fig. 16.4. Recovery of acid-insoluble DNA after irradiation of two variants of *E. coli* B. Loss of acid-insoluble material shows the extent of degradation. (After Ahnstrøm, George & Cramp, 1978.)

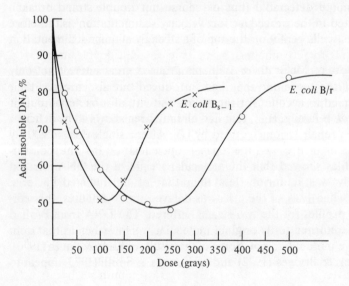

of McGrath & Williams may have been of about the right magnitude to result in almost maximum degradation of DNA in the irradiated B/r organisms, whereas in B_{s-1} the dose would have been high enough so that little degradation occurred. As explained in Chapter 2 and p. 228), misleading conclusions may be drawn about the effects of a process that depends both on dose and on time, if observations are made only after a single dose, and at a fixed time. Whatever the reason may be, the original results of McGrath & Williams could not be repeated by Drs W. A. Cramp & G. Ahnström (personal communication).

In the course of many experiments in which the McGrath & Williams technique has been used to observe single-stranded breaks several problems have been encountered.

Some of the earlier results obtained by lysing cells on top of alkaline-sucrose gradients may have been affected by an artefact pointed out by Elkind (1971) and by Sedgwick & Bridges (1972): if, during lysis, the cells are exposed even to quite dim light, results may be irreproducible because additional breaks are then produced.

A less easily avoided problem of the technique is the 'anomalous sedimentation' of cell contents unless doses of the order of 120 Gy or more are delivered to the cells before they are lysed. If lower doses are used, the DNA sediments rapidly as a complex (e.g. Elkind & Kamper, 1970; Ormerod & Lehmann, 1971) so formation and repair of single-strand breaks in mammalian cells can be investigated only after doses far beyond the range within which survival can be observed.

Accurate measurement of those processes depends also on complete denaturation of the DNA. Ahnström & Erixon (1973) and Ahnström & Edvardsson (1974) found that strand separation depended on the magnitude of the dose delivered to the cells as well as on the time of exposure to alkaline solution. After treatment of DNA from unirradiated cells by strong alkali for 20 minutes, most of it was still double-stranded. A different method was devised by Ahnström & Edvardsson (1974) to study the induction and repair of single-strand breaks induced in the DNA of irradiated cells. It depends on the initiation of unwinding in DNA, when breaks occur; treatment with mild alkali (NaOH, 2×10^{-2} mol dm^{-3}) stops the strands from reassociating. The suspension is neutralized and disrupted by sonication. The single- and double-stranded fragments are then separated by hydroxylapatite chromatography. The yield of single-strand breaks is calculated from the fraction of total DNA remaining in double-stranded form. Figure 16.5 shows their results on the induction and repair of breaks in V79 hamster cells by fast neutrons and X-rays. For induction of the same number of breaks, the dose of neutrons was larger by a factor of 4.4; but when the cells were incubated for 2 hours before they were treated by alkali, no residual breaks were detected in the X-irradiated cells, whereas some of these remained unrepaired after neutron irradiation (p. 225).

With minor modifications, that method of measuring single-strand breaks was used by Ahnstrøm, George & Cramp (1978) to re-examine induction and repair of breaks in *E. coli* B/r and B_{s-1}, and by Hesslewood (1978) to make comparable observations on a sensitive and comparatively resistant variant of the mouse lymphoma cell line L5178Y. Full dose–effect curves were used for calculating o.e.r.'s and other ratios. Hesslewood compared also the effects of fast neutron and X-irradiation. Some of the results are summarized in table 16.3.

Some points common to both sets of observations merit special attention:

(1) The comparative radiosensitivity of the more sensitive strains was not associated with residual breaks.
(2) O.e.r.'s for induction of breaks were considerably higher than for cell killing; and higher than several previous estimates.
(3) The fraction of residual breaks after maximum repair was higher in cells that had been anoxic during irradiation. This observation conforms with those of Modig, Edgren & Révész (1974) and Koch & Painter (1975). Table 16.3 shows that o.e.r.'s for residual breaks failed to match those for cell killing. However, by using bacterial mutants differing in specific repair pathways Town, Smith & Kaplan (1972, 1973) were able to distinguish three types of repair of single-strand breaks. One of these, 'Type I', was more effective after irradia-

Fig. 16.5. Residual single-strand breaks in the DNA of irradiated Chinese hamster cells exposed to 40 grays of neutrons or 9 grays of X-rays, as a function of time of incubation at 37 °C (after Ahnstrøm & Edvardsson, 1974.)

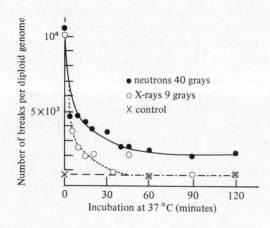

tion under anoxia. They concluded that this was a factor, but not the only one, in oxygen sensitization.

With values of o.e.r. for induction of strand breaks as high as those shown in table 16.3, it is tempting to speculate that primary events in cell membranes contribute, and that strand breaks induced by that mechanism are readily repaired. However, an analogous process would then have to be presumed to result from the association of virus DNA with its outer coat, since Palcic & Skarsgard (1975), using velocity sedimentation in alkaline-sucrose gradients, reported equal o.e.r.'s, about 3.3, for induction of single-strand breaks in the DNA of mammalian cells and of adenovirus. Substantial values of o.e.r. were reported also for induction of double-strand breaks in bacteriophage DNA, although oxygen did not enhance inactivation (Frey & Hagen, 1974).

Oxygen is not the only agent that differentially modifies biological inactivation and the induction of breaks in DNA. Hotz (1974) re-examined a phenomenon that had previously been reported: the presence of cysteine or cysteamine protects bacteriophage DNA against inactivation but concomitantly increases the ratio of double-strand breaks to lethal events. Hotz found that the increase in numbers of breaks was roughly the same

Table 16.3. *Comparisons of o.e.r.'s and values of RBE for cell killing and for initial and residual single-strand breaks in DNA of resistant and closely related sensitive strains, bacteria and mammalian cells. Strand breaks measured by 'DNA unwinding' and hydroxylapatite chromatography*

	Bacteria		Mammalian cells	
1. *o.e.r. values*	*E. coli* B/r	*E. coli* B_{s-1}	L5178Y/R	L5178Y/S
X-rays, killing	4.0	2.0	3.0	2.7
X-rays, initial strand breaks	6 to 7	6 to 7	7.8	7.8
X-rays, residual strand breaks	2.6	2.6	2.0	2.0
Neutrons, killing			1.5	1.5
Neutrons, initial strand breaks			4.7	4.7
Neutrons, residual strand breaks			2.7	2.7

2. *RBE values*, neutrons/X-rays (from slopes of dose–effect curves, irradiation in air)

Killing	2.1	1.3
Initial strand breaks	0.37	0.37
Residual strand breaks	1.3	1.3

After Ahnstrøm, George & Cramp (1978) and Hesslewood (1978).

for phage DNA treated by cysteamine after and before irradiation. He pointed out that no method had yet been devised for ascertaining that breaks had occurred in cellular or virus DNA before its extraction and subsequent handling. He suggested that irradiation might weaken bonds that could subsequently be attacked by secondary reactions during extraction and exposure to chemical environments of a profoundly different nature from those present at the time of irradiation. That suggestion is supported by the observation of Ahnstrøm *et al.* (1978) that residual breaks could be detected, after incubation in growth medium, when the solution in which the bacteria were lysed contained NaCl at 0.15 mol dm^{-3}, but not when the concentration was 0.05 mol dm^{-3}.

The contribution of unrepaired single-strand breaks to the killing of repair-deficient bacteria may perhaps be regarded as an open question, in view of the conflicting reports on this topic. There seem to be more uniform conclusions regarding the only exceptionally radiosensitive mammalian cells available for testing. The results of Hesslewood (1978) on L5178Y/R and S agreed with those of Fox & Fox (1973) although a different technique was used. Taylor *et al.* (1975) and Paterson *et al.* (1976) found no difference in the rate and extent of annealing of single-strand breaks in skin fibroblasts cultured from normal humans and those suffering from ataxia telangiectasia (AT cells).

Much less has been done on the induction and repair of double-strand breaks in the DNA of irradiated cells. This has been measured by observing the distribution of molecular weight of the DNA after velocity sedimentation in neutral sucrose gradient. It has been proposed that such breaks must be irreparable, and lethal, the reasoning being that an intact strand is required to act as templet for repair of a break, and that a double-strand break may be expected to lead to 'collapse' of the molecule. That view fails to take into account the attachment of the DNA to membrane. In any event, evidence of the repair of double-strand breaks has been found in bacteria (e.g. Burrell, Feldschreiber & Dean, 1971), mammalian cells (e.g. Corry & Cole, 1973) and yeast (Resnick & Martin, 1976). Lehmann & Stevens (1977) reported not only that double-strand breaks were repaired in skin fibroblasts, but also that the rate and extent of repair were the same for normal and AT cells.

Another mode of repair studied in cells after exposure to ionizing radiation is one that involves excision of regions of DNA, followed by resynthesis, the detection and measurement of which necessitates methods allowing this 'unscheduled' or 'repair' synthesis to be resolved from normal replication of DNA. This mode of repair can be observed after doses considerably less than those required for investigating DNA strand breaks by velocity sedimentation. Experiments aimed at correlating mammalian cell survival with capacity for this mode of repair have posed the question in different ways. For example, Shaeffer & Merz (1971) used seven mammalian cell lines, originating in humans, mice and a marsupial, to test a possible correlation between the extent of repair synthesis and relative radiosensitivity. No correlation was found. In case the lack of correlation was attributable to the method used by Shaeffer & Merz for measuring repair synthesis, Byfield, Lee & Kulhanian (1976) used two different methods to establish whether or not a correlation of that sort could be found, again with negative results. Scott, Fox & Fox (1974) compared closely related cell lines with different sensitivities, namely radiosensitive and resistant variants of L5187Y, and found repair synthesis to be more extensive in the former.

Excision of damaged DNA, followed by repair synthesis, was first demonstrated in repair-proficient bacteria that had been exposed to germicidal UV; and the sensitivity to UV of some mutant strains has been reasonably well correlated with deficiencies in one or more of the biochemical pathways involved in the process. The results of Scott *et al.* may therefore be relevant to the observation that the strain which is more sensitive to ionizing radiation is in fact much the more resistant to UV (Beer *et al.*, 1973; table 15.2).

From that point of view, it seems remarkable that Paterson *et al.* (1976) did find γ-irradiated AT cells to be defective in removing endonuclease-sensitive sites and in performing repair synthesis. Such defects could be expected to render the cells exceptionally sensitive to germicidal UV; but values of D_0 for UV survival curves have been about the same for normal skin fibroblasts and for AT cells from several different patients (table 15.3). Lehmann *et al.* (1977) investigated the extent of excision repair after UV irradiation in many lines of skin fibroblasts cultured from human patients, but did not report on that mode of repair in AT cells.

Observation of defective repair capacity in AT cells involved their being exposed to γ-irradiation while they were anoxic (Paterson *et al.*, 1976). This is an important point that has not attracted comment. Presumably that condition was used because previous results had shown it to be required for the induction of endonuclease-sensitive sites by ionizing radiation (Setlow & Carrier, 1972; Paterson & Setlow, 1972; Setlow, Faulcon & Regan, 1976; p. 214). According to the model discussed in Chapter 15, lethal damage attributable to primary energy deposition in DNA should play a relatively lesser role when cells are oxygenated during irradiation, so a search for correlation between cell survival and one or other mode of repair to DNA might well prove more successful if cells were made anoxic during exposure.

Repair or bypass?

Effectively all effort to account for genetic differences in cellular radiosensitivity, and to modification thereof, has gone into research into the repair of lesions in DNA. The success of that approach must be limited if a significant fraction of lethal damage is attributable to primary events in bacterial or nuclear membranes. Attempts to correlate P repair always with a specific mode of repair of DNA will also fail if, in some instances, cells can originate colonies despite the persistence of un-repaired lesions. An example suggesting bypass of damage, rather than repair, is afforded by the great variability in the response of *E. coli* strain B and some other strains possessing the same property. They have been characterized by a gene named 'fil' or 'lon', thought to confer a tendency to grow into filaments long enough to reach tens or hundreds of times their diameter. Survival after irradiation of *E. coli* B, and other 'fil$^+$' strains, is increased, in general, by conditions in which growth is slowed, and, in particular, by specific inhibition of protein synthesis (Gillies & Alper, 1959; Alper & Gillies, 1960; Gillies 1961; Forage & Gillies, 1964). Post-irradiation treatments that might be interpreted as evidence of 'repair' may operate by inhibiting filament formation, inhibiting the lysis of fila-ments (Brown & Gillies, 1972) or promoting the formation of cross-septa within filaments (Adler & Hardigree, 1964).

Under UV irradiation the lesion leading to filament formation in *E. coli* B was identified by Deering (1958) as being in nucleic acid; and, from the similarity in the modes of modification of survival by post-irradiation treatment, it was inferred that this is also a result of exposure to ionizing radiation, classified as type N damage (Alper, 1963b). But the observations of Brown & Gillies in particular suggest that relatively high survival is achieved if the lysis of filaments is inhibited. Such a phenomenon could be regarded as evidence that the effects of a lesion can be bypassed; or, alternatively, that the extent to which some lesions are lethal depends on post-irradiation conditions. In other words, some forms of damage require to be biochemically 'fixed' if cell death is to result.

The details of that mode of killing of *E. coli* B and other filamentous strains do not apply even to closely related ones (Alper & Gillies, 1958; Adler & Hardigree, 1965). But they suggest a possible interpretation of 'repair of potentially lethal damage' also in some higher cells, as was indeed proposed by Belli & Shelton (1969); in which case a correlation between increase in survival and a positive process of repair of DNA would be difficult to establish.

Other than chromosomal aberrations, morphological changes con-sequent on the irradiation of mammalian cells have received compara-tively little attention. Their growth into 'giant cells' was noted by Puck &

Marcus (1956) and properties of such cells were investigated by Whitmore *et al.* (1958) and Tolmach & Marcus (1960), but the doses given were too large to leave detectable viable cells, i.e. cells that might give rise to viable daughters despite morphological distortion. However, observations of Dettor *et al.* (1972) by electron microscopy led them to suggest that the sensitizing action of hypertonic treatment on Chinese hamster cells was attributable to 'confirmation' of damage that would otherwise not be lethal.

Korogodin *et al.* (1967) could detect no changes in yeast cells during the period of liquid holding recovery. These appeared when they were plated subsequently, and viable daughters were seen to come from cells that at the same time manifested gross morphological distortion. However, it was not made clear whether the proportion of abnormal-appearing but still viable cells was greater in the population that had recovered than in those plated immediately.

Stimulation of repair by irradiation

As mentioned on· p. 170, survivors of an exposure to a conditioning dose, followed by a radiation-free interval, had an increased capacity for repair, in experiments with HeLa cells and three species of algae. A similar effect was reported also for bacteria by Pollard & Achey (1975), in experiments based on a different approach. Pollard and his co-workers had previously found that ionizing radiation, UV and certain chemical agents could induce a factor controlling the extent of DNA degradation, and it was inferred that the agent was a protein. Using several variants of *E. coli* K12, Pollard & Achey set out to test whether a conditioning dose followed by incubation in nutrient medium would induce a factor that would render the survivors more resistant to ionizing radiation with respect to their colony-forming ability. The prediction was proved correct, except in the case of three UV-sensitive strains that were deficient in specific pathways for repair to DNA. The repair-promoting factor was induced by exposure to nalidixic acid, which inhibits DNA synthesis, as well as to γ-rays and UV irradiation, the latter being used for most of the results reported. Additional capacity for repair was manifested both by decreased terminal slopes to the survival curves and, to a somewhat greater extent, increased extrapolation numbers.

Additional capacity for repair in *Chlamydomonas reinhardii* is demonstrated by an increase in the value of D_0 for survivors of a conditioning dose, after an interval for recovery. Bryant (1975) found that this change in response could be prevented by treating the cells with inhibitors of protein synthesis during the recovery period. He also observed that the development of the decrease in sensitivity depended on the magnitude of the first dose and on time, the maximum effect requiring 5 hours (Bryant, 1976). An important observation was that the effect of a given dose was the same whether oxygen was present or not (fig. 16.6). The o.e.r. for cell killing was the same for cells that had been rendered more resistant as for previously unirradiated cells.

Additional repair capacity induced in two other species of algae was observed as increased capacity for Q repair: shoulders of survival curves were considerably wider for the survivors of a conditioning dose than for previously unirradiated cells of *Oedogonium cardiacum* (Horsley & Laszlo, 1971) and *Clostiferum moniliferum* (Howard & Cowie, 1975). In both cases the values of D_Q for the survival curves were increased by factors 1.5 to 2, and this 'over-repair' was suppressed when the cells were treated by an inhibitor of protein synthesis during the interval after the conditioning dose (Horsley & Laszlo, 1973; Howard & Cowie, 1976). Inhibition of protein synthesis did not affect the extent of normal Elkind recovery in the survivors of a conditioning dose to *Oedogonium* (Horsley & Laszlo, 1973), as had been observed also with mammalian cells (Berry, 1966; Elkind, Moses & Sutton-Gilbert, 1967a). Howard & Cowie (1976) observed that the process leading to 'over-repair' could be switched on also by a conditioning dose of UV.

Fig. 16.6. Increase in value of D_0, survival curves for *Chlamydomonas reinhardii*, as a function of time after a conditioning dose of 150 grays given to aerobic or anoxic cells. Vertical bars show 95% confidence intervals on estimates of D_0. (After Bryant, 1976.)

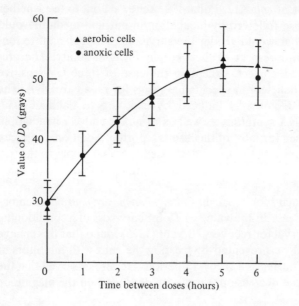

To the extent that the phenomenon has been investigated, there is a measure of uniformity in the results of investigations with *E. coli* and the three species of algae, and they suggest a possible relationship with the initiation of synthesis of 'protein X' in *E. coli*, which has been found to be switched on when DNA synthesis is inhibited by a variety of agents, including UV (Inouye & Pardee, 1970; Gudas & Pardee, 1976). That protein was not synthesized in one of the strains which Pollard & Achey had found non-inducible for increased radioresistance. Ionizing radiation is well known also temporarily to inhibit DNA synthesis. However, there may be a difference in the process of over-repair in *Chlamydomonas*, since Bryant & Parker (1977) found Elkind recovery to be inhibited if cells were exposed to UV just before a conditioning dose of ionizing radiation. It may be, of course, that the inhibition of normal Elkind recovery masked a process that would have caused the terminal slope of the survival curve to decrease. Since the switching on of over-repair by ionizing radiation depended on dose, it might be expected that its occurrence after UV, if it occurred at all, would likewise depend on the dose.

Not many mammalian cell lines have been studied as exhaustively as Chinese hamster cells in respect of the parameters of secondary survival curves after a conditioning dose. The phenomenon of over-repair seems worthy of further investigation not only in the classes of cell in which it has been most easily demonstrated, but also in HeLa; and the possibility that it occurs also in other neoplastic cell lines warrants consideration, because of the important implications for radiotherapy.

17

Some aspects of cellular radiobiology
in vivo

Only a few selected topics are dealt with here, since several text-books and reviews have focussed attention primarily on the relevance of experimental radiobiology to medical radiology, in particular to radiotherapy.

As will be evident from preceding chapters, no generalizations can be made about the details of radiation responses of cells, even within the restricted class of mammalian cells; and no features have yet been detected that distinguish between neoplastic and normal cells in that respect. Some confidence may be felt in ascribing early effects on normal tissues to the death of clonogenic cells, but such correlations as exist depend on observations of the survival of those cells *in situ*; there is limited, but convincing, evidence of important differences when cells originating in normal tissues are grown and irradiated *in vitro*.

Survival curves *in situ* and *in vitro* for cells from normal tissues

As shown by Field (1969) and Field, Morgan & Morrison (1976), there are grounds for the assumption that observations on mouse skin reflect the responses to radiation of human skin; and the group of results most suitable for a direct comparison of cell survival *in situ* and *in vitro* are those made on mouse skin *in situ* and on freshly cultured human fibroblasts.

Survival curves for the latter have repeatedly been shown to be exponential, or very nearly so (Puck *et al.*, 1957; Norris & Hood, 1962; Cox & Masson, 1974; Weichselbaum *et al.*, 1976). This means that no Elkind recovery is expected, and indeed none was observed by Cox & Masson. But it has long been the experience of radiotherapists that administering doses in fractions separated by one or more days spares skin greatly, and this has been shown to be true also of reactions to radiation by the skin of pigs (Fowler *et al.*, 1963), rats (Field, Jones & Thomlinson, 1967) and mice (Fowler *et al.*, 1965). In the direct measurement of survival of stem cells *in situ* in mouse skin, doses given in two fractions separated by some hours were substantially larger than the single dose to give the same effect: the value of $D_2 - D_1$ (p. 170) for a 24-hour recovery interval was 3.75 Gy, according to Withers (1967c). Slopes of survival curves constructed by Emery *et al.* (1970) were about the same for single-shot and D_2 curves when the fractionation interval was 7 hours; $D_2 - D_1$ was about 4 Gy. Fowler *et al.* (1965) constructed dose–effect curves from their observations on skin reactions in mice after irradiation by 15-MeV electrons and calculated values of $D_2 - D_1$ ranging from 2.9 to 8.4 Gy for a 24-hour fractionation interval. Obviously there can be no comparable experimental evidence with respect to skin reactions in humans; but some observations were made on the effects of comparatively low doses on small areas of the skin of volunteers in order to establish the RBE of the neutron beam from the Medical Research Council cyclotron, before neutron radiotherapy was undertaken. From comparisons of the dose–effect curves for single doses to two volunteers and six fractions to another volunteer, it may be deduced that $D_2 - D_1$ was not less than 3 Gy (Field *et al.*, 1976).

Survival curves for clonogenic cells *in situ*, in other organized tissues, may also be deduced to have comparatively large values of D_Q, and therefore of extrapolation number, as shown in table 17.1. Values of *n* are higher than for survival curves on most established cell lines *in vitro*.

It is probably not coincidence that survival curves have also been exponential, or nearly so, for cells freshly cultured from organized tissues other than skin (Puck *et al.*, 1957; Norris & Hood, 1962). As noted on p. 198, Durand & Sutherland (1972) found that cells grown *in vitro*, as spheroids, had a greater capacity for Q repair than single cells, and they gradually lost this capacity after separation. It is plausible that capacity for Q repair should be affected when cells from organized tissues are subjected to the much more drastic procedures necessary to get them growing *in vitro*. This capacity may partly be restored as cultures become established (Cox & Masson, 1974).

A small, but significant, difference in the extrapolation numbers of the survival curves for stem cells in haemopoietic tissues was observed when one was constructed from counts of clones on the spleens of irradiated mice, arising from endogenous surviving cells, while the other involved irradiation of haemopoietic tissue *in vitro*, followed by injection into mice previously irradiated with a dose large enough so that there were virtually no endogenous colony-formers left (Till & McCulloch, 1963; fig. 17.1). It would appear that the comparatively small capacity for Q repair in those cells can be reduced by taking tissues from an animal, keeping it *in vitro* for a period, then injecting it into another. Survival curves for stem cells in haemopoietic tissue irradiated *in vitro* and scored

Table 17.1. *Extrapolation number (n) to survival curves on clonogenic cells in situ, radiation of low LET. Calculation of n: (1) Best estimate of D_0, and maximum value of $D_2 - D_1$, for fractionation intervals up to 24 hours, assumed to give D_Q: $n = e^{D_Q/D_0}$. D_2 is the sum of two doses, separated by a recovery interval, required to leave the same surviving fraction as the single shot D_1. (2) n given by the maximum ratio of survivors of the same dose fractionated (by intervals not more than 24 hours) and given as a single shot.*

Tissue	D_0 (Gy)	Interval between fractions (hours)	D_Q ($D_2 - D_1$) (Gy)	n (survival ratio)	$n(e^{D_Q/D_0})$	Reference
Growing cartilage, rat leg	1.6	24	4	15	12	Kember (1967)
Epidermal cells, mouse	1.25	24	3.5	20	16	Withers (1967c)
Epidermal cells, mouse	1.3	9, 24 7	4.2		25	Emery et al. (1970)
Crypt stem cells, mouse jejunum	1.1	24	3.3	20	20	Withers, Brennan & Elkind (1970)
Crypt stem cells, mouse jejunum	1.2 (a) 1.6 (b)	24 24	5.2 (a) 6.9 (b)		76 75	Hornsey (1973)
Spermatogenic stem cells, mouse testes	1.8	4 4	2.7 (c)	5 (c)	4.5 (c)	Withers et al. (1974)

(a) 7-MeV electrons, 60 Gy min^{-1}.
(b) 8-MV X-rays, 1 Gy min^{-1}.
(c) Neither $D_2 - D_1$ nor the survival ratio reached a constant value. Recovery increased with increasing dose per fraction.

for colony-formers in recipient animals have sometimes been exponential (Silini & Maruyama, 1965).

Thus the limited amount of information available on survival curves for cells left *in situ* after irradiation allows comparisons to be made with those constructed *in vitro*, or *in vivo* after being handled *in vitro*; and it must be concluded that survival curves *in situ* are not then adequately represented. In addition to the artefacts that may be introduced into such assays by the act of separating cells, there is also the possibility of misleading results introduced by changes in temperature and in the nutritional environment after irradiation, as exemplified by some of the results cited in Chapter 16. It is unlikely that all possible artefacts could be avoided, even with meticulous care taken to standardize procedures in every detail. It cannot be assumed, therefore, that survival curve parameters for clonogenic cells left *in situ* are reflected by those pertaining to the same cells taken from excised tumours that have been irradiated in the animal.

The initial regions of survival curves for normal tissues *in vivo*

Cell survival curve parameters cited in preceding chapters have mostly referred to terminal regions. But if those regions are not approached until doses of several grays have been delivered to the cells, the information will be of little practical value in the evaluation of radiation hazards, and of limited value in its application to radiotherapy, since in most radiotherapy regimens dose fractions are of the order of 2 or 3 grays. Measurement of parameters like the enhancement ratio for

Fig. 17.1. Split-dose timing curves for colony-forming cells in haemopoietic tissue, assayed by counting spleen nodules (adapted from Till & McCulloch, 1963).

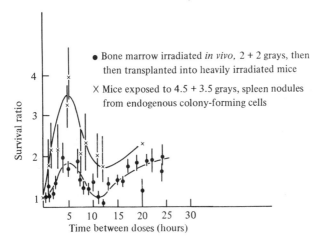

a sensitizing drug, or the o.e.r., has usually been based on comparisons between terminal regions of survival curves. If a modifying treatment is precisely dose-multiplying such measurements are valid also for the initial regions of dose–effect curves: but there is some evidence of different multiplication factors in initial and terminal regions, for example in the sensitizing action of oxygen (e.g. Chapman *et al.*, 1975; McNally & de Ronde, 1976). As shown in Chapter 14, survival curves may have a common extrapolation number even if dose-multiplication factors for initial and final slopes are different. When the effects of neutrons and X-rays are compared, differences in initial slopes may be much larger than in terminal ones (fig. 9.12; Withers, Brennan & Elkind, 1970).

As explained on p. 16, clones originating from surviving stem cells in organized tissues cannot be counted until the majority have been killed. For example, doses of the order of 10 Gy must be given before skin clones can be counted in mice (Withers, 1967a). The radiation responses of some normal tissues have been quantified only in terms of doses needed for the achievement of a given degree of injury, and dose–effect relationship are then assumed to reflect the killing of the relevant clonogenic cells. In that kind of experiment, too, single doses of the order of 10 Gy or more may be required before any response is detectable. This is true when skin reactions are used to assess effects of radiation on normal tissues, a guide frequently used since Fowler *et al.* (1963) devised a method for quantifying responses on the basis of subjective judgment of the severity of skin reactions.

The only method devised so far for 'measuring' initial slopes of survival curves *in vivo* has been the observation of effects of small dose fractions, each followed by an interval considered sufficient for full Elkind recovery. If the total accumulated dose is not sufficient for the realization of a detectable response, these may be followed by a large enough single 'test dose'. In such experiments it is assumed that doses are in the region of an initial defined slope if there is no detectable difference in the effect of a single dose and that same dose split into two or more fractions separated by intervals for recovery (e.g. Wambersie *et al.*, 1974).

The administration of a few small dose fractions followed by a single test dose, to make the overall effect detectable, has been used in experiments in which measurements were made of total doses to give a defined amount of tissue injury (e.g. Wambersie *et al.*, 1974; Dutreix & Wambersie, 1975; Denekemp & Harris, 1975b; Field & Hornsey, 1975); and the parameter evaluated was the ratio of the 'final' to the 'initial' slope (see Elkind, 1975, for summary). The same general experimental design was also used by Withers and his co-workers, whose measurements involved the counting of clones originating from surviving cells in normal tissues (e.g. Chen & Withers, 1972; Withers, 1975a; Withers *et al.*, 1975). Since the value of D_0 for the terminal region of a survival curve *in situ* can be directly measured by that technique, multi-fraction experiments should yield the value $_1D_0$, for the initial slope, given the validity of certain assumptions, as discussed below. From the survival curve for jejunum crypt cells, as reconstructed by Withers *et al.* (1975), it may be seen that the value of $_1D_0$ was of the order of 4 Gy.

A somewhat different experimental design was used by Douglas & Fowler (1976) who examined the relationship between the number and magnitude of repeated dose fractions, down to 1.07 Gy per fraction, required to achieve a given level of injury to the skin of mice. Their method of establishing parameters for the initial region of the survival curve, assumed to be reflected by the skin injury, was analogous with that shown

Fig. 17.2. A plot of the function F_e (inversely proportional to the product of dose per fraction and fraction number) for achievement of a given level of reaction in the skins of mice. Open symbols refer to irradiations given to anaesthetized mice; closed symbols to unanaesthetized mice. All fractions administered within 8 days. (After Douglas & Fowler, 1976.)

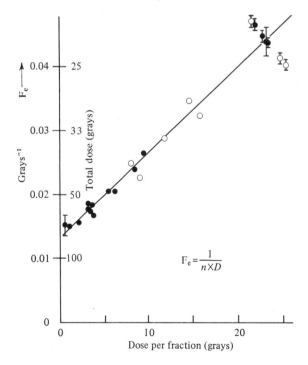

$$F_e = \frac{1}{n \times D}$$

in fig. 14.3. They posited that fractional survival, s, is given by

$$\ln s = F_e \cdot D \tag{17.1}$$

where D is dose and F_e is a function of dose to be determined. Thus F_e may be established from the relationship between $\ln s/D$ and dose. It is assumed that equation 17.1 holds for each of n fractions of magnitude D, and that the predetermined level of injury is achieved when fractional survival has been reduced to S. If n fractions are required for the achievement of the end-point, $S = s^n$, and $\ln S = n \cdot \ln s = n \cdot F_e D$. Since S is constant, F_e may be established by plotting $1/nD$ as a function of dose per fraction. Douglas & Fowler's results, when plotted that way, yielded a straight line, from which it could be concluded that F_e was given by $-(\alpha + \beta D)$, so that equation 17.1 was of the form $\ln s = -(\alpha D + \beta D^2)$ (fig. 17.2). Their plot of $1/nD$ against dose per fraction was linear up to the value of the single dose needed for the end-point, about 22 Gy. However, as previously explained (p. 194), fig. 14.4), the appearance of linearity is not a dependable guide to the range of doses over which a survival curve is 'continuously bending'. If the skin injury reflects the killing of clonogenic cells, observable by the technique devised by Withers (1967a), the dose–effect curve constructed by Douglas & Fowler might have been expected to approximate to a terminal exponential with doses of 9 Gy or more, since the same strain of mice was used in both series of experiments. In his observations on Elkind recovery in the clonogenic cells, Withers (1967c) showed the survival curves for single shots and two dose fractions to be parallel, and, as explained on p. 192, this result cannot be realized in such experiments if the single-shot curve is continuously bending. Emery et al. (1970), who used a different strain of mice, and found slopes to be about the same for the single-shot survival curve and for the two-fraction curve with a 7-hour recovery interval between them, fitted those curves to counts of clones formed after total doses of 13.2 Gy and more.

It is possible, however, that a survival curve reconstructed from observations of the effects of many small fractions may be continuously bending although the single-shot curve is exponential in the higher dose region investigated. Several assumptions are involved in the delineation of survival curves, or the initial regions thereof, from experiments requiring many dose fractions. One of these is that cells surviving one or more fractions always respond to further doses as if they had not previously been irradiated. Specific examples were quoted in Chapters 13 and 16 showing that this is not necessarily so. While there are not many such examples, there are perhaps even fewer recorded tests of the assumption: the early demonstration of its validity over five dose fractions, for Chinese hamster cells (Elkind & Sutton, 1960) has been adopted as the basis for interpreting results of multi-fraction experiments. But the response of V79 Chinese hamster cells irradiated in stationary phase was found to depart from that pattern when a larger number of dose fractions was given (McNally & de Ronde, 1976). They observed a diminution in Q repair, after the cells had been exposed to many dose fractions; this would cause a reconstructed survival curve to bend continuously.

There are several other assumptions involved in the deduction of survival curve parameters from experiments in which many, or even only two dose fractions are used. The interval between successive fractions must be long enough to ensure full Elkind recovery, but not so long that some viable cells will divide. As pointed out by Withers (1975b) there is also the possibility that an initially asynchronous population may become synchronized to a certain extent during a fractionation regimen, because cells in the more sensitive stages of the cycle will be selectively killed by each fraction. The extent of the distortion that 'redistribution' might introduce would of course depend on many factors.

Experiments involving the administration of many fractions to experimental animals are extremely laborious and, for reasons given above, the results are difficult to interpret. But they are indispensable as a means of discovering how normal tissues respond to dose fractions of a size comparable with those commonly given in radiotherapy schedules. Experiments with that aim have been designed in different ways, and have sometimes led to different conclusions (see e.g. Withers *et al.*, 1975, for discussion). Perhaps the identification of the reasons for these differences would in itself contribute to our insight into the modes of response of normal tissues to low doses of radiation.

Some questions of interpretation may arise even in results of experiments in which only two dose fractions are used, to deduce a value of D_Q from $D_2 - D_1$. If results can be plotted in the manner of fig. 13.6A, it is plausible that $D_2 - D_1$ represents a real value of D_Q, although it will not be the true value for the single-shot curve if the recovery interval has not been long enough. But some two-dose-fraction experiments on tumours as well as on normal tissues have shown $D_2 - D_1$ to increase with increasing D_1, no plateau being reached. (e.g. Field, Jones & Thomlinson, 1968; Withers *et al.*, 1974a). One interpretation is that the largest practicable conditioning dose is not large enough for survival to be 'off the shoulder' of the curve. But if there is no independent information about the single-shot survival curve, failure of $D_2 - D_1$ to reach a maximum value might equally well suggest that 'over-repair' was occurring, as with HeLa cells (Lockhart, Elkind & Moses, 1961; see Chapter 16).

Oxygen as a radiosensitizer *in vivo*: normal tissues

The enhancing action of oxygen in killing clonogenic cells in animal tissues can obviously not be examined in as much detail as *in vitro*. Conventional measurement of o.e.r. requires the comparison of dose–effect curves taken with partial pressures of oxygen near zero and high enough to confer 'full oxygen sensitization'. A method often used for rendering tissues anoxic is to occlude the supply of blood, for example to the limbs or tails of experimental animals, or to grafted tumours if these are in a suitable location. Mice can tolerate asphyxia for 40 to 50 seconds, so, if sufficiently high dose rates are available, a dose can be delivered within the last few seconds of that period (Wright & Batchelor, 1959). In some experiments irradiation *in vivo* is required, although the survival of the animal is not. It is common practice in that case to kill the animals before irradiation, the presumption being that tissue oxygen partial pressures will rapidly fall to zero.

It is easier to achieve 'radiobiological anoxia' by such means than to be sure that the P_{O_2} in the tissues of animals breathing normally is high enough to confer full oxygen sensitivity. The radiosensitivity of several tissues has been found to increase, if animals breathe oxygen at one atmosphere, or higher pressures, instead of air during irradiation. This may be because the tissue is 'radiobiologically hypoxic' as a whole, or because there is normally considerable variation in P_{O_2} among the cells, so that at any instant a minor fraction will be markedly hypoxic. The former case should be recognizable by a dose-multiplying relationship between effects on air- and oxygen-breathing animals, the latter by dose–effect curves that diverge with increasing dose. However, the usable range of doses is not always sufficient for this distinction to be drawn. Where it can be, this is indicated in table 17.2, a summary of some relevant results.

Anaesthesia can evidently reduce tissue P_{O_2} (Lindop & Rotblat, 1963), but, as shown by Hornsey, Myers & Andreozzi (1977), not in the tails (or, probably, other extremities) of rodents, which have been used by several groups investigating normal tissue reactions. Hornsey *et al.* found that skin reactions in mouse-tails were more severe for oxygen- than air-breathing mice when they were unanaesthetized; but little difference for anaesthetized mice. They attributed this result to the function of rodents' tails in temperature regulation, the mechanism for which is suppressed by anaesthesia. The temperature of the animals falls partly because the blood-vessels in the tail dilate; and the increased blood-flow evidently results in a higher P_{O_2}. Hornsey *et al.* suggested that this phenomenon might also affect observations of tissue reaction in rodents' feet.

Some tissues are regarded as uniformly hypoxic, like cartilage, and this is thought to demonstrate a P_{O_2} lower than that of other tissues which do not demonstrate a comparable increase. As regards growing cartilage, Kember (1967) pointed out that this assumption is not necessarily correct: even if blood-vessels are comparatively remote from clonogenic cells, oxygen may diffuse to them through non-metabolizing tissue. An alternative, or additional explanation may, however, be found in the relationship between radiosensitivity and P_{O_2} expressed by equation 6.1, and the probability that the constants differ as between clonogenic cells in different tissues. This has indeed been demonstrated for the parameter m (overall o.e.r.). Hornsey (1971) reported that the o.e.r. for haemopoietic tissue was about 2.1. This was based both on the counting of endogenous spleen clones and on the LD_{50} at 30 days; the 95% confidence interval for both tests together was 1.8 to 2.3. For mice irradiated in the same conditions, namely under anoxia produced by asphyxiation for 35 seconds, or while breathing oxygen at one atmosphere, the o.e.r. was 2.7 both for death at 4 days and for the killing of stem cells in the crypts of Lieberkühn (Alper & Hornsey, 1968; Hornsey, 1970).

No measurement has been made of a K value for mammalian cells *in vivo*, but it would be unrealistic to expect it to have the same value for clonogenic cells in different normal tissues, since experiments on mammalian cells *in vitro* have shown that the value of K depends not only on the cell line, but also on how cells are irradiated, or how they are grown before irradiation (table 6.4). Values of K reported up to the present for mammalian cells *in vitro* range from about 2 mm Hg (Cullen., 1976) to 12 mm Hg (Moore, Pritchard & Smith, 1972), but there are no grounds for supposing that values of K for cells *in vivo* cannot lie outside this range. Clearly more data are desirable, and it would be helpful to have them even for cells *in vitro*, irradiated otherwise than singly and while in exponential growth. Some of the results quoted in table 17.2 might well be accounted for by a relevant K value of the order of 10 to 15 mm Hg.

It has been suspected for several decades that some human tumours may be difficult to sterilize by radiation because they contain hypoxic but viable cells (e.g. Mottram, 1935). This has been, and still is, a focal point of attack in researches aimed directly at increasing the success of radiotherapy in controlling human tumours. A method that has been used clinically for some years has been irradiation of patients while they breathe oxygen at high pressure, in a specially constructed chamber. The rationale is that an increase in P_{O_2} of all tissues will raise that of the hypoxic tumour cells; whereas normal tissues, at one time thought to be 'at full oxygen sensitivity', would suffer little or no increase. That view was based on early observations relating sensitivity to P_{O_2}, and these indicated a linear rise in sensitivity followed by an absolute plateau (cf. Gray *et al.*, 1953). However, there is no such plateau in the relationship

260

Table 17.2. *Investigations bearing on extent of radiobiological hypoxia in normal tissues (radiation all of low LET). 'Sensitivity ratios' are calculated as the inverse of the ratio of doses to give an effect when the respired gas is oxygen, at one or three atmospheres, and the dose to give the same effect when air is respired. Results calculated for effects at dose D, air breathing*

Tissue	Test of damage	Dose rate (Gy/min)	Ratio O₂ (1 atmos)	at dose D, grays	Ratio O₂ (3 atmos)	at dose D, grays	Ref.
1. No anaesthesia							
Growing cartilage, mouse tail	stunted growth	0.9	1.30	5 to 15 DM (a)	1.36	5 to 15 DM	1
Skin, other tissues, mouse tail	tail necrosis after 6 months	4.3	1.3 / 1.4	49 (b) / 47			2
Skin, mouse tail	skin reaction 12–85 days after irrad.	1.7	1.0 / 1.4	20 to 30 (c) / 35 to 47			3
Haemopoietic tissue, mouse	50% deaths, 6–30 days	0.5			1.13	5.45 }	4
Intestinal mucosa, mouse	50% deaths in 5 days	0.5			1.14	10.3	4
2. Under anesthesia							
Skin, human patients	faint to moderate erythema, 20–100 days after irrad.	3			1.33 / 1.75	2.5 / 12.5	5
Skin, mouse leg	reaction, average over 30 days	240 to 360	1	14	1.4	20 }	6
Skin, mouse leg	death of clonogenic cells	2	1.9	40	2.0	63	7
Skin, mouse leg, two strains of mice	death of melanocytes	2.3	1.28 (d)	4 to 16, DM (d)	1.08	10 }	8
Skin, mouse tail	skin reaction, 12–85	1.7	<1.1	25 to 35	1.24	20	3

Growing cartilage, rat knee	death of clonogenic cells	2.0	1.1	16 to 20 ~DM (e) 4 to 10, DM	1.13	16 to 20, DM (e) 4 to 10, DM	9
Growing cartilage, rat tail	stunted growth	3.5	1.09		1.22		10
Intestinal mucosa, mouse	death of clonogenic cells	5 60	1 2.7 (f)	21 27			11

Notes:
(a) DM = dose multiplying.
(b) mice at ambient temperature. ⎱ Hendry *et al.* (1976).
(c) mice at 37°C. ⎰
(d) Potten & Howard's results for the two strains differed in respect of survival curve parameters, but not in respect of the ratio. The figure given is the mean for both strains.
(e) The best survival curves fitted by Kember had different extrapolation numbers for air-breathing and anoxia; but the difference was not significant.
(f) Curves for air- and oxygen-breathing coincident to 27 grays; with higher doses the curve for air-breathing and nitrogen asphyxia had the same slope.

References: 1, Howard-Flanders & Wright, 1957; 2, Hendry *et al.*, 1976; 3, Hornsey *et al.*, 1977; 4, Christensen *et al.*, 1969; 5, van den Brenk *et al.*, 1965; 6, Fowler *et al.*, 1965; 7, Withers, 1967b; 8, Potten & Howard, 1969; 9, Kember, 1967; 10, Dixon, 1968; 11, Hornsey, 1970.

described by equation 6.1. If that relationship holds *in vivo*, the effect of irradiating a patient breathing oxygen at high pressure will be to increase the effective dose to a normal tissue in a manner which depends on the particular values of m and K relevant to the clonogenic cells in that tissue. Figure 17.3 shows the effective increase in dose to normal human tissues caused by irradiation under high-pressure oxygen, depending on the relevant values of m and K. Calculations were based on the assumption that 'full oxygen sensitivity' will be achieved under hyperbaric oxygen, and that the P_{O_2} in the tissues of humans breathing air is 40 mm Hg.

Early clinical use of high-pressure oxygen radiotherapy demonstrated increased serious damage to cartilaginous structures (Churchill-Davidson *et al.*, 1966); this may reflect the comparative radiobiological hypoxia of stem cells in cartilage, first reported by Howard-Flanders & Wright (1955). Observations of increased morbidity in other tissues (e.g. Dische, 1978) can, however, not so convincingly be correlated with evidence from experiments on animals. This may well be because most of the observations quoted in table 17.2 refer to early effects, rather than the 'late' damage which is the hazard of concern to the radiotherapist.

Dose–response curves for tumours left *in situ*

Direct observation of cell survival *in situ* in irradiated tumours has not been achieved, but two methods have commonly been used to

Fig. 17.3. To show how the values of K and m for cells in normal tissues will determine the effective increase in dose when oxygen at 3 atmospheres is respired. Calculations are based on the assumption that the P_{O_2} of oxygen normally present is that of venous blood (~40 mm Hg). It is also assumed that 'maximum oxygen sensitivity' will be attained for all cells in normal tissues when oxygen is breathed at 3 atmospheres. Arrows on abscissa of left-hand panel show the range of K values so far reported for mammalian cells *in vitro*.

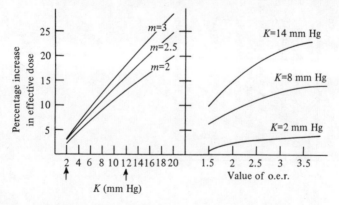

construct dose–effect curves. One of them is to give graded doses to groups of grafted tumours of about the same size, and relate the radiation dose to the fraction of tumours that fail to regrow after an arbitrarily chosen period. The parameter used to describe the results is often the 'TCD50', i.e. the dose required to control, or sterilize, 50% of tumours irradiated. Another method, developed by Thomlinson (1960), is to implant small volumes of tumours just below the dermis in a site in which they will grow into spherical tumours not firmly attached to underlying tissue. Diameters in three dimensions are then measured regularly. After doses that will leave some viable clonogenic cells the tumours will regrow; the smaller the surviving fraction, the longer it will take for the growing tumour to reach a given size (fig. 17.4). Delay in reaching an arbitrarily chosen size was plotted by Thomlinson & Craddock (1967) as a function of dose (fig. 17.5). The dose–effect curve pertaining to the clamped, and therefore uniformly hypoxic, tumours shows a smooth increase of effect with dose. The curve for animals breathing normally demonstrates an initial region with a steeper slope, then a region which

Fig. 17.4. The response of the transplantable sarcoma RIB_5 in the rat to single doses of X-rays. All tumours were treated on day O when growing in the same site and at the same size. Curve A, untreated control. Curves B to H, the response to radiation doses of 5, 10, 20, 30, 40, 50 and 60 grays respectively, administered to rats breathing air. Curve K, the response of tumours in animals that were cured. Curve X, 4000 rads given to anoxic tumours for comparison with curve D, 2000 rads given to unclamped tumours while animals breathed air. (After Thomlinson, 1977.)

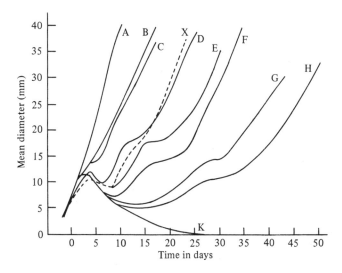

breaks away and becomes parallel with that for anoxic tumours. Thomlinson & Craddock attributed the regrowth of tumours after the lower doses to repopulation mainly by surviving cells that had been well oxygenated during irradiation, whereas those cells would all be killed by higher doses and regrowth would depend on survival of the more resistant hypoxic fraction. Dose–effect curves of that form, often referred to as 'biphasic', are often observed when tumours are irradiated while the animals breathe normally, and are interpreted as demonstrating that tumours contain both oxygenated and hypoxic cells. But it is important to note that growth-delay curves may not be detectably biphasic with very low or very high hypoxic cell fractions, as may indeed be true when survival curves are taken directly on mixed populations. The hypothetical curves shown in fig. 17.6 have been drawn with commonly observed parameters: extrapolation number = 5, o.e.r. = 2.5, ratio of final to initial slope = 4. If the population contained 40% completely hypoxic cells, the remainder being fully oxygenated, the point of inflection in the overall survival curve would hardly be detectable. Even if the indirect test of growth delay reflected only cell killing, it could not be expected to reveal the presence of the 60% of well-oxygenated cells by a biphasic appearance. Furthermore, if a treatment were used that selectively sensitized the hypoxic cells

Fig. 17.5. Radiation-induced delay in the growth of the rat sarcoma RIB$_5$. Dose–effect curves based on time of growth from 9 to 25 mm diameter. (Adapted from Thomlinson & Craddock, 1967.)

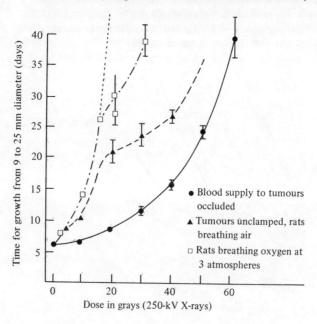

• Blood supply to tumours occluded

▲ Tumours unclamped, rats breathing air

□ Rats breathing oxygen at 3 atmospheres

in the dose-multiplying mode, ratios of doses to give the same effects, in the modified and unmodified conditions, would vary with the level of effect (fig. 17.6).

Shrinkage or 'regression' of tumours

After irradiation some experimental tumours shrink considerably before they start to regrow (fig. 17.4) and rates of 'regression' of human tumours are usually noted by radiotherapists during a course of treatment. Trott *et al.* (1977) could find no correlation between rates of tumour control when they compared effects of both single-doses and fractionated regimens on three transplantable mouse tumours with different biological characteristics. Their results support the view of Thomlinson (1978), who has argued that the rate and extent of shrinkage of treated tumours depend on their biological characteristics, not on the nature or 'dose' of treatment that results in shrinkage. His thesis is supported by measurements made on a variety of human tumours after

Fig. 17.6. Hypothetical survival curves for a population consisting of 60% oxygenated cells, 40% anoxic, o.e.r. = 2.5.

A: survival curve, oxygenated cells $\Big\}$ $n = 5$, $_1D_0 = 4 \times D_0$
B: survival curve, anoxic cells

It could barely be detected that A + B, the resultant curve for the whole population, is 'biphasic'. Curve B' (dashed) represents the the effect of a dose-reducing agent for hypoxic cells, e.g. a radio-sensitizer which is accurately dose-modifying by the factor 2.0. The new resultant curve is A + B', (dotted) which is not dose-modifying for A + B. At the fractional survival level 10^{-2} the dose-reducing factor is significantly less for the combined population than for B alone.

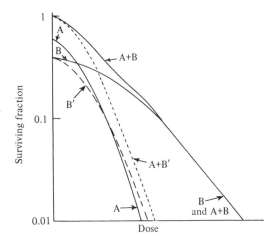

different forms of therapy. In some cases, when tumours had shrunk after one mode of treatment, then started to grow again, a different treatment was given. Thomlinson's measurements show that the rates of shrinkage were then identical with those observed after the first treatment, as exemplified by fig. 17.7. This is an important point, since the rate at which human tumours regress during and after a course of radiotherapy is often regarded as an indication of the adequacy of the treatment.

Survival curves for clonogenic cells from irradiated tumours

Cell cultures *in vitro* made from experimental tumours have frequently been used to construct cell survival curves. The usual procedure is to make a single-cell suspension, then to make viable counts either by counting numbers of tumours growing in recipient animals (p. 15) or by growing the cells *in vitro*. That may be done by the normal plating techniques, or alternatively by mixing aliquots of cell suspensions with molten soft agar, at a temperature just above gel point, together with

Fig. 17.7. Shrinkage of a primary carcinoma of the breast after treatment, first by cytotoxic drugs, and, after it had started to regrow, by radiation. The slopes of the exponential curves relating reduction in volume to time after the start of treatment are not significantly different, if account is taken of the tailing-off in shrinkage and commencement of detectable regrowth at about 60 days after commencement of the first treatment. In contrast, slopes of shrinkage curves differed over a range of nearly 40, as between tumours in different patients, although the same treatments had been given. (From measurements by Dr R. H. Thomlinson, who provided the figure.)

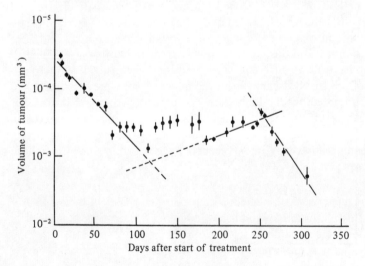

required nutrients. This mixture, which sometimes includes heavily irradiated 'feeder' cells, is then cast on a bed of slightly stiffer agar. As cells divide the progeny remain in cohesion, forming colonies in the agar, so there is no need for attachment to a surface.

An important parameter in such experiments is the 'plating efficiency', i.e. the ratio of colony-forming cells to the total plated. Some experimenters base their assessment of plating efficiency not on total cell counts, but on counts of cells that exclude certain dyes (p. 14). Clearly high plating efficiencies are desirable. Some neoplastic cell lines have been specially selected as yielding proportions of colony-formers *in vitro* as high as 30 to 40%. This has been achieved by making many alternate passages *in vivo* and *in vitro*, and good agreement has been found between survival curves on cells grown *in vitro* before irradiation and on those from tumours irradiated *in vivo*, then excised and plated (Barendsen & Broerse, 1969; Rockwell, Kallman & Fajardo, 1972). The tumour line EMT6 was originally selected by Rockwell *et al.* for just those properties, from several sub-lines that had originated in the same tumour. Nevertheless, Fu *et al.* (1975b) found that survival curves were not the same for cells grown and irradiated *in vitro* and those irradiated as tumours; moreover, survival curves differed, depending on whether tumours had grown in the flanks or lungs of mice.

Good agreement was found by Thomson & Rauth (1974) between survival curves for KHT tumour cells grown *in vitro* and for those from tumours irradiated *in vivo*, the tumour of origin not having been previously selected for good growth *in vitro*. In contrast, the radiation response of a Visking rat sarcoma was found by Dawson *et al.* (1973) to be strongly dependent both on method of growth before irradiation and on the method of assay. Survival curves were somewhat different for freshly made cultures irradiated and assayed *in vitro* and for established cultures. Much bigger differences were found when tumours had been irradiated *in vivo* and the assays made *in vitro*; irradiation *in vivo* followed by assay *in vivo* was different again, and the form, as well as the parameters of the survival curve from the assay *in vivo* were changed when heavily irradiated 'feeder' cells were inoculated together with the tumour cells. Such results suggest that assays of cell survival may depend also on the site in the recipient animal at which tumour formation is observed, but no comparative experiments seem to have been done other than those of Hornsey (1963) (p. 218).

Comparisons of methods for assessing responses of tumour cells

It is important to establish whether the two methods of measuring radiation effects on tumours left *in situ* yield essentially the same information; and perhaps even more important to know whether survival curves from the cells of excised tumours can be used as a basis for predicting the most effective method of sterilizing the tumours: that is, do such survival curves reflect survival of clonogenic cells left *in situ?* These problems may be tackled by observing how modifying procedures change dose–effect curves constructed concomitantly by different methods; but at the present time information bearing on this point is very sparse. Similar relationships between dose–effect curves for delay in growth and tumour control were found when effects of fast neutrons and X-rays were compared by Field *et al.* (1967) and Fowler *et al.* (1972) in experiments respectively on a rat fibrosarcoma and a mouse mammary carcinoma. The factor by which misonidazole sensitized mouse mammary tumours to single doses of radiation was judged to be about the same for growth delay and tumour control (Sheldon, Foster & Fowler, 1974). However McNally & Sheldon (1977), using a different transplantable mouse tumour, failed to achieve as high an enhancement ratio by that compound for delay in tumour growth as for control of the tumours. By the latter test, the values of TCD50 after single doses delivered to tumours in the presence and absence of misonidazole were respectively 38 and 78 Gy, giving an enhancement ratio of 2.1. Dose–effect curves based on growth delay gave values of enhancement ratio that increased with dose. The highest dose usable when drug was present was 30 Gy, and at that level the enhancement ratio was about 1.7. This illustrates a difficulty inherent in comparisons of the effects of modifying treatments on dose–effect curves for growth delay and tumour control, when modification depends on cellular oxygen content, since most animal tumours are heterogeneous in that respect. Doses used for measuring growth delay are of necessity in a lower range than is used for measuring the TCD50, so modifying factors for the two end-points may be unequal unless the experimental design includes provision for effective abolition of variations in intracellular P_{O_2} within the tumour (see fig. 17.6).

If delay in the growth of tumours, after irradiation, is attributable wholly or mainly to the death of clonogenic cells, oxygen enhancement ratios for the two effects should be the same. With the rat fibrosarcoma RIB$_5$, Thomlinson & Craddock (1967) observed the o.e.r. for growth delay to be 3.4. McNally (1972) constructed survival curves for cells in suspensions from those tumours irradiated in the appropriate conditions *in vivo*, and found the o.e.r. for cell killing to be about 2, whether viability was assessed by colony formation *in vitro* or by measuring the numbers of cells required to originate tumours in the brains of recipient rats. The o.e.r. was about the same when the cells were irradiated *in vitro*, about 4 hours after the cells had been plated. The plating efficiency was about 10 to 15%, which is about the middle of the range that has been reported for cells from solid tumours not specifically selected for high plating efficiency. In experiments with a sub-line of RIB$_5$ that had been so selected, McNally (1975) found a discrepancy also when a change of radiation quality was used as modifying procedure, comparative effects of neutrons and X-rays being used for observations both on growth delay and on the survival of cells *in vitro* from tumours irradiated *in vivo*. For equal doses of neutrons, values of RBE were significantly greater according to the latter test.

If it can be accepted that growth delay and tumour control are well correlated, discrepancies of the kind reported by McNally could be interpreted in two ways: either tumour sterilization depends to a significant extent on radiation effects other than the killing of clonogenic cells; or removing irradiated cells from tumours and placing them in a new environment causes significant changes in the survival curve parameters that would apply *in situ*. The latter possibility is suggested both by some of the results summarized in Chapter 16 and by the differences in cell survival curves for normal tissue cells *in vitro* and *in vivo*, discussed above.

Perhaps both factors play a part. But in the event, if such discrepancies prove to be common, assays *in vitro* or *in vivo* of cells from excised tumours, or parts thereof, cannot be relied upon for guidance as to the most effective method of sterilizing tumours of the type under consideration.

One aspect of such assays has already suggested a reason for some discrepancy, namely 'recovery from potentially lethal damage', the phrase used to describe the observation that some tumour cells show higher survival when the irradiated tumours are left *in situ* for a few hours, rather than excised immediately. The phenomenon has been regarded as analogous with the increase in survival often achieved by delay in plating mammalian cells *in vitro* (Hahn & Little, 1972) which, as described in Chapter 16, is manifested by a decrease in the slope of the survival curve. However, the changes in survival observed in some experiments of this type with tumours are better described as demonstrating an increase in the extent of Q repair. This is illustrated by results of Courtenay *et al.* (1976) on a tumour that had originated as a human metastatic pancreatic carcinoma, and was propagated as a xenograft in mice. The hypoxic fraction was estimated as about 25% of the clonogenic population, so tumours irradiated in mice breathing normally yielded survival curves relating mainly to that fraction, as in most experiments of this kind. The increase in survival consequent on leaving the tumours *in situ* for 18 hours was constant at all dose levels, suggesting an increase in Q, but not in P repair. Survival curves for cells from Lewis lung carcinomata growing in the flanks of mice were the same by techniques *in vivo* (lung colony) and by techniques *in vitro* (soft agar colony) (Shipley *et al.*, 1975). For tumour cells from mice breathing normally during irradiation survival curves as drawn reflected only the response of the 'chronically hypoxic' fraction. Maximal increases in survival were seen when tumours were left *in situ* for 4 to 8 hours after γ-irradiation. According to results presented graphically, those increases were almost constant for all doses, suggesting an increase in Q repair, although the slope of the regression line for the cells that had recovered was marginally less than for those plated immediately after irradiation.

Urano *et al.* (1976) found no difference in the extent to which survival of tumour cells increased by being left for a few hours *in situ*, whether this was measured for the naturally or 'chronically' hypoxic tumour fraction or for the total population, made 'acutely' hypoxic by killing the mice before irradiation. In contrast, McNally & Sheldon (1977) observed an increase in survival only in the chronically hypoxic cells, which they estimated to constitute 50% or more of the tumour they used. Maximum increase in survival was observed for tumour cells left for 6 hours *in situ*, and was dose-dependent. It was estimated that D_0 for the survival curve was then greater than for the immediately plated cells by a factor of about 1.3.

'Recovery from potentially lethal damage' in chronically hypoxic cells accounted only in part for the discrepancies in the results of McNally & Sheldon, as between observations on tumour control and survival of tumour cells *in vitro*. It could not account for the failure of the survival curve parameters *in vitro* to predict the TCD50, nor for the differences in response conferred by the presence of misonidazole during irradiation. The TCD50 dose was reduced by the factor 2.1, but the enhancement ratio as observed from the survival curves was only 1.6, that difference being independent of recovery from potentially lethal damage.

Some aspects of hypoxic cell sensitizers used *in vivo*

In the first experiments showing that metronidazole and misonidazole sensitized mouse tumours to radiation, it was observed that their growth was delayed when the mice were treated by those compounds independently of, or after, irradiation (Begg, Sheldon & Foster, 1974; Denekemp & Harris, 1975a). Those effects appear to be attributable to cytotoxic action of electron-affinic compounds that is specific for hypoxic cells (Hall & Roizin-Towle, 1975; Sridhar, Koch & Sutherland, 1976). Such cytotoxic action might be regarded as an additional benefit of the use of electron-affinic sensitizers as an adjuvant to radiotherapy. Furthermore, Stratford & Adams (1977) showed the cytotoxic action of misonidazole to increase sharply with increasing temperature. Currently, there is vigorous research into the possibility that advantage is to be gained from treating cancers by hyperthermia, or hyperthermia combined with radiation, and from that point of view the concomitant use of a com-

pound with those properties could be very favourable. All these considerations are based, of course, on the supposition that hypoxic cells
in some human tumours afford a reason for the failure of radiotherapy
to sterilize them.

However, preliminary clinical trials of misonidazole have given evidence of toxicity that is unlikely to be associated with intracellular
hypoxia, namely the induction of peripheral neuropathies when rather
high doses are given. This effect limits the total dose that can be administered to patients (Saunders *et al.*, 1978; Dische, Saunders & Flockhart,
1978). Brown (1977) observed that misonidazole acted as a toxic agent
for tumour cells only when some of them were hypoxic, but that neigh-
bouring well-oxygenated cells were then killed; and Olive & Durand
(1978) suggested that enzymic reduction in hypoxic cells of nitrohetero-
cyclic compounds might engender products that could damage aerobic
cells. Cellular hypoxia within solid tumours is considered to result from
consumption of oxygen by the actively respiring cells nearest to blood
vessels (Thomlinson & Gray, 1955) so drugs that increase cellular respiration rates could steepen oxygen gradients through tumours and thereby
increase the fraction of hypoxic cells. Using multicellular spheroids (p.
180) as models of tumours, Biaglow & Durand (1976) investigated effects
of certain electron-affinic compounds in changing the size of the fraction of hypoxic cells, and therefore the overall sensitivity of the spheroids
to radiation. They concluded that the property of stimulating cell respiration might offset potential advantages of some compounds found to be
very effective sensitizers of hypoxic mammalian cells *in vitro*.

It has been suggested that the use of hypoxic cell sensitizers would avoid
one disadvantage of hyperbaric oxygen radiotherapy, namely the increased sensitivity conferred also on some normal tissues (p. 262). As a
result, radiotherapists have come to reduce the total doses delivered in
hyperbaric oxygen regimens, which may offset some of the potential gain
from increased oxygenation of the tumours. Some compounds have been
shown to be almost as effective as oxygen in sensitizing hypoxic cells; but
at the time of writing there is little evidence as to whether such compounds
would increase the lethal effect of radiation on cells in which partial pressures of oxygen were somewhat too low for maximum sensitization: in
other words, whether the effects of hypoxic cell sensitizers and oxygen are
additive.

The results of McNally & de Ronde (1978) on two sensitizing compounds showed that they were not completely additive, when tested on
mammalian cells *in vitro* (p. 84). The o.e.r. was 2.8, but when misonidazole at 10^{-4} mol dm^{-3} and oxygen at P_{O_2} about 8 mm Hg were combined,
the factor for sensitization was 2.5, just the same as for the oxygen alone,
although misonidazole in that concentration, without oxygen, gave an

enhancement ratio of 1.25. However, it may be deduced from their results that the K value (equation 6.1) was rather low, probably less than 4 mm Hg, and, as suggested above, K values for some clonogenic cells *in vivo* may be much higher. Results reported to date of investigations *in vivo* with misonidazole do not give a clear answer. Denekamp, Michael & Harris (1974) and Dische, Gray & Zanelli (1976) found that drug to sensitize hypoxic skin respectively of mice and of human patients. No sensitizing effect was observed by Denekamp *et al.* on the skins of mice breathing pure oxygen, and with oxygen flowing over the skin, and Dische *et al.*, in their investigation of the effect of misonidazole on 'oxic' skin, passed a stream of oxygen over the areas under irradiation, so the skin P_{O_2} may have been higher than normal. Hendry (1978) reported that misonidazole had no sensitizing effect for the killing of clonogenic cells in haemopoietic tissue, or the crypt of intestinal mucosa, in mice breathing normally; but there was a considerable increase in the extent of necrosis seen in the tails. The target tissue for that form of damage is considered to be radio-biologically hypoxic in mice breathing normally.

Necrosis in the tails of rodents can perhaps be regarded as a form of 'late damage' to a normal tissue, in contrast with those effects for which the enhancing action of oxygen, and, to a lesser extent, of hypoxic cell sensitizers, have been explored. Since clinical trials of radiosensitizing drugs are planned in many centres, at the time of writing, it may well be that clinical observation will provide information as to the occurrence or non-occurrence of increased late damage to normal tissues, from the use of radiosensitizing drugs, before laboratory evidence is adequate for the purposes of prediction.

Concluding remarks

Great variability in the responses of mammalian cells to radiation has been demonstrated in every aspect dealt with in preceding chapters; but almost nothing is known about the subtle biological and biochemical differences underlying the variability. Thus, *in vitro* or *in vivo*, there is no way of predicting in detail how cells of a given kind will respond to radiation, or to any modifying procedure. The protective effect of hypoxia remains the radiobiological phenomenon of widest generality, and this one possible reason for failure in the local control of tumours by radiation has received considerable attention. Apart from techniques already mentioned for dealing with this possibility, namely hyperbaric oxygen radiotherapy, or the use of hypoxic cell sensitizers, others have been canvassed, like the use of radioprotective agents which, as shown in Chapter 8, act preferentially on well-oxygenated cells; or a combination of protective and sensitizing compounds (e.g. Harris, 1976). Reduction in the extent of protection by hypoxia against radiations of high LET has likewise usually been the rationale suggested for their use in radiotherapy.

Measures already being taken, or proposed, to deal with the problem presented by hypoxic cells in tumours afford the prospect of therapeutic gain, since they are selected on the basis that well-oxygenated cells in normal tissues will be much less affected; or, conversely, that those cells will be protected to a greater degree by chemical agents than hypoxic cells. At the same time, it is generally recognized that there are limitations on the gain to be expected from the use of such measures. It was predicted by Thomlinson (1961) that, after a dose of radiation to a tumour, some of the viable cells, having been protected by their hypoxia, might then acquire access to a better oxygen supply because dead cells would be removed. Thus the tumour might 're-oxygenate', a process that has been observed to occur in several experimental tumours; and its occurrence has been deduced also from the responses of some animal tumours to radiation given in several fractions. It seems that methods developed by radiotherapists on the basis of clinical experience over many decades, including the administration of radiation in rather small dose fractions, may already be adequate to deal with hypoxic cells in some human tumours; and in such cases little or no advantage is to be expected from techniques aimed specifically at the differential sensitization of hypoxic cells.

Reasons for the failure in local control of tumours, other than hypoxia, have not received so much attention, perhaps because methods for dealing with them are difficult to devise. A considerable problem may be presented if there is a greater capacity for repair in the clonogenic tumour cells than in those of the normal tissues at risk, added to which is the disquieting possibility suggested in Chapter 16, that an initial dose of radiation might, by inhibiting DNA synthesis, itself induce increased repair capacity in some neoplastic cell lines.

Even if the causes of 'radioresistance' of some human tumours are identified, this does not necessarily mean that remedies will be immediately obvious; but at least the application of a technique that is expensive and effort-consuming, yet unsuitable, could be avoided. With the wide choice of methods of treatment now open to cancer therapists, there is clearly as great a requirement as there ever has been for vigorous research in cellular radiobiology at a fundamental level.

REFERENCES AND AUTHOR INDEX

Authors' sole publications are listed chronologically. Where a first author has more than one paper with co-authors, these are listed alphabetically according to the second author's name and so on. Abbreviations of names of journals are from *The World List of Scientific Periodicals*. The figures in square brackets at the end of references show the pages on which the papers are cited.

Adams, G. E. (1972). Radiation chemical mechanisms in radiation biology. *Adv. Radiation Chem.* **3**, 125–208. [26, 78, 93]

Adams, G. E. (1977). Hypoxic cell sensitizers for radiotherapy. In *Cancer: A Comprehensive Treatise*, ed. F. F. Becker, vol. 6, pp. 181–223. New York: Plenum Press. [78]

Adams, G. E., Agnew, D. A., Stratford, I. J. & Wardman, P. (1976). Applications of pulse radiolysis and cellular fast-mixing techniques to the study or radiation damage in cells. In *Proceedings Fifth Symposium on Microdosimetry*, Euratom 5452 d–e–f, pp. 16–36. [80, 81]

Adams, G. E., Asquith, J. C., Dewey, D. L., Foster, J. L., Michael, B. D. & Willson, R. L. (1971a). Electron affinic sensitization Part II: para-nitroacetophenone: A radiosensitizer for anoxic bacterial and mammalian cells. *Int. J. Radiat. Biol.* **19**, 575–85. [81]

Adams, G. E., Baverstock, K. F., Cundall, R. B. & Redpath, J. L. (1973). Radiation effects on α-chymotrypsin in aqueous solution: pulse radiolysis and inactivation studies. *Radiat. Res.* **54**, 375–87 [27]

Adams, G. E., Cooke, M. S. & Michael, B. D. (1968). Rapid mixing in radiobiology. *Nature, Lond.* **219**, 1368–9 [72, 75]

Adams, G. E. & Dewey, D. L. (1963). Hydrated electrons and radiobiological sensitization. *Biochem. biophys. Res. Comm.* **12**, 473–7 [78]

Adams, G. E., Michael, B. D., Asquith, J. C., Shenoy, M. A., Watts, M. E. & Whillans, D. W. (1975). Rapid-mixing studies on the time-scale of radiation damage in cells. In *Radiation Research. Biomedical, Chemical and Physical Perspectives*, ed. O. F. Nygaard, H. I. Adler and W. K. Sinclair, pp. 478–92. New York and London: Academic Press. [77, 83, 84]

Adams, G. E., Willson, R. L., Bisby, R. H. & Cundall, R. B. (1971b). On the mechanism of the radiation-induced inactivation of ribonuclease in dilute aqueous solution. *Int. J. Radiat. Biol.* **20**, 405–15. [27]

Adler, M. I. & Hardigree, A. A. (1964). Analysis of a gene controlling cell division and sensitivity to radiation in *Escherichia coli*. *J. Bacteriol.* **87**, 720–26. [246]

274

Adler, H. I. & Hardigree, A. A. (1965). Postirridiation growth, division and recovery in bacteria. *Radiat. Res.* **25**, 92–102. [246]

Ahnstrøm, G. & Edvardsson, K-A. (1974). Radiation-induced single-strand breaks in DNA determined by rate of alkaline strand separation and hydroxylapatite chromatography: an alternative to velocity sedimentation. *Int. J. Radiat. Biol.* **26**, 493–7 [225, 241, 242]

Ahnstrøm, G. & Erixon, K. (1973). Radiation induced strand breakage in DNA from mammalian cells. Strand separation in alkaline solution. *Int. J. Radiat. Biol.* **23**, 285–9. [241]

Ahnstrøm, G., George, A. M. & Cramp, W. A. (1978). Extensive and equivalent repair in both radiation resistant and radiation sensitive *E. coli* determined by a DNA unwinding technique. *Int. J. Radiat. Biol.* **36**, 317–27 [240, 242, 243, 244]

Alexander, P. (1957). Effect of oxygen on inactivation of trypsin by the direct action of electrons and alpha particles. *Radiat. Res.* **6**, 653–60. [25]

Alexander, P. (1961). Mouse lymphoma cells with different radiosensitivities. *Nature, Lond.* **192**, 572–3. [122, 217]

Alexander, P. (1963). Chemical protection in chemical systems. In *Radiation Effects in Physics, Chemistry and Biology*, ed. M. Ebert and A. Howard, pp. 254–71. Amsterdam: North-Holland Publishing Co. [88]

Alexander, P. & Charlesby, A. (1955). Physico-chemical methods of protection against ionizing radiations. In *Radiobiology Symposium, 1954*, ed. Z. M. Bacq and P. Alexander, pp. 49–60. London: Butterworth. [25, 74]

Alexander, P., Lett, J. T. & Dean, C. J. (1965). The role of post-irradiation repair processes in chemical protection and sensitization. *Prog. biochem. Pharmacol.* **1**, 22–40. [89]

Alper, T. (1955). Bacteriophage as indicator in radiation chemistry. *Radiat. Res.* **2**, 119–34. [27, 183]

Alper, T.(1956). The modification of damage caused by primary ionization of biological targets. *Radiat. Res.* **5**, 573–86. [25, 46, 64, 71, 74, 134, 139]

Alper, T. (1959). Variability in the oxygen effect observed with microorganisms. Part 1. Haploid yeast: single and budding cells. *Int. J. Radiat. Biol.* **1**, 414–19, [150]

Alper, T. (1961a). Variability in the oxygen effect observed with microorganisms. Part II. *Escherichia coli* B. *Int. J. Radiat. Biol.* **3**, 369–77, [60, 93]

Alper, T. (1961b). Effects on subcellular units and free-living cells. In *Mechanisms in Radiobiology*, ed. M. Errera and A. Forssberg, vol. 1, pp. 353–417. New York and London: Academic Press. [227]

Alper, T. (1962). The dependence of chemical protective action on oxygen, as studied with bacteria. *Br. J. Radiol.* **35**, 361. [88]

Alper, T. (1963a). Chemical protection of various bacteria and its involvement with the oxygen effect. Reported by E. A. Wright. In *Radiation Effects in Physics, Chemistry and Biology*, ed. M. Ebert and A. Howard, pp. 276–89. Amsterdam: North-Holland Publishing Co. [89, 91, 92]

Alper, T. (1963b). Lethal mutations and cell death. *Physics Med. Biol.* **8**, 365–85. [208, 246]

Alper, T. (1963c). Comparison between the oxygen enhancement ratios for neutrons and X-rays, as observed with *Escherichia coli* B. *Br. J. Radiol.* **36**, 97–101. [238]

Alper, T. (1967). A characteristic of the lethal effect of ionizing radiation on 'Hcr⁻' bacterial strains. *Mutation Res.* **4**, 15–20. [52]

Alper, T. (1970). Mechanisms of lethal radiation damage to cells. In *Proceedings of the Second Symposium on Microdosimetry*, ed. H. G. Ebert pp. 5–36. Euratom, 14452 d–f–e. [159]

Alper, T. (1971). Cell death and its modification: the roles of primary lesions in membranes and DNA. In *Biophysical Aspects of Radiation Quality*, Ink. Atomic Energy Agency, Vienna, pp. 171–83, London: HMSO. [112, 122]

Alper, T. (1974). Radiobiological support for radiotherapy. What kind of research? *Lancet*, I June 29, 1328–30. [199]

Alper. T. (ed.) (1975). *Cell Survival after Low Doses of Radiation*. London: Institute of Physics and John Wiley & Sons. [39, 117]

Alper, T. (1976a). Another method for testing the applicability of the hyperbolic oxygen equation and for calculating K. *Int. J. Radiat. Biol.* **30**, 389–92. [55, 57]

Alper, T. (1976b). Modern trends and creeds in radiobiology. In *Radiation and Cellular Control Processes*. ed. J. Kiefer, pp. 307–18 Berlin, Heidelberg, New York: Springer-Verlag. [222, 225]

Alper, T., Bewley, D. K. & Fowler, J. F. (1962). Chemical protection against alpha particle irradiation. *Nature, Lond.* **194**, 1245–7. [127, 129]

Alper, T. & Bryant, P. E. (1974). Reduction in oxygen enhancement ratio with increase in LET: tests of two hypotheses. *Int. J. Radiat. Biol.* **26**, 203–8 [134, 135, 136, 140, 142]

Alper, T., Fowler, J. F., Morgan, R. L., Vonberg, D. D., Ellis, F. & Oliver, R. (1962). The characterisation of the 'Type C' survival curve. *Br. J. Radiol.* **35**, 722–3. [38]

Alper, T. & Gillies, N. E. (1958). 'Restoration' of *Escherichia coli* strain B after irradiation: its dependence on suboptimal growth conditions. *J. gen. Microbiol.* **18**, 461–72. [238, 246]

Alper, T. & Gillies, N. E. (1960). The relationship between growth and survival after irradiation of *Escherichia coli* strain B and two resistant mutants. *J. gen. Microbiol.* **22**, 113–28. [246]

Alper, T., Gillies, N. E. & Elkind, M. M. (1960). The sigmoid survival curve in radiobiology. *Nature, Lond.* **186**, 1062–3. [37]

Alper, T. & Haig, D. A. (1968). Protection by anoxia of the scrapie agent and some DNA and RNA viruses irradiated as dry preparations. *J. gen. Virol.* **3**, 157–66. [25, 26, 29]

Alper, T. & Hornsey, S. (1968). The effect of hypoxia during irradiation on four-day death in mice given single and split doses of electrons. *Br. J. Radiol.* **41**, 375–80. [259]

Alper, T. & Howard-Flanders P. (1956). The role of oxygen in modifying the radiosensitivity of *E. coli* B. *Nature, Lond.* **178**, 978–9 [53]

Alper, T. & Moore, J. L. (1967). The interdependence of oxygen enhancement ratios for 250 kVp X rays and fast neutrons. *Br. J. Radiol.* **40**, 843–8. [112, 122, 128, 141, 215]

Alper, T., Moore, J. L. & Bewley, D. K. (1967a). LET as a determinant of bacterial radio-sensitivity, and its modification by anoxia and glycerol. *Radiat. Res.* **32**, 277–93. [61, 112, 123, 124, 128, 129, 130]

Alper, T., Moore, J. L. & Smith, P. (1967b). The role of dose rate, irradiation technique and LET in determining radiosensitivities at low oxygen concentrations. *Radiat. Res.* **32**, 780–91. [55, 56, 61, 62, 140]

Anderson, R. F. & Patel, K. B. (1977). Radiosensitization of *Serratia marcescens* by bipyridinium compounds. *Int. J. Radiat. Biol.* **32**, 471–9. [79, 82]

Anderson, R. F., Patel, K. B. & Smithen, C. E. (1978). Radiosensitization of *Serratia marcescens* by nitropyridinium compounds. *Br. J. Cancer* **37**, suppl. 3, 103–6. [79, 82]

Anderson, R. S. & Turkowitz, H. (1941). The experimental modification of the sensitivity of yeast to roentgen rays. *Am. J. Roentg.* **46**, 537–41. [50]

Arlett, C. F. (1970). Influence of post-irradiation conditions on the survival of Chinese hamster cells after γ-irradiation. *Int. J. Radiat. Biol.* **17**, 515–26. [178, 196, 232]

Ashwood-Smith, M. J. (1961). The radioprotective action of dimethyl sulphoxide and various other sulphoxides. *Int. J. Radiat. Biol.* **3**, 41–8. [96]

Asquith, J. C., Watts, M. E., Patel, K., Smithen, C. E. & Adams, G. E. (1974). Electron affinic sensitization V. Radiosensitization of hypoxic bacteria and mammalian cells *in vitro* by some nitroimidazoles and nitropyrazoles. *Radiat. Res.* **60**, 108–18. [161]

Attix, F. H. & Roesch, W. C. (eds.) (1966–1972). *Radiation Dosimetry*, vol. 2, 1966; vol. 1, 1968; vol. 3, 1969; supplement no. 1, 1972. New York and London: Academic Press. [6]

Averbeck, D. & Ebert, M. (1973). Diploidy and repair of radiation damage in *Saccharomyces* mutants. *Mutation Res.* **19**, 305–12. [52, 238]

Avery, O. T., MacLeod, C. M. & McCarty, M. (1944). Studies on the chemical nature of the substance inducing transformation of pneumococcal types. Induction of transformation by a desoxyribonucleic acid fraction isolated from *Pneumococcus* Type III. *J. exp. Med.* **79**, 137–57. [206]

Bacchetti, S. & Mauro, F. (1965). Recovery from sublethal X-ray damage in surviving yeast cells. *Radiat. Res.* **25**, 103–14. [192]

Bacq, Z. M. & Alexander, P. (1955). *Fundamentals of Radiobiology* (1st edition) London: Butterworth. [64, 225]

Baker, M. L., Dalrymple, G. V., Sanders, J. L. & Moss, A. J. Jr (1970). Effects of radiation on asynchronous and synchronized L cells under energy deprivation. *Radiat. Res.* **42**, 320–30. [236, 237]

Barendsen, G. W. (1968). Responses of cultured cells, tumours and normal tissues to radiation of different linear energy transfer. In *Current Topics in Radiation Research*, ed. M. Ebert and A. Howard vol. 4, pp. 295–356. Amsterdam: North-Holland Publishing Co. [128]

Barendsen, G. W. & Broerse, J. J. (1969). Experimental radiotherapy of a rat rhabdomyosarcoma with 15 McV neutrons and 300 KV X-rays. *Europ. J. Cancer* **5**, 373–91. [267]

Barendsen, G. W., Koot, C. J., van Kersen, G. R., Bewley, D. K., Field, S. B. & Parnell, C. J. (1966). The effect of oxygen on impairment of the proliferative capacity of human cells in culture by ionizing radiations of different LET. *Int. J. Radiat. Biol.* **10**, 317–27. [114, 115, 116, 117, 124, 128, 135, 136]

Baverstock, K. F. & Burns, W. G. (1976). Primary production of oxygen from irradiated water as an explanation for decreased radiobiological oxygen enhancement at high LET. *Nature, Lond.* **260**, 316–18. [142]

Beam, C. A., Mortimer, R. K., Wolfe, R. G. & Tobias, C. A. (1954). The relation of radioresistance to budding in *Saccharomyces cerevisiae*. *Arch. Biochem. Biophys.* **49**, 110–22. [150]

Bedford, J. S. & Hall, E. J. (1966). On the shape of the dose-response curve for HeLa cells cultured *in vitro* and exposed to γ-radiation. *Nature. Lond.* **209**, 1363–4. [16]

Beer, J. Z., Lett, T. P. & Alexander, P. (1963). Influence of temperature and medium on the X-ray sensitivities of leukaemia cells *in vitro*. *Nature, Lond.* **199**, 193–4. [232]

Beer, J. Z., Szumiel, I. & Walicka, M. (1973). Cross-sensitivities to UV light and X-rays of two strains of murine lymphoma L5178Y cells *in vitro*. *Studia Biophysica* **36/37**, 175–82. [217, 218, 245]

Begg, A. C., Sheldon, P. W. & Foster, J. L. (1974). Demonstration of radio-sensitization of hypoxic cells in solid tumours by metronidazole. *Br. J. Radiol.* **47**, 399–404. [78, 270]

Belli, J. A. & Bonte, F. (1963). Influence of temperature on the radiation response of mammalian cells in tissue culture. *Radiat. Res.* **18**, 272–6. [232]

Belli, J. A. & Shelton, M. (1969). Potentially lethal radiation damage: repair by mammalian cells in culture. *Science* **165**, 490–2. [232, 233, 235, 246]

Bender, M. A. & Gooch, P. C. (1962). The kinetics of X-ray survival of mammalian cells *in vitro*. *Int. J. Radiat. Biol.* **5**, 133–45. [16, 40]

Ben-Hur, E., Elkind, M. M. & Bronk, B. W. (1974). Thermally enhanced radioresponse of cultured Chinese hamster cells: inhibition of repair of sublethal damage and enhancement of lethal damage. *Radiat. Res.* **58**, 38–51. [232]

Berry, R. J. (1964). On the shape of X-ray dose–response curves for the reproductive survival of mammalian cells. *Br. J. Radiol.* **37**, 948–51. [16]

Berry, R. J. (1965). A reduced oxygen enhancement ratio for X-ray survival of HeLa cells *in vitro*, after treatment with 'Methotrexate'. *Nature. Lond.* **208**, 1108–10. [218]

Berry, R. J. (1966). Effects of some metabolic inhibitors on X-ray dose–response curves for the survival of mammalian cells *in vitro*, and on early recovery between fractional X-ray doses. *Br. J. Radiol.* **39**, 458–63. [178, 236, 248]

Berry, R. J., Bewley, D. K. & Parnell, C. J. (1965). Reproductive capacity of mammalian tumour cells irradiated *in vivo* with cyclotron-produced fast neutrons. *Br. J. Radiol.* **38**, 613–17. [128, 136]

Berry, R. J., Hall, E. J. & Cavanagh, J. (1970). Radiosensitivity and the oxygen effect for mammalian cells cultured *in vitro* in stationary phase. *Br. J. Radiol.* **43**, 81–90. [148, 149]

Bertsche, U. & Liesem, H. (1976). Survival of yeast after heavy iron irradia-tion. In *Radiation and Cellular Control Processes*, ed. J. Kiefer, pp. 105–10. Berlin, Heidelberg, New York: Springer-Verlag. [238]

Bewley, D. K. (1968). Fast neutrons—LET distributions and the response of mammalian cells. In *Biophysical Aspects of Radiation Quality*, Int. Atomic Energy Agency, Vienna, pp. 65–85. London: HMSO. [124]

Bewley, D. K. (1972). Radiation quality and its influence on biological response. *Br. med. Bull.* **29**, 7–11. [153]

Bewley, D. K., McNally, N. J. & Page, B. C. (1974). Effect of the secondary charged-particle spectrum on cellular response to fast neutrons. *Radiat. Res.* **58**, 111–21. [119, 126, 127, 128, 136, 140]

Biaglow, J. E. & Durand, R. E. (1976). The effects of nitrobenzene derivatives on oxygen utilization and radiation response of an *in vitro* tumor model. *Radiat. Res.* **65**, 529–39. [271]

Biebl, R. (1963). Chemical protection against the effects of α-rays and of thermal neutrons in plant cells. Reported by E. A. Wright, In *Radiation Effects in Physics, Chemistry and Biology*, ed. M. Ebert and A. Howard, pp. 276–89. Amsterdam: North-Holland Publishing Co. [127]

Bird, R. & Burki, J. (1975). Survival of synchronized Chinese hamster cells exposed to radiation of different linear energy transfer. *Int. J. Radiat. Biol.* **27**, 105–20. [116, 157, 158, 162]

Blum, E. & Alper, T. (1971). Radiation-target molecular weights of urease and of L-glutamate dehydrogenase, and their relevance to the size of the functional subunits. *Biochem. J.* **122**, 677–80. [24, 29]

Boag, J. W. (1975). The statistical treatment of cell survival data. In *Cell Survival After Low Doses of Radiation*, ed. T. Alper, pp. 40–51. London: Institute of Physics and John Wiley & Sons. [16]

Bond, V. P., Fliedner, T. M. & Archambeau, J. O. (1965). *Mammalian Radiation Lethality: a Disturbance in Cellular Kinetics.* New York and London: Academic Press. [3]

Braams, R. (1960). Changes in the radiation sensitivity of some enzymes and the possibility of protection against the direct action of ionizing particles *Radiat. Res.* **12**, 113–19. [25, 93]

Bridges, B. A. (1960). Sensitization of *Escherichia coli* to gamma radiation by N-ethyl-maleimide. *Nature, Lond.* **188**, 415. [74]

Bridges, B. A. (1962a). Protection of *Pseudomonas* sp. against gamma-radiation by dimethyl sulphoxide. *Int. J. Radiat. Biol.* **5**, 101–4. [88, 89, 97]

Bridges, B. A. (1962b). The chemical protection of *Pseudomonas* species against ionizing radiation. *Radiat. Res.* **17**, 801–8. [95, 97]

Bridges, B. A. (1969). Sensitization of organisms to radiation by sulphydryl-binding agents. *Adv. Radiation Biol.* **3**, 123–76. [76]

Broerse, J. J., Barendsen, G. W. & van Kersen, G. R. (1968). Survival of cultured human cells after irradiation with fast neutrons of different energies in hypoxic and oxygenated conditions. *Int. J. Radiat. Biol.* **13**, 559–72. [119, 126, 127, 140]

Brown, D. & Gillies, N. E. (1972). The relationship between filaments, killing and restoration in irradiated *Escherichia coli* Strain B. *J. gen. Microbiol.* **70**, 461–70. [246]

Brown, J. M. (1977). Cytotoxic effects of the hypoxic cell radiosensitizer Ro7–0582 to tumor cells *in vivo*. *Radiat. Res.* **72**, 469–86. [271]

Brustad, T. (1960). Study of the radiosensitivity of dry preparations of lyso-zyme, trypsin, and desoxyribonuclease, exposed to accelerated nuclei of hydrogen, helium, carbon, oxygen and neon. *Radiat. Res. suppl.* **2**, 65–74. [32]

Brustad, T. (1961). Molecular and cellular effect of fast charged particles. *Radiat. Res.* **15**, 139–58. [112]

Brustad, T. (1967). On the mechanisms of radiation inactivation of enzymes in dilute solution. In *Radiation Research*, ed. G. Silini, pp. 384–96. Amsterdam: North-Holland Publishing Co. [27]

Brustad, T. (1968). Rapid-mixing and pulse radiolysis equipment as tools in radiobiological research. *Scand. J. clin. Lab. Invest.* **22**, 31–40. [85]

Brustad, T. & Singsaas, B. (1971). On the time-scale for radio-protection by and permeation of glycerol in anoxic cells of *Escherichia coli* B. *Radiat. Res.* **45**, 94–109. [97]

Bryant, P. E. (1968). Survival after fractionated doses of radiation: modi-fication by anoxia of the response of *Chlamydomonas*. *Nature, Lond.* **219**, 75–77. [174]

Bryant, P. E. (1970). The effect of hypoxia on recovery from sublethal damage in *Chlamydomonas*. *Int. J. Radiat. Biol.* **17**, 527–32. [174, 175]

Bryant, P. E. (1973). LET as a determinant of oxygen enhancement ratio and shape of survival curve for *Chlamydomonas*. *Int. J. Radiat. Biol.* **23**, 217–26. [16, 38, 62, 123, 124, 136, 222]

Bryant, P. E. (1974). Change in sensitivity of cells after split dose recovery. A further test of the repair hypothesis. *Int. J. Radiat. Biol.* **26**, 499–504. [170]

Bryant, P. E. (1975). Decrease in sensitivity of cells after split-dose recovery: evidence for the involvement of protein synthesis. *Int. J. Radiat. Biol.* **27**, 95–102. [170, 247]

Bryant, P. E. (1976). Absence of oxygen effect for induction of resistance to ionising radiation. *Nature, Lond.* **261**, 588–90. [170, 248]

Bryant, P. E. & Lansley, I. (1975). Survival-curve shape and oxygen enhancement ratio at low doses of *Chlamydomonas*. In *Cell Survival after Low Doses of Radiation*, ed. T. Alper, pp. 107–13. London: Institute of Physics and John Wiley & Sons. [16, 189, 190]

Bryant, P. E. & Parker, J. (1977). Evidence for location of the site of accumulation for sub-lethal damage in *Chlamydomonas*. *Int. J. Radiat. Biol.* **32**, 237–46. [179, 249]

Burrell, A. D., Feldschreiber, P. & Dean, C. J. (1971). DNA–membrane association and the repair of double breaks in X-irradiated *Micrococcus radiodurans*. *Biochim. biophys. Acta* **247**, 38–53. [244]

Butts, J. J. & Katz, R. (1967). Theory of RBE for heavy ion bombardment of dry enzymes and viruses. *Radiat. Res.* **30**, 855–71. [107]

Byfield, J. E., Lee, Y. C. & Kulhanian, F. (1976). X-ray excision repair replication and radiation survival in placental mammalian cells. *Int. J. Radiat. Oncol. Biol. Phys.* **1**, 937–43. [245]

Caldwell, W. L., Lamerton, L. F. & Bewley, D. K. (1965). Increased sensitivity *in vitro* of murine leukaemia cells to fractionated X-rays and fast neutrons. *Nature, Lond.* **208**, 168–70. [36, 144]

Catcheside, D. G. & Lea, D. (1943). The effect of ionization distribution on chromosome breakage by X-rays. *J. Genet.* **45**, 186–96. [134, 137]

Chadwick, K. H. & Leenhouts, H. P. (1973). A molecular theory of cell survival. *Physics Med. Biol.* **18**, 78–87. [42, 188, 192]

Chapman, J. D., Dugle, D. L., Reuyers, A. P., Meeker, B. E. & Borsa, J. (1974). Studies on the radiosensitizing effect of oxygen in Chinese hamster cells. *Int. J. Radiat. Biol.* **26**, 383–9. [63]

Chapman, J. D., Gillespie, C. J., Reuvers, A. P. & Dugle, D. L. (1975). Radio-protectors, radiosensitizers, and the shape of the mammalian cell survival curve. In *Cell Survival after Low Doses of Radiation*, ed. T. Alper, pp. 135–40. London: Institute of Physics and John Wiley & Sons. [254]

Chapman, J. D., Reuvers, A. P., Borsa, J. & Greenstock, C. L. (1973). Chemical radioprotection and radiosensitization of mammalian cells growing *in vitro*. *Radiat. Res.* **56**, 291–306. [88, 89, 95, 98, 99, 213]

Chapman, J. D., Reuvers, A. P. & Gillespie, C. J. (1976). Radiation chemical probes in the study of mammalian cell inactivation and their influence on radiobiological effectiveness. In *Proceedings of the Fifth Symposium on Microdosimetry*, ed. J. Booz, H. G. Ebert and B. G. R. Smith, pp. 775–93, EUR 5452 d–e–f, Commission of the European Communities. [99]

Chapman, J. D., Sturrock, J., Boag, J. W. & Crookall, J. O. (1970). Factors affecting the oxygen tension around cells growing in plastic Petri dishes. *Int. J. Radiat. Biol.* **17**, 305–28. [54, 58, 158]

Chapman, J. D., Urtasun, R. C., Blakely, E. A., Smith, K. C. & Tobias, C. A.

(1978). Hypoxic cell sensitizers and heavy charged-particle radiations. *Br. J. Cancer* **37**, suppl. 3, 184–8. [131]

Chapman, J. D., Webb, R. G. & Borsa, J. (1971). Radiosensitization of mammalian cells by *p*-nitroacetophenone I. Characterization in asynchronous and synchronous populations. *Int. J. Radiat. Biol.* **19**, 561–73. [81, 161]

Chen, K. Y. & Withers, H. R. (1972). Survival characteristics of stem cells of gastric mucosa in C₃H mice subjected to localized gamma irradiation. *Int. J. Radiat. Biol.* **21**, 521–34. [255]

Christensen, G. M., Dahlke, L. W., Griffin, J. T., Moutvic, J. C. & Jackson, K. L. (1969). The effect of high-pressure oxygen on acute radiation mortality in mice. *Radiat. Res.* **37**, 283–86. [261]

Churchill-Davidson, I., Foster, C. A., Wiernik, G., Collins, C. D., Pizey, N. C. D., Skeggs, D. B. L. & Purser, P. R. (1966). The place of oxygen in radiotherapy. *Br. J. Radiol.* **39**, 321–31. [262]

Conger, A. D. (1956). The effect of oxygen on the radiosensitivity of mammalian cells. *Radiology* **66**, 63–8 [220]

Conger, A. D. & Randolph, M. L. (1959). Magnetic centres (free radicals) produced in cereal embryos by ionizing radiation. *Radiat. Res.* **11**, 54–66. [64]

Corry, P. M. & Cole, A. (1973). Double strand rejoining in mammalian DNA. *Nature New Biol.* **245**, 100–1. [244]

Courtenay, V. D. (1969). Radioresistant mutants of L5178Y cells. *Radiat. Res.* **38**, 186–203. [217]

Courtenay, V. D., Smith, I. E., Peckham, M. J. & Steel, G. G. (1976). *In vitro* and *in vivo* radiosensitivity of human tumour cells obtained from a pancreatic carcinoma xenograft. *Nature, Lond.* **263**, 771–2. [269]

Cox, R. & Masson, W. K. (1974). Changes in radiosensitivity during the *in vitro* growth of diploid human fibroblasts. *Int. J. Radiat. Biol.* **26**, 193–6. [168, 250, 251]

Cox, R., Thacker, J. & Goodhead, D. T. (1977a). Inactivation and mutation of cultured mammalian cells by aluminium characteristic ultrasoft X-rays. *Int. J. Radiat. Biol.* **31**, 561–76. [36, 143, 144]

Cox, R., Thacker, J., Goodhead, D. T. & Munson, R. J. (1977b). Mutation and inactivation of mammalian cells by various ionising radiations. *Nature, Lond.* **267**, 415–27. [114, 116, 117, 118]

Cramp, W. A. (1966). Radiation protection of *Shigella flexneri* by compounds containing the thiourea molecular structure. In *Energy Transfer in Radiation Processes* ed. G. O. Phillips, pp. 153–9. Amsterdam: Elsevier. [89, 95]

Cramp, W. A. (1967). The toxic action on bacteria of irradiated solutions of copper compounds. *Radiat. Res.* **30**, 221–36. [71]

Cramp, W. A. (1969). Radiation protection of *Shigella flexneri* by ethanol, β-mercaptoethanol and several polyhydric alcohols. *Int. J. Radiat. Biol.* **15**, 227–32. [96, 98]

Cramp, W. A. (1970). Radiosensitization of *Shigella flexneri* and *Escherichia coli* by methyl hydrazine, indane trione and acriflavine. *Int. J. Radiat. Res.* **17**, 559–67. [179]

Cramp, W. A. & Bryant, P. E. (1975). The effects of rifampicin on electron and neutron irradiated *E. coli* B/r and B$_{s-1}$. Survival, DNA degradation and DNA synthesis by membrane fragments. *Int. J. Radiat. Biol.* **27**, 143–56. [128]

Cramp, W. A. & Petrusek, R. (1974). The synthesis of DNA by membrane–DNA complexes from *E. coli* B/r and *E. coli* B$_{s-1}$ after exposure to UV light; a comparison with the effects of ionizing radiation. *Int. J. Radiat. Biol.* **26**, 277–84. [211]

Cramp, W. A. & Walker, A. (1974). The nature of the new DNA synthesized by DNA–membrane complexes isolated from irradiated *E. coli. Int. J. Radiat. Biol.* **25**, 175–87. [210]

Cramp, W. A. & Watkins, D. K. (1970). The modification of post-irradiation DNA degradation in *Escherichia coli* B/r. *Radiat. Res.* **41**, 312–25. [240]

Cramp, W. A., Watkins, D. K. & Collins, J. (1972). The synthesis of DNA by membrane–DNA complexes from *E. coli* B/r and B_{s-1} after exposure to fast electrons and neutrons: the measurement of oxygen-enhancement ratios and RBE values. *Int. J. Radiat. Biol.* **22**, 379–87. [209, 215]

Cullen, B. M. (1976). Pre-irradiation culture conditions as a determinant of the radiosensitivity of mammalian cells at various oxygen concentrations. Ph. D. Thesis, University of London. [60, 61, 128, 259]

Cullen, B. M. & Lansley, I. (1974). The effect of pre-irradiation growth conditions on the relative radiosensitivities of mammalian cells at low oxygen concentrations. *Int. J. Radiat. Biol.* **26**, 579–88. [57, 60, 61, 62, 159]

Curtis, S. B. (1970). The effect of track structure on OER at high LET. In *Charged Particle Tracks in Solids and Liquids* ed. G. E. Adams, D. K. Bewley and J. W. Boag, pp. 140–2. Bristol: The Institute of Physics and the Physical Society. [124]

Dale, W. M., Davies, J. V. & Russell, C. (1961). Nitric oxide as a modifier of radiation effects on *Shigella flexneri. Int. J. Radiat. Biol.* **4**, 1–13. [72]

Dalrymple, G. V., Sanders, J. L. & Baker, M. L. (1967). Dinitrophenol decreases the radiation sensitivity of L cells. *Nature, Lond.* **216**, 708–09. [236]

Dalrymple, G. V., Sanders, J. L., Baker, M. L. & Wilkinson, K. P. (1969). The effect of 2,4-dinitrophenol on the repair of radiation injury in cells. *Radiat. Res.* **37**, 90–102. [236]

Davies, D. R. (1963). Radiation-induced chromosome aberrations and loss of reproductive integrity in *Tradescantia. Radiat. Res.* **20**, 726–40. [219]

Davies, D. R. (1967). Absence of an oxygen effect and the genetically determined repair process. *Nature, Lond.* **215**, 829–32. [52]

Davies, H. G. & Haynes, M. E. (1975). Light- and electron-microscope observations on certain leukocytes in a teleost fish and a comparison of the envelope-limited monolayers of chromatin structural units in different species. *J. Cell. Sci.* **17**, 263–85. [18, 19, 20]

Dawson, K. B., Madoc-Jones, H., Mauro, F. & Peacock, J. H. (1973). Studies on the radiobiology of a rat sarcoma treated *in situ* and assayed *in vitro. Eur. J. Cancer* **9**, 59–68. [267]

Deering, R. A. (1958). Studies on division inhibition and filament formation of *Escherichia coli* by ultraviolet light. *J. Bacteriol.* **76**, 123–30. [246]

Deering, R. A. & Rice, R. (1962). Heavy ion irradiation of HeLa cells. *Radiat. Res.* **17**, 774–86. [113, 116]

Deering, R. A., Smith, M. S., Thompson, B. K. & Adolf, A. C. (1970). Gamma-ray-resistant and -sensitive strains of slime mold (*Dictyostelium discoideum*). *Radiat. Res.* **43**, 711–28. [56]

DeFilippes, F. M. & Guild, W. R. (1959). Irradiation of solutions of transforming DNA. *Radiat. Res.* **11**, 38–53. [27]

Denekamp, J., Emery, E. W. & Field, S. B. (1971). Response of mouse epidermal cells to single and divided doses of fast neutrons. *Radiat. Res.* **45**, 80–4. [182]

Denekamp, J., Fowler, J. F., Kragt, K., Parnell, C. J. & Field, S. B. (1966). Recovery and repopulation in mouse skin after irradiation with cyclotron neutrons as compared with 250-Kv X-rays or 15-Mev electrons. *Radiat. Res.* **29**, 71–84. [182]

Denekamp, J. & Harris, S. R. (1975a). Tests of two electron-affinic radiosensitizers *in vivo* using regrowth of an experimental carcinoma. *Radiat. Res.* **61**, 191–203. [78, 270]

Denekamp, J. & Harris, S. R. (1975b). The response of mouse skin to multiple small doses of radiation. In *Cell Survival after Low Doses of Radiation*, ed. T. Alper, pp. 342–50. London: Institute of Physics and John Wiley & Sons. [254]

Denekamp, J., Michael, B. D. & Harris, S. R. (1974). Hypoxic cell radiosensitizers: comparative tests of some electron affinic compounds using epidermal cell survival *in vivo*. *Radiat. Res.* **60**, 119–32. [272]

Dennis, J. A. (1977). Hit and target theories and the molecular theory of radiation action: notes on the influence of radiation quality. *National Radiological Protection Board: Report NRPB-R61* Harwell, Oxfordshire OX11 ORQ. [115, 137]

Deschner, E. & Gray, L. H. (1959). Influence of oxygen tension on X-ray-induced chromosomal damage in Ehrlich ascites tumor cells irradiated *in vitro* and *in vivo*. *Radiat. Res.* **11**, 115–46. [62, 220]

Dettor, C. M., Dewey, W. C., Winans, L. F. & Noel, J. S. (1972). Enhancement of X-ray damage in synchronous Chinese hamster cells by hypertonic treatments. *Radiat. Res.* **52**, 352–72. [247]

Dewey, D. L. (1960). Effect of glycerine on the X-ray sensitivity of *Serratia marcescens*. *Nature, Lond.* **187**, 1008–10. [97]

Dewey, D. L. (1963). The X-ray sensitivity of *Serratia marcescens*. *Radiat. Res.* **19**, 64–87. [96]

Dewey, D. L. (1965). Cysteine and the radiosensitivity of bacteria. *Prog. biochem. Pharmacol.* **1**, 59–64. [75]

Dewey, D. L. & Boag, J. W. (1959). Modification of the oxygen effect when bacteria are given large pulses of radiation. *Nature, Lond.* **183**, 1450–1. [55]

Dewey, D. L. & Michael, B. D. (1965). The mechanisms of radiosensitization by iodoacetamide. *Biochem. biophys. Res. Commun.* **21**, 392–6. [72]

Dewey, D. L. & Stein, G. (1970). The action of atomic hydrogen, hydrated electrons, and ionizing radiation on bacteriophage T_7 in aqueous solution. *Radiat. Res.* **44**, 345–58. [85]

Dische, S. (1978). Hyperbaric oxygen. The Medical Research Council trials and their clinical significance. *Br. J. Radiol.* **51**, 888–94. [262]

Dische, S., Gray, A. J. & Zanelli, G. D. (1976). Clinical testing of the radiosensitizer Ro-07-0582. II. Radiosensitization of normal and hypoxic skin. *Clin. Radiol.* **27**, 159–166. [272]

Dische, S., Saunders, M. I. & Flockhart, I. R. (1978). The optimum regime for the administration of misonidazole and the establishment of multi-centre clinical trials. *Br. J. Cancer* **37**, suppl. 3, 318–21. [271]

Dittrich, W. (1960). Treffermischkurven. *Z. Naturf.* **15b**, 261–6. [22]

Dixon, B. (1968). The effect of radiation on the growth of vertebrae in the tails of rats. I. Single doses of X-rays and the effect of oxygen. *Int. J. Radiat. Biol.* **13**, 355–68. [261]

Dixon, B. (1969). The effect of radiation on the growth of vertebrae in the tails of rats. III. the response to cyclotron neutrons. *Int. J. Radiat. Biol.* **15**, 541–8. [182]

Djordjević, B. & Tolmach, L. J. (1967). Responses of synchronous populations of HeLa cells to ultraviolet irradiation at selected stages of the generation cycle. *Radiat. Res.* **32**, 327–46. [155, 156]

Doniach, I. & Logothetopoulos, J. H. (1955). Effects of radioactive iodine on the rat thyroid's function, regeneration and response to goitrogens. *Br. J. Cancer* **9**, 117–27. [12]

Dooley, D. C. & Nester, E. W. (1973). Deoxyribonucleic acid–membrane complexes in the *Bacillus subtilis* transformation system. *J. Bacteriol.* **114**. 711–22. [206]

Douglas, B. G. & Fowler, J. F. (1976). The effect of multiple small doses of X-rays on skin reactions in the mouse and a basic interpretation. *Radiat. Res.* **66**, 401–26. [194, 255]

Durand, R. E. & Biaglow, J. E. (1974). Modification of the radiation response of an *in vitro* tumour model by control of cellular respiration. *Int. J. Radiat. Biol.* **26**, 597–601. [234]

Durand, R. E. & Olive, P. L. (1977). Fast neutron effects on multi-cell spheroids. *Br. J. Radiol.* **50**, 423–9. [180, 181]

Durand, R. E. & Sutherland, R. M. (1972). Effects of intercellular contact on repair of radiation damage. *Expl Cell Res.* **71**, 75–80. [180, 198, 199, 251]

Dutreix, J. & Wambersie, A. (1975). Cell survival curves deduced from non-quantitative reactions of skin, intestinal mucosa and lung. In *Cell Survival after Low Doses of Radiation*, ed. T. Alper, pp. 335–40. London: Institute of Physics and John Wiley & Sons. [254]

Ebert, M. & Alper, T. (1954). Influence of dissolved gases on hydrogen peroxide formation and bacteriophage inactivation by radiation. *Nature, Lond.* **173**, 987–9. [85]

Edwards, A. A. & Dennis, J. A. (1975). The calculation of charged particle fluence and LET spectra for the irradiation of biologically significant materials by neutrons. *Physics Med. Biol.* **20**, 395–409. [107]

Ehmann, U. K., Nagasawa, H., Petersen, D. F. & Lett, J. T. (1974). Symptoms of X-ray damage to radiosensitive mouse leukaemic cells: asynchronous populations. *Radiat. Res.* **60**, 453–72. [36]

Ehret, C. F., Smaller, B., Powers, E. L. & Webb, R. B. (1960). Thermal annealment and nitric oxide effects on free radicals in X-irradiated cells. *Science* **132**, 1768–9. [64]

Ekert, B. & Grunberg-Manago, M. (1966). Effets des rayons γ sur l'efficacité de quelques polyribonucléotides en tant que messagers. *C. R. Acad. Sc. Paris* **263**, 1762–5. [27]

Ekert, B. & Latarjet, M-F. (1971). Inactivation par les rayons γ des propriétés fonctionelles des RNA de transfert d'*E. coli* (phényl-alanine et lysine). *Int. J. Radiat. Biol.* **20**, 521–40. [27]

Elkind, M. M. (1967). Sublethal X-ray damage and its repair in mammalian cells. In *Radiation Research*, ed. G. Silini, pp. 558–86. Amsterdam: North-Holland Publishing Co. [171]

Elkind, M. M. (1970). Damage and repair processes relative to neutron (and charged particle) irradiation. *Curr. Topics Radiat. Res.* **7**, 1–44. [171]

Elkind, M. M. (1971). Sedimentation of DNA released from Chinese hamster cells. *Biophys. J.* **11**, 502–20. [241]

Elkind, M. M. (1975). A summary and review of the conference. In *Cell Survival after Low Doses of Radiation*, ed. T. Alper, pp. 376–88. London: Institute of Physics & John Wiley & Sons. [254]

Elkind, M. M. & Kamper, C. (1970). Two forms of repair of DNA in mammalian cells following irradiation. *Byophys. J.* **10**, 237–45. [241]

Elkind, M. M., Moses, W. B. & Sutton-Gilbert, H. (1967a). Radiation responses of mammalian cells grown in culture. VI. Protein, DNA and RNA inhibition during the repair of X-ray damage. *Radiat. Res.* **31**, 156–73. [174, 178, 248]

Elkind, M. M. & Redpath, J. L. (1977). Molecular and cellular biology of radiation lethality. In *Cancer: A Comprehensive Treatise*, ed. F. F. Becker, vol. 6, pp. 51–99. New York: Plenum Press. [171, 179]

Elkind, M. M. & Sinclair, W. K. (1965). Recovery in X-irradiated mammalian cells. *Curr. Topics Radiat. Res.* **1**, 165–220. [171]

Elkind, M. M. & Sutton, H. (1959). X-ray damage and recovery in mammalian cells in culture. *Nature, Lond.* **184**, 1293–5. [164, 165]

Elkind, M. M. & Sutton, H. (1960). Radiation response of mammalian cells grown in culture. I. Repair of X-ray damage in surviving Chinese hamster cells. *Radiat. Res.* **13**, 556–93. [139, 165, 171, 172, 173, 256]

Elkind, M. M., Sutton-Gilbert, H., Moses, W. B. & Kamper, C. (1967b). Sublethal and lethal radiation damage. *Nature, Lond.* **214**, 1088–92. [166, 178, 192, 196]

Elkind, M. M., Swain, R. W., Alescio, T., Sutton, H. & Moses, W. B. (1965). Oxygen, nitrogen, recovery and radiation therapy. In *Cellular Radiation Biology*, pp. 442–61. Baltimore: Williams & Wilkins Company. [174]

Elkind, M. M. & Whitmore, G. F. (1967). *The Radiobiology of Cultured Mammalian Cells.* New York: Gordon & Breach. [38, 151, 166, 171, 219]

Elkind, M. M., Whitmore, G. F. & Alescio, T. (1964). Actinomycin D: suppression of recovery in X-irradiated mammalian cells. *Science* **143**, 1454–7. [178]

Emery, E. W., Denekamp, J, Ball, M. M. & Field, S. B. (1970). Survival of mouse skin epithelial cells following single and divided doses of X-rays. *Radiat. Res.* **41**, 450–66. [251, 252, 256]

Emmerson, P. T. (1967). Enhancement of the sensitivity of anoxic *Escherichia coli* B / r to X-rays by triacetoneamine N-oxyl *Radiat. Res.* **30**, 841–9. [73]

Emmerson, P. T. (1968). Sensitization of anoxic recombination-deficient mutants of *Escherichia coli* K12 to X-rays by triacetoneamine N-oxyl. *Radiat. Res.* **36**, 410–17. [73]

Emmerson, P. T. & Howard-Flanders, P. (1964). Sensitization of anoxic bacteria to X-rays by di-t-butyl nitroxide and analogues. *Nature, Lond.* **204**, 1005–6. [73]

Emmerson, P. T. & Howard-Flanders, P. (1965). Preferential sensitization of anoxic bacteria to X-rays by organic nitroxide-free radicals. *Radiat. Res.* **26**, 54–62. [73]

Ephrussi-Taylor & Latarjet, R. (1955). Inactivation, par les rayons X, d'un facteur transformant du Pneumococque. *Biochim. biophys. Acta* **16**, 183–97. [206]

Epp, E. R., Weiss, H., Kessaris, N. D., Santomasso, A., Heslin, J. & Ling, C. C. (1973). Oxygen diffusion times in bacterial cells irradiated with high-intensity pulsed electrons: new upper limit to the lifetime of oxygen-sensitive species suspected to be induced at critical sites in bacterial cells. *Radiat. Res.* **54**, 171–80. [65]

Epp, E. R., Weiss, H. & Santomasso, A. (1968). The oxygen effect in bacterial cells irradiated with high-intensity pulsed electrons. *Radiat. Res.* **34**, 320–5. [55, 65]

Erikson, R. L. & Szybalski, W. (1961). Molecular radiobiology of human cell lines. I. Comparative sensitivity to X-rays and ultraviolet light of cells containing halogen-substituted DNA. *Biochem. biophys. Res. Commun.* **4**, 258–61. [96]

Evans, H. J. (1967). Repair and recovery at chromosome and cellular levels: Similarities and differences. In *Recovery and Repair Mechanisms in Radiobiology. Brookhaven Symp. Biol*, no. 20, pp. 111–31. [219, 221]

Evans, H. J. & Neary, G. J. (1959). The influence of oxygen on the sensitivity of *Tradescentia* pollen tube chromosomes to X-rays. *Radiat. Res.* **11**, 636–41. [62]

Evans, R. G., Bagshaw, M. A., Gordon, L. F., Kurkjian, S. D. & Hahn, G. M. (1974). Modification of recovery from potentially lethal X-ray damage in plateau phase Chinese hamster cells. *Radiat. Res.* **59**, 597–605. [233]

Field, S. B. (1969). The relative biological effectiveness of fast neutrons for mammalian tissues. *Radiology* **93**, 915–20. [250]

Field, S. B. & Hornsey, S. (1975). The response of mouse skin and lung to fractionated X-rays. In *Cell Survival after Low Doses of Radiation*, ed. T. Alper, pp. 362–8. London: Institute of Physics and John Wiley & Sons. [254]

Field, S. B., Jones, T. & Thomlinson, R. H. (1967). The relative effects of fast neutrons and X-rays on tumour and normal tissue in the rat. 1. Single doses. *Br. J. Radiol.* **40**, 834–41. [250, 268]

Field, S. B., Jones, T. & Thomlinson, R. H. (1968). The relative effects of fast neutrons and X-rays on tumour and normal tissue in the rat. Part 2. Fractionation: recovery and reoxygenation. *Br. J. Radiol.* **41**, 597–607. [182, 257]

Field, S. B., Morgan, R. L. & Morrison, R. (1976). The response of human skin to irradiation with X-rays or fast neutrons. *Int. J. Radiat. Oncol. Biol. Physics* **1**, 481–6. [250, 251]

Fielden, E. M., Ewing, D. & Roberts, P. B. (1974). Additive effects in the radiosensitization of *B. megaterium* spores by p-nitroacetophenone and norpseudopelletierine *N*-oxyl. *Radiat. Res.* **58**, 489–97. [81]

Fisher, G. J., Watts, M. E., Patel, K. B. & Adams, G. E. (1978). Sensitization of ultraviolet damage in bacteria and mammalian cells. *Br. J. Cancer* **37**, suppl. 3, 111–14. [71]

Forage, A. J. (1971). The dependence of the oxygen enhancement ratio on the test of damage in irradiated bacteria. *Int. J. Radiat. Biol.* **20**, 427–36. [123]

Forage, A. J. & Alper, T. (1970). A decrease in anoxic sensitivity, and resultant increase in oxygen enhancement ratio, with decrease in energy of fast electrons. *Int. J. Radiat. Biol.* **17**, 527–32. [52, 124, 125]

Forage, A. J. & Alper, T. (1973). Evidence for differing modes of interaction of acriflavine with ultraviolet-induced lesions in an Hcr⁺ bacterial strain. *Molec. gen. Genet.* **122**, 89–100. [179]

Forage, A. J. & Gillies, N. E. (1964). Restoration of *Escherichia coli* strain B after γ-irradiation. *J. gen. Microbiol.* **37**, 33–9. [246]

Ford, C. E., Hamerton, J. L., Barnes, D. W. H. & Loutit, J. F. (1956). Cytological identification of radiation-chimaeras. *Nature, Lond.* **177**, 452–4. [3]

Foster, C. J., Malone, J., Orr, J. S. & MacFarlane, D. E. (1971). The recovery of the survival curve shoulder after protracted hypoxia. *Br. J. Radiol.* **44**, 540–5. [201]

Fowler, J. F. (1964). Differences in survival curve shapes for formal multi-target and multi-hit models. *Physics Med. Biol.* **9**, 177–88. [41]

Fowler, J. F., Denekamp, J., Page, H. N. C., Begg, A. C., Field, S. B. & Butler, K. R. (1972). Fractionation with X-rays and neutrons: A comparison of tumour and normal tissue effects in C_3H mice. *Br. J. Radiol.* **45**, 237–49. [268]

Fowler, J. F., Kragt, K., Ellis, R. E., Lindop, P. J. & Berry, R. J. (1965). The effect of divided doses of 15 Mev electrons on the skin response of mice. *Int. J. Radiat. Biol.* **9**, 241–52. [250, 261]

Fowler, J. F., Morgan, R. L., Silvester, J. A., Bewley, D. K. & Turner, B. A. (1963). Experiments with fractionated X-ray treatment of the skin of pigs. *Br. J. Radiol.* **36**, 188–96. [250, 254]

Fox, B. W. & Fox, M. (1973). DNA single-strand rejoining in two pairs of cell-lines showing the same and different sensitivities to X-rays. *Int. J. Radiat. Biol.* **24**, 127–35. [244]

Frey, R. & Hagen, U. (1974). Oxygen effect on irradiated DNA. *Radiat. environm. Biophys.* **11**, 125–33. [243]

Fried, V. A. & Novick, A. (1973). Organic solvents as probes for the structure and function of the bacterial membrane: effects of ethanol on the wild type and ethanol-resistant mutant of *Escherichia coli* K12. *J. Bacteriol.* **114**, 239–48. [212]

Fu, K. K., Phillips, T. L., Kane, L. J. & Smith, V. (1975a). Tumor and normal tissue response to irradiation *in vivo*: Variation with decreasing dose rates. *Radiology* **114**, 709–16. [184, 185]

Fu, K. K., Phillips, T. L., Wharam, M. D. & Kane, L. J. (1975b). The influence of growth and irradiation conditions on the radiation response of the EMT6 tumour. In *Cell Survival after Low Doses of Radiation*, ed. T. Alper, pp. 251–8. London: Institute of Physics and John Wiley & Sons. [267]

George, K. C., Shenoy, M. A., Joshi, D. S., Bhatt, B. Y., Singh, B. B. & Gopal-Ayengar, A. R. (1975). Modification of radiation effects on cells by membrane-binding agents – procaine HCl. *Br. J. Radiol.* **48**, 611–14 [79]

Gillespie, C. J., Chapman, J. D., Reuvers, A. P. & Dugle, D. L. (1975). The inactivation of Chinese hamster cells by X-rays: synchronized and exponential cell populations. *Radiat. Res.* **64**, 353–64. [157, 188]

Gillies, N. E. (1961). The use of auxotrophic mutants to study restoration in *Escherichia coli* B after ultra-violet irradiation. *Int. J. Radiat. Biol.* **3**, 379–87. [246]

Gillies, N. E. & Alper, T. (1959). Reduction in the lethal effects of radiations on *Escherichia coli* B by treatment with chloramphenicol. *Nature, Lond.* **183**, 237–8. [210, 246]

Gillies, N. E., Obioha, F. I. & Ratnajothi, N. H. (1979). An oxygen-dependent X-ray lesion in the cell wall of *E. coli* B/r sensitive to penicillin. *Int. J. Radiat. Biol.* (in press). [222, 223]

Ginoza, W. (1967). The effects of ionizing radiations on nucleic acids of bacteriophages and bacterial cells. *A. Rev. Microbiol.* **21**, 325–62. [30]

Ginoza, W. & Norman, A. (1957). Radiosensitive molecular weight of tobacco mosaic virus nucleic acid. *Nature, Lond.* **179**, 520–1. [25, 93]

Goodhead, D. T. (1977). Inactivation and mutation of cultured mammalian cells by aluminium characteristic ultrasoft X-rays. III. Implications for theory of dual radiation action. *Int. J. Radiat. Biol.* **32**, 43–70. [144]

Goodhead, D. T. & Thacker, J. (1977). Inactivation and mutation of cultured mammalian cells by aluminium characteristic ultrasoft X-rays. I. Properties of aluminium X-rays and preliminary experiments with Chinese hamster cells. *Int. J. Radiat. Biol.* **31**, 541–59. [143]

Gordy, W., Ard, W. B. & Shields, H. (1955). Microwave spectroscopy of biological substances. I. Paramagnetic resonance in X-irradiated amino acids and proteins. *Proc. natn. Acad. Sci. U.S.A.* **41**, 983–1009. [64]

Gordy, W. & Miyagawa, I. (1960). Electron spin resonance studies of mechanisms for chemical protection from ionizing radiation. *Radiat. Res.* **12**, 211–29. [25]

Goscin, S. A. & Fridovich, I. (1973). Superoxide dismutase and the oxygen effect. *Radiat. Res.* **56**, 565–9. [67]

Gray, L. H. (1954). Some characteristics of biological damage induced by ionizing radiations. *Radiat. Res.* **1**, 189–213. [64, 134]

Gray, L. H. (1956). A method of oxygen assay applied to a study of the removal of dissolved oxygen by cysteine and cysteamine. In *Progress in Radiobiology*, ed. J. S. Mitchell, B. E. Holmes and C. L. Smith, pp. 267–74. Edinburgh: Oliver & Boyd. [88]

Gray, L. H., Conger, A. D., Ebert, M., Hornsey, S. & Scott, O. C. A. (1953). The concentration of oxygen dissolved in tissues at the time of irradiation as a factor in radiotherapy. *Br. J. Radiol.* **26**, 638–48. [259]

Gray, L. H., Green, F. O. & Hawes, C. A. (1958). Effect of nitric oxide on the radiosensitivity of tumour cells. *Nature, Lond.* **182**, 952–3. [72]

Gudas, L. J. & Pardee, A. B. (1976). DNA synthesis inhibition and the induction of protein X in *Escherichia coli. J. molec. Biol.* **101**, 459–77. [249]

Hahn, G. M., Bagshaw, M. A., Evans, R. G. & Gordon, L. F. (1973). Repair of potentially lethal lesions in X-irradiated, density-inhibited Chinese hamster cells: metabolic effects and hypoxia. *Radiat. Res.* **55**, 280–90. [233, 235]

Hahn, G. M. & Little, B. (1972). Plateau-phase cultures of mammalian cells: an *in vitro* model for human cancer. *Curr. Topics Radiat. Res.* **8**, 39–83. [149, 229, 235, 269]

Hall, E. J. (1972). Radiation dose-rate: a factor of importance in radiobiology and radiotherapy. *Br. J. Radiol.* **45**, 81–97. [185]

Hall, E. J. (1975). Biological problems in the measurement of survival at low doses. In *Cell Survival after Low Doses of Radiation*, ed. T. Alper, pp. 13–24. London: Institute of Physics and John Wiley & Sons. [16, 190, 191]

Hall, E. J. & Roizin-Towle, L. (1975). Hypoxic cell sensitizers: radiobiological studies at the cellular level. *Radiology* **117**, 453–7. [270]

Hall, E. J., Roizin-Towle, L. Theus, R. B. & August, L. S. (1975). Radiobiological properties of high-energy cyclotron-produced neutrons used for radiotherapy. *Radiology* **117**, 173–8. [131]

Han, A. & Elkind, M. M. (1977). Additive action of ionizing and non-ionizing radiations throughout the Chinese hamster cell-cycle. *Int. J. Radiat. Biol.* **31**, 275–82. [155, 156]

Han, A., Sinclair, W. K. & Kimler, B. F. (1976). The effect of N-ethylmaleimide on the response to X-rays of synchronized HeLa cells. *Radiat. Res.* **65**, 337–50. [158, 161, 162]

Hanawalt, P. C. & Setlow, R. B. (eds.) (1975). *Molecular Mechanisms for the Repair of DNA.* New York and London: Plenum Press. [238]

Hannan, R. S. & Shepherd, H. J. (1954). Some after-effects in fats irradiated with high-energy electrons and X-rays. *Br. J. Radiol.* **27**, 36–42. [221]

Harris, J. W. (1976). Radiation modifiers. An evaluation of recent research and clinical potential. In *Modification of Radiosensitivity of Biological Systems*, Int. Atomic Energy Agency, Vienna, pp. 11–28. London: HMSO. [272]

Harris, J. W. & Power, J. A. (1973). Diamide: A new radiosensitizer for anoxic cells. *Radiat. Res.* **56**, 97–109. [76, 77, 81]

Harris, J. W., Power, J. A. & Koch, C. J. (1975). Radiosensitization of hypoxic mammalian cells by diamide. I. Effects of experimental conditions on survival. *Radiat. Res.* **64**, 270–80. [196]

Harrop, H. A., Maughan, R. L., Michael, B. D. & Rupp, W. D. (1978). Dose rate effects in *Escherichia coli* K-12 and repair deficient mutants. *Br. J. Radiol.* **51**, 559. [186, 213]

Hawes, C., Howard, A. & Gray, L. H. (1966). Induction of chromosome structural damage in Ehrlich ascites tumour cells. *Mutation Res.* **3**, 79–89. [62]

Haynes, R. H. (1966). The interpretation of microbial inactivation and recovery phenomena. *Radiat. Res.* Suppl. 6, 1–29. [44]

Hendry, J. H. (1978). Sensitization of hypoxic normal tissue. *Br. J. Cancer* **37**, suppl. 3, 232–4. [272]

Hendry, J. H., Rosenberg, I., Greene, D. & Stewart, J. G. (1976). Tolerance of rodent tails to necrosis after 'daily' fractionated X-rays or D-T neutrons. *Br. J. Radiol.* **49**, 690–9. [261]

Hershey, A. D. & Chase, M. (1952). Independent functions of viral protein and nucleic acid in growth of bacteriophage. *J. gen. Physiol.* **36**, 39–56. [206]

Hesslewood, I. P. (1978). DNA strand breaks in resistant and sensitive murine lymphoma cells detected by the hydroxylapatite chromatographic technique. *Int. J. Radiat. Biol.* **34**, 461–9. [215, 225, 242, 243, 244]

Hetzel, F. W., Kruuv, J. & Frey, H. E. (1976). Repair of potentially lethal damage in X-irradiated V79 cells. *Radiat, Res.* **68**, 308–19. [235]

Hewitt, H. B. (1958). Studies of the dissemination and quantitative transplantation of a lymphocytic leukaemia of CBA mice. *Br. J. Cancer* **12**, 378–401. [15]

Hewitt, H. B. & Read, J. (1950). Search for an effect of oxygen on the direct X-ray inactivation of bacteriophage. *Br. J. Radiol.* **23**, 416–23. [206]

Hewitt, H. B. & Wilson, C. W. (1959). A survival curve for mammalian leukaemia cells irradiated *in vivo*. *Br. J. Cancer* **13**, 69–75. [15, 148]

Hewitt, H. B. & Wilson, C. W. (1961). Survival curves for tumour cells irradiated *in vivo*. *Ann. N. Y. Acad. Sci.* **95**, 818–27. [39]

Hillová, J. & Drášil, V. (1967). The inhibitory effect of iodoacetamide on recovery from sub-lethal damage in *Chlamydomonas rheinhardii. Int. J. Radiat. Biol.* **12**, 201–8. [170]

Hollaender, A. (ed.) (1960). *Radiation Protection and Recovery*, Oxford: Pergamon Press. [227]

Hornsey, S. (1963). A comparison of the survival curves of Ehrlich ascites tumour cells after X-irradiation produced by two different techniques. In *Radiation Effects in Physics, Chemistry and Biology*, reported by H. B. Hewitt, ed. M. Ebert and A. Howard, pp. 234–249. Amsterdam: North-Holland Publishing Co. [218, 267]

Hornsey, S. (1970). The effect of hypoxia on the sensitivity of the epithelial cells of the jejunum. *Int. J. Radiat. Biol.* **18**, 539–6. [259, 261]

Hornsey, S. (1971). Differences in the effect of hypoxia on the radiation sensitivity of the bone marrow and the intestine in mice. *Br. J. Radiol.* **44**, 357–60. [259]

Hornsey, S. (1973a). The radiosensitivity of the intestine. In *Strahlenempfindlichkeit von Organen und Organensystemen der Säugtiere und des*

Menschen, ed. H. Braun, F. Heuck, H-A. Ladner, O. Messerschmidt, K. Musshoff, and C. Streffer, pp. 78–88. Stuttgart: Georg Thieme Verlag. [4]

Hornsey, S. (1973b). The effectiveness of fast neutrons compared with low LET radiation on cell survival measured in the mouse jejunum. *Radiat. Res.* **55**, 58–68. [182, 252]

Hornsey, S. & Alper, T. (1966). An unexpected dose-rate effect in the killing of mice by radiation. *Nature, Lond.* **210**, 212–13. [185]

Hornsey, S. & Bewley, D. K. (1971). Hypoxia in mouse intestine induced by electron irradiation at high dose-rates. *Int. J. Radiat. Biol.* **19**, 479–83. [185]

Hornsey, S., Kutsutani, Y. & Field, S. B. (1975). Damage to mouse lung with fractionated neutrons and X-rays. *Radiology* **116**, 171–4. [182]

Hornsey, S., Myers, R. & Andreozzi, U. (1977). Differences in the effects of anaesthesia on hypoxia in normal tissues. *Int. J. Radiat. Biol.* **32**, 609–12. [258, 261]

Hornsey, S. & Silini, G. (1961). Studies on cell-survival of irradiated Ehrlich ascites tumour. II. Dose–effect curves for X-ray and neutron irradiations. *Int. J. Radiat. Biol.* **4**, 135–41. [119, 128, 136]

Hornsey, S. & Silini, G. (1962). Recovery of tumour cells cultured *in vivo* after X-ray and neutron irradiations. *Radiat. Res.* **16**, 712–22. [119, 167, 168]

Horowitz, I. A., Norwint, H. & Hall, E. J. (1975). Conditioned medium from plateau-phase cells. Effect on growth of proliferative cells and on repair of potentially lethal radiation damage. *Radiology* **114**, 723–6. [233, 235]

Horsley, R. J. & Laszlo, A. (1971). Unexpected additional recovery following a first X-ray dose to a synchronized cell culture. *Int. J. Radiat Biol.* **20**, 593–6. [248]

Horsley, R. J. & Laszlo, A. (1973). Additional recovery in X-irradiated *Oedogonium cardiacum* can be suppressed by cycloheximide. *Int. J. Radiat. Biol.* **23**, 201–4. [248]

Hotz, G. (1966). Untersuchungen über den Mechanismus des Strahlenschutzes von Thiol- und Disulfidverbindungen am Modell biologischer Elementarein-heiten. *Z. Naturf.* **21b**, 148–52. [94]

Hotz, G. (1968). Der Einfluss von Sauerstoff auf den direkten Effekt von ^{60}Co-Gammastrahlung bei der Inaktivierung von Phagen-nuklein-säure. *Studia Biophysica* **12**, 49–57. [23, 26]

Hotz, G. (1974). Infectious DNA from coliphage T_1. V. Detection of gamma-radiation-induced latent double-strand breaks by cysteamine. *Radiat. environm. Biophys.* **11**, 157–64. [243]

Hotz, G. & Müller, A. (1962). Der Einfluss von Cystein- und Cysteamin-Konzentration auf die Inaktivierung röntgenbestrahlter T-Phagen. *Z. Naturf.* **17b**, 34–7. [94]

Howard, A. (1968). The oxygen requirement for recovery in split-dose experi-ments with *Oedogonium*. *Int. J. Radiat. Biol.* **14**, 341–50. [174]

Howard, A. & Cowie, F. G. (1975). Survival-curve characteristics in a desmid. In *Cell Survival after Low Doses of Radiation*. ed. T. Alper, pp. 3–10. London: Institute of Physics and John Wiley & Sons. [248]

Howard, A. & Cowie, F. G. (1976). Over-repair in *Closterium*: increased radioresistance caused by an earlier exposure to radiation. In *Radiation and Cellular Control Processes*, ed. J. Kiefer, pp. 188–95, Heidelberg and New York: Springer-Verlag. [248]

Howard, A. & Pelc, S. (1953). Synthesis of desoxyribonucleic acid in normal and irradiated cells and its relation to chromosome breakage. Supplement, *Heredity, Lond.* **6**, 261–73. [151]

Howard-Flanders, P. (1957). Effect of nitric oxide on the radiosensitivity of bacteria. *Nature, Lond.* **180**, 1191–2. [72]

Howard-Flanders, P. (1958). Physical and chemical mechanisms in the injury of cells by ionizing radiations. *Adv. biol. med. Phys.* **6**, 553–603. [59, 74, 134]

Howard-Flanders, P. (1960). Effect of oxygen on the radiosensitivity of bacteriophage in the presence of sulphydryl compounds. *Nature, Lond.* **186**, 485–7. [26, 93]

Howard-Flanders, P. & Jockey, P. (1960). Similarities in the effects of oxygen and nitric oxide on the rate of inactivation of vegetative bacteria by X-rays. *Radiat. Res.* **13**, 466–78. [72]

Howard-Flanders, P., Levin, J. & Theriot, L. (1963). Reactions of deoxyribonucleic acid radicals with sulfhydryl compounds in X-irradiated bacteriophage systems. *Radiat. Res.* **18**, 593–606. [94, 95]

Howard-Flanders, P. & Moore, D. (1958). The time interval after pulsed irradiation within which injury to bacteria can be modified by dissolved oxygen. I. A search for an effect of oxygen 0.02 second after pulsed irradiation. *Radiat. Res.* **9**, 422–37. [64]

Howard-Flanders, P. & Wright, E. A. (1955). Effect of oxygen on the radiosensitivity of growing bone and a possible danger in the use of oxygen during radiotherapy. *Nature, Lond.* **175**, 418–29. [262]

Howard-Flanders, P. & Wright, E. A. (1957). The effect of oxygen on the radiosensitivity of growing bone in the tail of the mouse. *Br. J. Radiobiol.* **30**, 593–9. [261]

Hutchinson, F. & Pollard, E. (1961a). Physical principles of radiation action. In *Mechanisms in Radiobiology*, ed. M. Errera and A. Forssberg, vol. 1, pp. 1–70. New York and London: Academic Press, [6, 28]

Hutchinson, F. & Pollard, E. (1961b). Target theory and radiation effects on biological molecules. In *Mechanisms in Radiobiology*, ed. M. Errera and A. Forssberg, vol. 1, pp. 71–91. New York and London: Academic Press. [28, 30]

Inouye, M. & Pardee, A. B. (1970). Changes of membrane proteins and their relation to deoxyribonucleic acid synthesis and cell division of *Escherichia coli. J. biol. Chem.* **245**, 5813–9. [249]

Jacobson, B. (1957). Evidence for recovery from X-ray damage in *Chlamydomonas. Radiat. Res.* **7**, 394–406. [164]

Jain, V. K. & Pohlit, W. (1972). Influence of energy metabolism on the repair of X-ray damage in living cells. I. Effect of respiratory inhibitors and glucose on the liquid-holding reactivation in yeast. *Biophysik* **8**, 254–63. [233, 237]

Jain, V. K. & Pohlit, W. (1973). Influence of energy metabolism on the repair of X-ray damage in living cells. II. Split-dose recovery, liquid-holding reactivation and division delay reversal in stationary populations of yeast. *Biophysik* **9**, 155–65. [177]

Jain, V. K., Pohlit, W. & Purohit, S. C. (1973). Influence of energy metabolism on the repair of X-ray damage in living cells. III. Effect of 2-deoxy-D-glucose on the liquid-holding reactivation in yeast. *Biophysik* **10**, 137–42. [234]

Jain, V. K., Pohlit, W. & Purohit, S. C. (1975). Influence of energy metabolism on the repair of X-ray damage in living cells. IV. Effects of 2-deoxy-D-glucose on the repair phenomena during fractionated irradiation of yeast. *Radiat. environm. Biophys.* **12**, 315–20. [177]

Johansen, I. (1974). Competition between tetramethylpiperidinol N-oxyl and oxygen in effects on single-strand breaks in episomal DNA and in killing after X-irradiation in *Escherichia coli. Radiat. Res.* **58**, 398–408. [73, 74, 82]

Johansen, I., Gulbrandsen, R., Fielden, E. M. & Sapora, O. (1977). Additive effects shown by combinations of nitroxyl and electron-affinic hypoxic cell sensitizers. *Radiat. Res.* **70**, 597–603. [81, 85]

Johansen, I., Gulbrandsen, R. & Pettersen, R. (1974). Effectiveness of oxygen in promoting X-ray-induced single-strand breaks in circular phage λ DNA and killing of radiosensitive mutants of *Escherichia coli. Radiat. Res.* **58**, 384–97. [63, 142]

Johansen, I. & Howard-Flanders, P. (1965). Macromolecular repair and free radical scavenging in the protection of bacteria against X-rays. *Radiat. Res.* **24**, 184–200.

Johns, H. E. & Cunningham, J. R. (1974). *The Physics of Radiology*, 3rd edition. Springfield, Illinois: Charles C. Thomas. [6]

Kapp, D. S. & Smith, K. C. (1970). Repair of radiation-induced damage in *Escherichia coli*. II. Effect of *rec* and *uvr* mutations on radiosensitivity, and repair of X-ray-induced single-strand breaks in deoxyribonucleic acid. *J. Bacteriol.* **103**, 49–54. [239]

Katz, R. (1970). RBE, LET, and $Z*/\beta$ Health Physics **18**, 175. [106]

Katz, R. (1973). Dosimetric implications of the one-hit detector. In *Proceedings of the Regional Conference on Radiation Protection*, pp. 38—46. Israel Atomic Energy Commission. [32]

Katz, R., Ackerson, B., Homayoonfar, M. & Sharma, S. C. (1971). Inactivation of cells by heavy ion bombardment. *Radiat. Res.* **47**, 402–25. [131]

Kellerer, A. M. (1975). Statical and biophysical aspects of the survival curve. In *Cell Survival after Low Doses of Radiation*, ed. T. Alper, pp. 69–85. London: Institute of Physics and John Wiley & Sons. [188]

Kellerer, A. M. & Rossi, H. H. (1972). The theory of dual radiation action. *Curr. Topics Radiat. Res.* **8**, 85–158. [41, 110, 137, 138, 143, 188, 192]

Kember, N. F. (1965). An *in vivo* cell survival system based on the recovery of rat growth cartilage from radiation injury. *Nature, Lond.* **207**, 501–3. [15]

Kember, N. F. (1967). Cell survival and radiation damage in growth cartilage. *Br. J. Radiol.* **40**, 496–505. [252, 259, 261]

Kember, N. F. (1969). Radiobiological investigations with fast neutrons using the cartilage clone system. *Br. J. Radiol.* **42**, 595–7. [182]

Kessaris, N. D., Weiss, H. & Epp, E. R. (1973). Diffusion of oxygen in bacterial cells after exposure to high intensity pulsed electrons: theoretical model and comparison with experiment. *Radiat. Res.* **54**, 181–91. [65]

Kiefer, J. (1968). Recovery from sub-lethal ultra-violet damage in diploid yeast. *Int. J. Radiat. Biol.* **13**, 339–400. [179]

Kiefer, J. (1971). The importance of cellular energy metabolism for the sparing effect of dose fractionation with electrons and ultra-violet light. *Int. J. Radiat. Biol.* **20**, 325–36. [176, 177]

Kiefer, J. (1973). The effect of some nucleic acid-binding metabolic inhibitors on split-dose sparing in UV- and electron-irradiated diploid yeast. *Int. J. Radiat. Biol.* **24**, 93–7. [178, 179]

Kihlman, B. A. (1958). The effect of oxygen, nitric oxide, and respiratory inhibitors on the production of chromosome aberrations by X-rays. *Expl. Cell Res.* **14**, 639–42. [72]

Kim, J. H., Eidinoff, M. L. & Laughlin, J. S. (1964). Recovery from sublethal X-ray damage of mammalian cells during inhibition of synthesis of deoxyribonucleic acid. *Nature, Lond.* **204**, 598–9. [178]

Kimler, B. F., Sinclair, W. K. & Elkind, M. M. (1977). N-ethylmaleimide sensitization of X-irradiated hypoxic Chinese hamster cells. *Radiat. Res.* **71**, 204—13. [159, 161]

Knippers, R. & Stratling, W. (1970). The DNA replicating capacity of isolated *E. coli* cell wall–membrane complexes. *Nature, Lond.* **226**, 713–7. [209]

Koch, C. J. & Burki, H. J. (1977). Enhancement of X-ray induced potentially lethal damage by low temperature storage of mammalian cells. *Br. J. Radiol.* **50**, 290–3. [232]

Koch, C. J. & Kruuv, J. (1971). The effect of extreme hypoxia on recovery after radiation by synchronized mammalian cells. *Radiat. Res.* **48**, 74–85. [176]

Koch, C. J., Kruuv, J. & Frey, H. E. (1973). Variation in radiation response as a function of oxygen tension. *Radiat. Res.* **53**, 33–42. [176]

Koch, C. J., Meneses, J. J. & Harris, J. W. (1977). The effect of extreme hypoxia and glucose on the repair of potentially lethal and sublethal radiation damage by mammalian cells. *Radiat. Res.* **70**, 542–51. [176, 234]

Koch, C. J. & Painter, R. B. (1975). The effect of extreme hypoxia on the repair of DNA single-strand breaks in mammalian cells. *Radiat. Res.* **64**, 256–69. [242]

Kohn, H. I. & Gunter, S. E. (1959). Factors influencing the radio-protective action of cysteine: effects in *Escherichia coli* due to drug concentration, temperature, time and pH. *Radiat. Res.* **11**, 732–44. [90]

Korogodin, V. I. & Malumina, T. S. (1959). The recovery of viability of irradiated yeast cells. *Priroda* October, 82–5.]233]

Korogodin, V. J., Meissel, M. N. & Remesova, T. (1967). Postirradiation recovery in yeast. In *Radiation Research*, ed. G. Silini, pp. 538–57. Amsterdam: North-Holland Publishing Co. [228, 229, 233, 247]

Kruuv, J. & Sinclair, W. K. (1968). X-ray sensitivity of synchronized Chinese hamster cells irradiated during hypoxia. *Radiat. Res.* **36**, 45–54. [159]

Kuppermann, A. (1974). Diffusion kinetics in radiation chemistry: an assessment. In *Physical Mechanisms in Radiation Biology*, ed. R. D. Cooper and R. W. Wood, pp. 155–76. Technical Information Centre, United States Atomic Energy Commission. [110, 111, 128, 130]

Laurie, J., Orr, J. S. & Foster, C. J. (1972). Repair processes and cell survival. *Br. J. Radiol.* **45**, 362–8. [199]

Lea, D. E. (1946). *Actions of Radiations on Living Cells*. Cambridge University Press. [18, 19, 25, 28, 31, 133, 183, 205]

Lea, D. E. & Catcheside, D. G. (1942). The mechanism of the induction by radiation of chromosome aberrations in *Tradescantia*. *J. Genet.* **44**, 216–45. [133, 134, 137]

Lea, D. E. & Haines, R. B. (1940). The bactericidal action of ultraviolet light. *J. Hyg. Camb.* **40**, 149–62. [102]

Legrys, G. A. & Hall, E. J. (1969). The oxygen effect and X-ray sensitivity in synchronously dividing cultures of Chinese hamster cells. *Radiat. Res.* **37**, 161–72. [159]

Lehmann, A. L. & Bridges, B. A. (1977). DNA repair. In *Essays in Biochemistry*, ed. P. N. Campbell and D. G. Greville, vol. 13, pp. 71–245. New York and London: Academic Press. [238]

Lehmann, A. R., Kirk-Bell, S., Arlett, C. F., Harcourt, S. A., de Weerd-

Kastelein, E. A., Keijzer, W. & Hall-Smith, P. (1977). Repair of ultraviolet light damage in a variety of human fibroblast cell lines. *Cancer Res.* **37**, 904–10. [245]

Lehmann, A. R. & Stevens, S. (1977). The production and repair of double strand breaks in cells from normal humans and from patients with ataxia telangiectasia. *Biochim. biophys. Acta.* **474**, 49–60. [244]

Leigh, B. (1968). The absence of an oxygen enhancement effect on induced chromosome loss. *Mutation Res.* **5**, 432–4. [220]

Lindop, P. & Rotblat, J. (1963). Dependence of radiation induced life shortening on dose-rate and anaesthetic. In *Cellular Basis and Aetiology of Late Somatic Effects of Ionizing Radiation*, ed. R. C. Harris, pp. 313–18. New York and London: Academic Press. [258]

Littbrand, B. & Révész, L. (1969). The effect of oxygen on cellular survival and recovery after radiation. *Br. J. Radiol.* **42**, 914–24. [50, 200]

Little, J. B. (1971). Repair of potentially lethal radiation damage in mammalian cells: enhancement by conditioned medium from stationary cultures. *Int. J. Radiat. Biol.* **20**, 87–92. [233]

Little, J. B. & Hahn, G. M. (1973). Life-cycle dependence of repair of potentially-lethal radiation damage. *Int. J. Radiat. Biol.* **23**, 401–7. [235]

Little, J. B., Hahn, G. M., Frindel, E. & Tubiana, M. (1973). Repair of potentially lethal damage *in vitro* and *in vivo*. *Radiology* **106**, 689–94. [234]

Lockhart, R. Z., Elkind, M. M. & Moses, W. R. (1961). Radiation response of mammalian cells grown in culture. II Survival and recovery characteristics of several sub-cultures of HeLa S₃ cells after X-irradiation. *J. natn. Cancer Inst.* **27**, 1393–404. [170, 171, 257]

Lohman, P. H. M., Bootsma, D. & Bridges, B. A. (1977). DNA repair mechanisms in mammalian cells. *Mutation Res.* **46**, 99–104. [238]

Longmuir, I. S., Knopp, J. A., Lee, T. -P., Benson, D. & Tang, A. (1977). The intracellular heterogeneity of oxygen concentrations as measured by ultraviolet television microscopy of P.B.A. fluorescence quenching by oxygen. Abstracts p. 14. *3rd Symposium, International Society for Oxygen Transport to Tissue.* [63]

Luzzati, V., Nicolaieff, A. & Masson, E. (1961). Structure de l'acide désoxyribonucléique en solution: Étude par diffusion des rayons X aux petits angles. *J. molec. Biol.* **3**, 185–201. [213]

Lyman, J. T. & Haynes, R. H. (1967). Recovery of yeast after exposure to densely ionizing radiations. *Radiat. Res.* Supplement 7, 222–30. [145, 238]

Lynn, K. R. & Raoult, A. P. D. (1976). γ-Irradiation of lima bean protease inhibitor in dilute aqueous solutions. *Radiat. Res.* **65**, 41–9. [27]

McCord, J. M. & Fridovich, I. (1969). Superoxide dismutase: an enzymatic function for erythrocuprein. *J. biol. Chem.* **244**, 6049–55. [67]

McGrath, R. A. & Williams, R. W. (1966). Reconstruction *in vivo* of irradiated *Escherichia coli* deoxyribonucleic acid; the rejoining of broken pieces. *Nature, Lond.* **212**, 534–5. [62, 239, 240, 241]

McNally, N. J. (1972). A low oxygen-enhancement ratio for tumour-cell survival as compared with that for tumour-growth delay. *Int. J. Radiat. Biol.* **22**, 407–10. [268]

McNally, . J. (1975). A comparison of the effects of radiation on tumor growth delay and cell survival. The effect of radiation quality. *Br. J. Radiol.* **48**, 141–5. [269]

McNally, N. J. (1976). The effect of a change in radiation quality on ability of electron affinic sensitizers to sensitize hypoxic cells. *Int. J. Radiat. Biol.* **29**, 191–6. [131]

McNally, N. J. & Bewley, D. K. (1969). A biological dosimeter using mammalian cells in tissue culture and its use in obtaining neutron depth dose curves. *Br. J. Radiol.* **42**, 289–94. [136]

McNally, N. J. & de Ronde, J. (1976). The effect of repeated small doses of radiation on recovery from sub-lethal damage by Chinese hamster cells irradiated in the plateau phase of growth. *Int. J. Radiat. Biol.* **29**, 221–34. [173, 254, 257]

McNally, N. J. & de Ronde, J. (1978). Interaction between electron-affinic sensitizers. *Br. J. Cancer* **37**, suppl. 3, 90–4. [84, 85, 271]

McNally, N. J. & Sheldon, P. W. (1977). The effect of radiation on tumour growth delay, cell survival and cure of the animal using a singe tumour system. *Br. J. Radiol.* **50**, 321–8. [268, 270]

Madoc-Jones, H. (1964). Variations in radiosensitivity of a mammalian cell line with phase of growth cycle. *Nature, Lond.* **203**, 983–4. [148, 149]

Manney, T. R., Brustad, T. B. & Tobias, C. A. (1963). Effects of glycerol and of anoxia on the radiosensitivity of haploid yeasts to densely ionizing particles. *Radiat. Res.* **18**, 374–88. [113, 123, 129, 144]

Marcovich, H. (1957). Sur le mécanisme de l'activité radioprotectrice de la cystéamine chez les bactéries. *Annls Inst. Pasteur, Paris* **93**, 456–62. [95]

Michael, B. D., Adams, G. E., Hewitt, H. B., Jones, W. B. G. & Watts, M. E. (1973). A posteffect of oxygen in irradiated bacteria: a submillisecond fast mixing study. *Radiat. Res.* **54**, 239–51. [65, 142]

Michael, B. D., Harrop, H. A., Maughan, R. L. & Patel, K. B. (1978). A fast kinetics study of the modes of action of some different radiosensitizers in bacteria. *Br. J. Cancer* **37**, suppl. 3, 29–33. [65, 84, 85, 224]

Miletić, B., Kućan, Ž., Drakulić, M. & Zajec, Lj. (1961). Effect of chloramphenicol on the biosynthesis of DNA in X-irradiated *Escherichia coli* B. *Biochem. biophys. Res. Commun.* **4**, 348–52. [240]

Miletić, B., Petrović, D., Han, A. & Šašel, L. (1964). Restoration of viability of X-irradiated L-strain cells by isologous and heterologous highly polymerized deoxyribonucleic acid. *Radiat. Res.* **23**, 94–103. [196]

Mitchison, J. M. (1972). Cell cycle markers provided by enzyme synthesis. *Curr. Topics Radiat. Res.* **7**, 244–7. [156, 199]

Modig, H. G., Edgren, M. & Révész, L. (1974). Dual effect of oxygen on the induction and repair of single-strand breaks in the DNA of X-irradiated mammalian cells. *Int. J. Radiat. Biol.* **26**, 341–53. [242]

Moore, J. L. (1965). An induced enzyme in X-irradiated *Escherichia coli*: comparison with lethal effects. *J. gen. Microbiol.* **41**, 119–26. [123]

Moore, J. L. (1966). The physiology of irradiated micro-organisms, with particular reference to enzyme induction. Ph.D. Thesis, University of Wales, University College, Cardiff. [212]

Moore, J. L., Pritchard, J. A. V. & Smith, C. W. (1972). Oxygen equilibration in the determination of K for HeLa S$_3$ (OXF). *Int. J. Radiat. Biol.* **22**, 149–58. [58, 259]

Mortimer, R., Brustad, T. & Cormack, D. V. (1965). Influence of linear energy transfer and oxygen tension on the effectiveness of ionizing radiations for induction of mutations and lethality in *Saccharomyces cerevisiae*. *Radiat. Res.* **26**, 465–82. [126]

Mosin, A. F. (1969). Influence of inhibitors of oxidative and substrate phosphorylation of ATP synthesis and the magnitude and rate of subsequent repair in yeasts subjected to gamma radiation. *Radiobiology* (translated from Russian 'Radiobiologiya') **9**, 25–33. [233, 237]

Moss, A. J. Jr, Dalrymple, G. V., Sanders, J. L., Wilkinson, K. P. & Nash, J.C. (1971). Dinitrophenol inhibits the rejoining of radiation-induced DNA breaks by L-cells. *Biophys. J.* **11**, 158–74. [237]

Mottram, J. C. (1935). On the alteration in the sensitivity of cells towards radiation produced by cold and by anaerobiosis. *Br. J. Radiol.* **8**, 32–9. [259]

Munson, R. J., Neary, G. J., Bridges, B. A. & Preston, R. J. (1968). The sensitivity of *Escherichia coli* to ionizing particles of different LET's. *Int. J. Radiat. Biol.* **13**, 205–24. [112, 215]

Nash, J. C., Dalrymple, G. V., Moss, A. J. Jr & Baker, M. L. (1974). Initial studies with a line of radioresistant rat tumour cells. *Radiat. Res.* **60**, 280–91. [236]

Neary, G. J. (1965). Chromosome aberrations and the theory of RBE. I. General considerations. *Int. J. Radiat. Biol.* **9**, 477–502. [134, 137, 139]

Neary, G. J. (1967). Chromosome aberrations, cell killing and the molecular basis of relative biological effectiveness of ionizing radiations. In *Radiation Research 1966*, ed. G. Silini, pp. 445–54. Amsterdam: North-Holland Publishing Co. [137]

Neary, G. J., Savage, J. R. K. & Evans, H. J. (1964). Chromatid aberrations in *Tradescantia* pollen tubes induced by monochromatic X-rays of quantum energy 3 and 1.5 KeV. *Int. J. Radiat. Biol.* **8**, 1–19. [137]

Ngo, F. Q. H., Han, A. & Elkind, M. M. (1977). On the repair of sublethal damage in V79 Chinese hamster cells resulting from irradiation with fast neutrons combined with X-rays. *Int. J. Radiat. Biol.* **32**, 507–11.

Nias, A. H. W. & Gilbert, C. W. (1975). Response of HeLa and Chinese hamster cells to low doses of photons and neutrons. In *Cell Survival after Low Doses of Radiation*, ed. T. Alper, pp. 93–9. London: Institute of Physics and John Wiley & Sons. [119]

Nias, A. H. W., Swallow, A. J., Keene, J. P. & Hodgson, B. W. (1973). Absence of a fractionation effect in irradiated HeLa cells. *Int. J. Radiat. Biol.* **23**, 559–69. [200]

Norman, A. & Ginoza, W. (1958). Molecular interactions in irradiated solids. *Radiat. Res.* **9**, 77–83. [25]

Norris, G. & Hood. S. L. (1962). Some problems in the culturing and radiation sensitivity of normal human cells. *Expl. Cell Res.* **27**, 48–62. [36, 250, 251]

Oberley, L. W., Lindgren, A. L., Baker, S. A. & Stevens, R. H. (1976). Superoxide ion as the cause of the oxygen effect. *Radiat. Res.* **68**, 320–8. [67]

Olive, P. L. & Durand, R. E. (1978). Activation of radiosensitizers by hypoxic cells. *Br. J. Cancer* **37**, suppl. 3, 124–8. [271]

Ormerod, M. G. & Alexander, P. (1963). On the mechanism of radiation protection by cysteamine: an investigation by means of electron spin resonance. *Radiat. Res.* **18**, 495–509. [25, 94]

Ormerod, M. G. & Lehmann, A. R. (1971). Release of high molecular weight DNA from the mammalian cell (L5178Y). Attachment of the DNA to the nuclear membrane. *Biochim. biophys. Acta* **228**, 331–43. [241]

Orr, J. S., Wakerley, S. E. & Stark, J. M. (1966). A metabolic theory of cell survival curves. *Physics Med. Biol.* **11**, 103–8. [199]

Palcic, B. & Skarsgard, L. D. (1975). Absence of ultrafast processes of repair of single-strand breaks in mammalian DNA. *Int. J. Radiat. Biol.* **27**, 121–33. [243]

Parker, L., Skarsgard, L. D. & Emmerson, P. T. (1969). Sensitization of anoxic mammalian cells by triacetoneamine *N*-oxyl. Survival and toxicity studies. *Radiat. Res.* **38**, 493–500. [73, 85]

Paterson, M. C., Boyle, J. M. & Setlow, R. B. (1971). Ultraviolet- and X-ray-induced responses of a deoxyribonucleic acid polymerase-deficient mutant of *Escherichia coli*. *J. Bacteriol.* 107, 61–7. [212]

Paterson, M. C. & Setlow, R. B. (1972). Endonucleolytic activity from *Micrococcus luteus* that acts on γ-ray-induced damage in plasmid DNA of *Escherichia coli* minicells. *Proc. natn. Acad. Sci. USA* **69**, 2927–31. [214, 245]

Paterson, M. C., Smith, B. P., Lohman, P. H. M., Anderson, A. K. & Fishman, L. (1976). Defective excision repair of γ-ray damaged DNA in human (ataxia telangiectasia) fibroblasts. *Nature, Lond.* **260**, 444–7. [244, 245]

Patrick, M. H. & Haynes, R. H. (1964). Dark recovery phenomena in yeast II. Conditions that modify the recovery process. *Radiat. Res.* **23**, 564–79. [236]

Patt, H. W., Tyree, E. B., Straube, R. L. & Smith, D. E. (1949). Cysteine protection against X irradiation. *Science* 110, 213–14. [90]

Pauly, H. (1959). X-Ray sensitivity and target volume of enzyme induction. *Nature, Lond.* 184, 1570. [123]

Petkau, A. & Chelack, W. S. (1974). Protection of *Acholeplasma laidlawii* B by superoxidedismutase. *Int. J. Radiat. Biol.* **26**, 421–6. [67, 221]

Phillips, R. A. & Tolmach, L. J. (1966). Repair of potentially lethal damage in X-irradiated HeLa cells. *Radiat. Res.* **29**, 413–32, [202, 227, 232, 235]

Pietronigro, D. D., Jones, W. B. G., Kalty, K. & Demopoulos, H. B. (1977). Interaction of DNA and liposomes as a model for membrane-mediated DNA damage. *Nature, Lond.* **267**, 78–9. [222, 224, 225]

Pollard, E. C. & Achey, P. M. (1975). Induction of radioresistance in *Escherichia coli*. *Biophys. J.* **15**, 1141–53. [247]

Potten, C. S. & Howard, A. (1969). Radiation depigmentation of mouse hair: the influence of local tissue oxygen tension on radiosensitivity. *Radiat. Res.* **38**, 65–81. [261]

Powers, E. L. (1962). Considerations of survival curves and target theory. *Physics Med. Biol.* **7**, 3–28. [42, 59, 199]

Powers, E. L. (1965). Some physico-chemical bases of radiation sensitivity in cells. In *Cellular Radiation Biology* pp. 286–304. Baltimore: Williams & Wilkins Company. [113]

Powers, E. L., Kaleta, B. F. & Webb, R. B. (1959). Nitric oxide protection against radiation damage. *Radiat. Res.* **11**, 461. [72]

Powers, W. E. & Tolmach, L. J. (1963). A multicomponent X-ray survival curve for mouse lymphosarcoma cells irradiated *in vivo*. *Nature, Lond.* **197**, 710–11. [45]

Puck, T. T. & Marcus, P. I. (1955). A rapid method for viable cell titration and clone production with HeLa cells in tissue culture: the use of X-irradiated cells to supply conditioning factors. *Proc. natn. Acad. Sci. USA* **41**, 432–7. [13, 33, 87]

Puck, T. T. & Marcus, P. I. (1956). Action of X-rays on mammalian cells. *J. exp. Med.* **103**, 653–66. [37, 246]

Puck, T. T., Morkovin, D., Marcus, P. I. & Cieciura, S. J. (1957). Action of X-rays on mammalian cells. II. Survival curves of cells from normal human tissues. *J. exp. Med.* **106**, 483–500. [36, 250]

Puga, A. & Tessman, I. (1973). Membrane binding of Phage S13 to messenger RNA. *Virology* **56**, 375–8. [205]

Quastler, II. (1945). Studies on Roentgen death in mice. *Am. J. Roentg.* **54**, 449–56. [2]

Quastler, H. (1956). The nature of intestinal radiation death. *Radiat. Res.* **4**, 303–20. [3]

Raleigh, J. A., Kremers, W. & Gaboury, B. (1977). Dose-rate and oxygen effects in models of lipid membranes: linoleic acid. *Int. J. Radiat. Biol.* **31**, 203–13. [185, 186]

Rasey, J. S., Nelson, N. J. & Carpenter, R. E. (1977). Damage repair in EMT-6 cells treated in vitro with X-rays or cyclotron neutrons. *Radiat. Res.* **70**, 642. [238]

Rauth, A. M. & Simpson, J. A. (1964). The energy loss of electrons in solids. *Radiat. Res.* **22**, 643–61. [28, 30]

Rauth, A. M. & Whitmore, G. F. (1966). The survival of synchronized L cells after ultraviolet irradiation. *Radiat. Res.* **28**, 84–95. [153, 155]

Read, J. (1952). Effect of ionizing radiations on the broad bean root: $\bar{x}$ Dependence of X-ray sensitivity on dissolved oxygen. *Br. J. Radiol.* **25**, 154–60. [123]

Redpath, J. L. (1978). The response of *E. coli* AB 2463 *recA* to fast neutron beams with mean energies in the range 4 to 25 MeV. *Brit. J. Radiol.* **51**, 524–7. [122]

Redpath, J. L. & Patterson, L. K. (1976). Radiosensitization of *Serratia marcescens* by cetylpyridinium chloride. *Radiology* **118**, 725. [79]

Redpath, J. L. & Patterson, L. K. (1978). The effect of membrane fatty acid composition on the radiosensitivity of *E. coli* K-1060. *Radiat. Res.* **75**, 443–7. [223]

Reinhard, R. D. & Pohlit (1976). Influence of intracellular adenosine-triphosphate concentration on survival of yeast cells following X-irradiation. In *Radiation and Cellular Control Processes*, ed. J. Kiefer, pp. 117–23. Heidelberg and New York: Springer-Verlag. [177]

Resnick, M. A. & Martin, P. (1976). The repair of double strand breaks in the nuclear DNA of *Saccharomyces cerevisiae* and its genetic control. *Molec. gen. Genet.* **143**, 119–29. [244]

Reuvers, A. P., Greenstock, C. L., Borsa, J. & Chapman, J. D. (1973). Studies on the mechanism of chemical protection by dimethyl sulphoxide. *Int. J. Radiat. Biol.* **24**, 533–6. [214]

Revell, S. H. (1959). The accurate estimation of chromatid breakage and its relevance to a new interpretation of chromatid aberrations induced by ionizing radiation. *Proc. R. Soc. B* **150**, 563–89. [221]

Ritter, M. A., Cleaver, J. E. & Tobias, C. A. (1977). High-LET radiations induce a large proportion of non-rejoining breaks. *Nature, Lond.* **266**, 653–5. [225]

Rockwell, S. C., Kallman, R. F. & Fajardo, L. F. (1972). Characteristics of a serially transplanted mouse mammary tumour and its tissue-culture adapted derivative. *J. natn. Cancer Inst.* **49**, 735–49. [267]

Roots, R. & Okada, S. (1972). Protection of DNA molecules of cultured mammalian cells growing *in vitro*. *Int. J. Radiat. Biol.* **21**, 329–42. [98]

Rossi, H. H. & Kellerer, A. M. (1974). Effects of spatial-temporal distribution of primary events. In *Physical Mechanisms in Radiation Biology*, ed. R. D. Cooper and R. W. Wood, pp. 224–243. Technical Information Center, United States Atomic Energy Commission. [138]

Ryter, A. & Jacob, F. (1966). Étude morphologique de la liaison du noyau a la membrane chez *E. coli* et chez les protoplastes de *B. subtilis*. *Annls Inst. Pasteur, Paris*, **110**, 801–12. [206]

Sanner, T. & Pihl, A. (1969). Significance and mechanism of the indirect effect in bacterial cells. The relative protective effect of added compounds in *Escherichia coli* B irradiated in liquid and frozen suspensions. *Radiat. Res.* **37**, 216–27. [97, 213]

Sapora, O., Fielden, E. M. & Loverock, P. S. (1977). A comparative study of the effect of two classes of radiosensitizer on the survival of several *E. coli* B and K12 mutants. *Radiat. Res.* **69**, 293–305. [52, 73, 75, 81]

Sapozink, M. D. (1977). Oxygen enhancement ratios in synchronous HeLa cells exposed to low-LET radiation. *Radiat. Res.* **69**, 27–39. [159, 160]

Sato, C., cited by S. Okada (1975). Repair studies at the molecular chromosomal and cellular levels: a review of current work in Japan. In *Radiation Research: Biomedical, Chemical and Physical Perspectives*, ed. O. F. Nygaard, H. I. Adler and W. K. Sinclair, pp. 694–702. New York and London: Academic Press. [36]

Saunders, M. I., Dische, S., Anderson, P. & Flockhart, I. (1978). The neurotoxicity of misonidazole and its relationship to dose, half-life and concentration in the serum. *Br. J. Cancer* **37**, suppl. 3, 268–70. [271]

Savage, J. R. K. & Miller, M. W. (1972). Some problems of chromosomal aberration studies in meristems. In *The Dynamics of Meristem Cell Populations*, ed. M. W. Miller and C. C. Kuehnert, pp. 211–24. New York: Plenum Publishing Corporation. [220]

Schambra, P. E. & Hutchinson, F. (1964). The action of fast heavy ions on biological material. II. Effects on T_1 and $\phi x-174$ bacteriophage and double-strand and single-strand DNA. *Radiat. Res.* **23**, 514–26. [32]

Schneider, E. & Kiefer, J. (1976). Delayed plating recovery in diploid yeast of different sensitivities after X-ray and alpha particle exposure. *Int. J. Radiat. Biol.* **29**, 77–84. [238]

Scott, D., Fox, M. & Fox, B. W. (1974). The relationship between chromosomal aberrations, survival and DNA repair in tumour cell lines of differential sensitivity to X-rays and sulphur mustard. *Mutation Res.* **22**, 207–21. [245]

Scott, O. C. A. (1963). Chemical protection in mammals. In *Radiation Effects in Physics, Chemistry and Biology*, ed. M. Ebert and A. Howard, pp. 294–304. Amsterdam: North-Holland Publishing Co. [88]

Sedgwick, S. G. & Bridges, B. A. (1972). Survival, mutation and capacity to repair single-strand DNA breaks after gamma irradiation in different Exr⁻ strains of *Escherichia coli*. *Molec. gen. Genet.* **119**, 93–102. [239, 241]

Setlow, R. B. & Carrier, W. L. (1972). Endonuclease activity toward DNA irradiated *in vitro* by gamma rays. *Nature New Biology* **241**, 170–2, [214, 245]

Setlow, R. B., Faulcon, F. M. & Regan, J. D. (1976). Defective repair of gamma-ray-induced DNA damage in xeroderma pigmentosum cells. *Int. J. Radiat. Biol.* **29**, 125–36. [214, 245]

Shaeffer, J. & Merz, T. (1971). A comparison of unscheduled DNA synthesis, D_0, cell recovery and chromosome number in X-irradiated mammalian cell lines. *Radiat. Res.* **47**, 426–36. [245]

Shalek, R. I. & Gillespie, T. L. (1960). The influence of oxygen upon the radiation damage of lysozyme. In *Radiation Biology and Cancer*, pp. 41–50. London: Peter Owen Ltd. [25]

Shekhtman, Ia L., Plokhoi, V. I. & Filippova, G. V. (1958). Effects of X- and α-rays on *E. coli* communis. Biofizika **3**, 479–86. [120]

Sheldon, P. W., Foster, J. L. & Fowler, J. F. (1974). Radiosensitization of C₃H mouse mammary tumours by a 2-nitroimidazole drug. *Br. J. Cancer* **30**, 560–5. [268]

Shenoy, M. A., Asquith, J. C., Adams, G. E., Michael, B. D. & Watts, M. E. (1975). Time-resolved oxygen effects in irradiated bacteria and mammalian cells: a rapid-mix study. *Radiat. Res.* **62**, 498–512. [84, 224]

Shenoy, M. A., Singh, B. B. & Gopal-Ayengar, A. R. (1974). Enhancement of radiation lethality of *E. coli* B/r by procaine hydrochloride. *Nature, Lond.* **248**, 415–6. [79]

Shipley, W. V., Stanley, J. A., Courtenay, D. & Field, S. B. (1975). Repair of radiation damage in Lewis lung carcinoma cells following in situ treatment with fast neutrons and γ-rays. *Cancer Res.* **35**, 932–8. [238, 269]

Silini, G. & Hornsey, S. (1961). Studies on cell-survival of irradiated Ehrlich ascites tumour. I. The effect of the host's age and the presence of non-viable cells on tumour takes. *Int. J. Radiat. Biol.* **4**, 127–34. [15]

Silini, G. & Maruyama, Y. (1965). X-ray and fast neutron survival response of 5-bromo-deoxycytidine-treated bone-marrow cells. *Int. J. Radiat. Biol.* **9**, 605–10. [36, 144, 253]

Sinclair, W. K. (1968). Cyclic X-ray responses in mammalian cells *in vitro*. *Radiat. Res.* **33**, 620–43. [151, 154]

Sinclair, W. K. (1969). Protection by cysteamine against lethal X-ray damage during the cell cycle of Chinese hamster cells. *Radiat. Res.* **39**, 135–54. [76, 161]

Sinclair, W. K. (1970). Dependence of radiosensitivity upon cell age. In *Time and Dose Relationships in Radiation Biology as Applied to Radiobiology*, Carmel Conference, pp. 97–107. Brookhaven Nat. Lab. 50203 (c–57). [151]

Sinclair, W. K. (1972). Cell cycle dependence of the lethal radiation response in mammalian cells. *Curr. Topics Radiat. Res.* **7**, 264–85. [156, 161, 199]

Sinclair, W. K. (1973). N-ethylmaleimide and the cyclic response to X-rays of synchronous Chinese hamster cells. *Radiat. Res.* **55**, 41–57. [76]

Sinclair, W. K. (1975). Mammalian cell sensitization, repair and the cell cycle. In *Radiation Research: Biomedical, Chemical and Physical Perspectives*, ed. O. F. Nygaard, H. I. Adler and W. K. Sinclair, pp. 742–9. New York and London: Academic Press. [162]

Sinclair, W. K. & Morton, R. A. (1964). Recovery following X-irradiation of synchronized Chinese hamster cells. *Nature, Lond.* **203**, 247–50. [166]

Sinclair, W. K. & Morton, R. A. (1966). X-ray sensitivity during the cell generation cycle of cultured Chinese hamster cells. *Radiat. Res.* **29**, 450–74. [154, 155, 157, 159]

Skarsgard, L. D. (1974). Complete survival curves measured at various stages of the cell cycle: X-rays and heavy ions. Unpublished: presented at the *Berkeley Workshop on High LET Radiation*, July 9–11, 1974. [162]

Skarsgard, L. D., Kihlman, B. A., Parker, L., Pujara, C. M. & Richardson, S. (1967). Survival, chromosome abnormalities, and recovery in heavy-ion- and X-irradiated mammalian cells. *Radiat. Res.* Suppl. **7**, 208–21. [116, 180, 219]

Smith, D. W., Schaller, H. E. & Bonhoeffer, F. J. (1970). DNA synthesis *in vitro*. *Nature, Lond.* **226**, 711–13. [206, 209]

Sparrman, B., Ehrenberg, L. & Ehrenberg, A. (1959). Scavenging of free radicals and radiation protection by nitric oxide in plant seeds. *Acta chem. scand.* **13**, 199–200. [72]

Sparvoli, E., Galli, M. G., Mosca, A. & Paris, G. (1976). Localization of DNA replicator sites near the nuclear membrane in plant cells. *Expl. Cell. Res.* **97**, 74–82. [206, 207]

Sridhar, R., Koch, C. & Sutherland, R. (1976). Cytoxicity of two nitroimidazole radiosensitizers in an *in vitro* tumour model. *Int. J. Radiat. Oncol. Biol. Physics*, **1**, 1149–57. [270]

Stapleton, G. E. (1955). The influence of pretreatments and posttreatments on bacterial inactivation by ionizing radiations. *Ann. N. Y. Acad. Sci.* **59**, 604–18. [147]

Stapleton, G. E., Billen, D. & Hollaender, A. (1953). Recovery of X-irradiated bacteria at sub-optimal incubation temperatures. *J. cell. comp. Physiol.* **41**, 345–58. [229, 230, 231]

Stratford, I. J. & Adams, G. E. (1977). Effect of hyperthermia on differential cytoxicity of a hypoxic cell radiosensitizer Ro–07–0582, on mammalian cells *in vitro*. *Br. J. Cancer* **35**, 307–13. [270]

Stratford, I. J., Maughan, R. L., Michael, B. D. & Tallentire, A. (1977). The decay of potentially lethal oxygen-dependent damage in fully hydrated *Bacillus megaterium* spores exposed to pulsed electron irradiation. *Int. J. Radiat. Biol.* **32**, 447–55. [66]

Stuy, J. H. (1961). Studies on the radiation inactivation of microorganisms. VII. Nature of the X-ray-induced breakdown of deoxyribonucleic acid in *Haemophilus influenzae*. *Radiat. Res.* **14**, 56–65. [240]

Swanson, C. P. (1955). Relative effect of qualitatively different ionizing radiations on the production of chromatid aberrations in air and nitrogen. *Genetics, Princeton* **40**, 195–203. [220]

Swanson, C. P. & Schwartz, D. (1953). Effect of X-rays on chromatid aberrations in air and in nitrogen. *Proc. natn. Acad. Sci. USA* **39**, 1241–50. [220]

Swenson, P. A. (1976). Physiological responses of *Escherichia coli* to far-ultraviolet radiation. In *Photochemical and Photobiological Reviews*, ed. K. C. Smith, vol. 1, pp. 269–387. New York: Plenum Publishing Co. [179]

Taylor, A. M. R., Harnden, D. G., Arlett, C. F., Harcourt, S. A., Lehmann, A. R., Stevens, S. & Bridges, B. A. (1975). Ataxia telangiectasia: a human mutation with abnormal radiosensitivity. *Nature, Lond.* **258**, 427–9. [218, 244]

Taylor, I. W. & Bleehen, N. M. (1977a). Changes in sensitivity to radiation and ICRF 159 during the life of monolayer cultures of EMT6 tumour line. *Br. J. Cancer* **35**, 587–94. [149, 196]

Taylor, I. W. & Bleehen, N. M. (1977b). Interaction of ICRF 159 with radiation and its effect on sublethal and potentially lethal radiation damage *in vitro*. *Br. J. Cancer* **36**, 493–500. [196, 197]

Terasima, T. & Tolmach, L. J. (1961). Changes in X-ray sensitivity of HeLa cells during the division cycle. *Nature, Lond.* **190**, 1210–11. [148, 151, 153]

Terasima, T. & Tolmach, L. J. (1963). Variations in several responses of HeLa cells to X-irradiation during the division cycle. *Biophys. J.* **3**, 11–33. [152, 153, 157]

Thoday, J. M. & Read, J. (1947). Effect of oxygen on the frequency of chromosome aberrations produced by X-rays. *Nature, Lond.* **160**, 608–9. [219]

Thoday, J. M. & Read, J. (1949). Effect of oxygen on the frequency of chromosome aberrations produced by alpha rays. *Nature, Lond.* **163**, 133–4. [123, 133, 220]

Thomlinson, R. H. (1960). An experimental method for comparing treatments of intact malignant tumours in animals and its application to the use of oxygen in radiotherapy. *Br. J. Cancer* **14**, 555–76. [263]

Thomlinson, R. H. (1961). The oxygen effect in mammals. In *Fundamental Aspects of Radiosensitivity: Brookhaven Symposium in Biology*, vol. 14, pp. 204–16. [273]

Thomlinson, R. H. (1977). Hypoxia and tumours. *J. clin. Pathol.* 30 Suppl. (Roy. Coll. Path.) 11, 105–13. [263]

Thomlinson, R. H. (1979). Measurement of the response of primary carcinoma of the breast to treatment. *Br. J. Radiol.* **52**, (in press). [265]

Thomlinson, R. H. & Craddock, A. (1967). The gross response of an experimental tumour to single doses of X-rays. *Br. J. Cancer* **21**, 108–23. [263, 264, 268]

Thomlinson, R. H. & Gray, L. H. (1955). The histological structure of some human lung cancers and the possible implications for radiotherapy. *Br. J. Cancer* **9**, 539–49. [271]

Thomson, J. E. & Rauth, A. M. (1974). An *in vitro* assay to measure the viability of KHT tumor cells not previously exposed to culture conditions. *Radiat. Res.* **58**, 262–76. [267]

Till, J. E. & McCulloch, E. A. (1961). A direct measurement of the radiation sensitivity of normal mouse bone marrow cells. *Radiat. Res.* **14**, 213–22. [15]

Till, J. E. & McCulloch, E. A. (1963). Early repair processes in marrow cells irradiated and proliferating *in vivo*. *Radiat. Res.* **18**, 96–105. [251, 253]

Timoféeff-Ressovsky, N. W. & Zimmer, K. G. (1947). *Das Trefferprinzip in der Biologie*. Leipzig: S. Hirzel Verlag. [18, 35, 133]

Todd, P. W. (1964). Reversible and irreversible effects of ionizing radiations in the reproductive integrity of mammalian cells cultured *in vitro*. Ph.D. Thesis, University of California. UCRL-11614. [114, 115, 116, 117, 124]

Todd, P. (1968). Fractionated heavy ion irradiation of cultured human cells. *Radiat. Res.* **34**, 378–89. [180]

Tolmach, L. J. & Marcus, P. I. (1960). Development of X-ray induced giant HeLa cells. *Expl. Cell. Res.* **20**, 350–60. [247]

Town, C. D., Smith, K. C. & Kaplan, H. S. (1972). Influence of ultrafast repair processes (independent of DNA polymerase-1) on the yield of DNA single-strand breaks in *Escherichia coli* K12 X-irradiated in the presence or absence of oxygen. *Radiat. Res.* **52**, 99–114. [242]

Town, C. D., Smith, K. C. & Kaplan, H. S. (1973). The repair of DNA single-strand breaks in *E. coli* K12 X-irradiated in the presence or absence of oxygen; the influence of repair on cell survival. *Radiat. Res.* **55**, 334–45. [242]

Tremblay, G. Y., Daniels, M. J. & Schaechter, M. (1969). Isolation of a cell membrane-DNA-nascent RNA complex from bacteria. *J. molec. Biol.* **40**, 65–76. [206]

Trott, K. R., Szczepanski, L. V., Kummermehr, J. & Hug, O. (1977). Tumour control probability and tumour regression rate after fractionated radiotherapy of two mouse tumours. In *Radiobiological Research and Radiotherapy*, vol. 1, pp. 29–42. Int. Atomic Energy Agency, Vienna. London: HMSO. [265]

Urano, M., Nesumi, N., Ando, K., Koike, S. & Ohnuma, N. (1976). Repair of potentially lethal damage in acute and chronically hypoxic cells *in vivo*. *Radiology* **118**, 447–51. [270]

Uretz, R. B. (1955). Additivity of X-rays and ultraviolet light in the inactivation of haploid and diploid yeast. *Radiat. Res.* **2**, 240–52. [16]

van den Brenk, H. A. S., Kerr, R. C., Richter, W. & Papworth, M. P. (1965). Enhancement of radiosensitivity of skin of patients by high pressure oxygen. *Br. J. Radiol.* **38**, 857–64. [261]

van den Brenk, H. A. S. & Moore, R. (1959). Effect of high oxygen pressure on the protective action of cystamine and 5-hydroxytryptamine in irradiated rats. *Nature , Lond.* **183**, 1530–1. [87]

van der Meer, C. & van Bekkum, D. W. (1959). The mechanism of radiation protection by histamine and other biological amines. *Int. J. Radiat. Biol.* **1**, 5–23. [87]

van der Schans, G. P. (1973). Cited by Blok, J. and Loman, H. The effects of γ-radiation in DNA. *Current Topics in Radiation Research* **9**, 166–245. [27]

Van der Schueren, E., Smith, K. C. & Kaplan, H. S. (1973). Modification of DNA repair and survival of X-irradiated *pol, rec* and *exr* mutants of *Escherichia coli* K12 by 2, 4-dinitrophenol. *Radiat. Res.* **55**, 346–55. [237]

Vergroesen, A. J., Budke, L. & Vos, O. (1967). Protection against X-irradiation by sulphydryl compounds. II. Studies on the relation between chemical structure and protective activity for tissue culture cells. *Int. J. Radiat. Biol.* **13**, 77–92. [91, 95]

Vos, O. & van Kaalen, M. C. A. C. (1962). Protection of tissue-culture cells against ionizing radiation. II. The activity of hypoxia, dimethyl sulphoxide, dimethyl sulphone, glycerol and cysteamine at room temperature and at $-196°C$. *Int. J. Radiat. Biol.* **5**, 609–21. [96]

Wambersie, A., Dutreix, J., Gueulette, J. & Lellouch, J. (1974). Early recovery for intestinal stem cells, as a function of dose per fraction, evaluated by survival rate after fractionated irradiation of the abdomen of mice. *Radiat. Res.* **58**, 498–515. [254]

Wardman, P. (1977). The use of nitroaromatic compounds as hypoxic cell sensitizers. *Curr. Topics Radiat. Res.* **11**, 347–8. [78]

Watkins, D. K. (1970). High o.e.r. for the release of enzymes from isolated mammalian lysosomes after ionizing radiation. *Adv. biol. med. Phys.* **13**, 289–306. [209]

Watkins, D. K. & Deacon, S. (1973). Comparative effects of electron and neutron irradiation on the release of enzymes from isolated rat-spleen lysosomes. *Int. J. Radiat. Biol.* **23**, 41–50. [32]

Watson, G. E. & Gillies, N. E. (1970). The oxygen enhancement ratio for X-ray induced chromosomal aberrations in cultured human lymphocytes. *Int. J. Radiat. Biol.* **17**, 279–83. [220]

Watts, M. E., Maughan, R. L. & Michael. B. D. (1978). Fast kinetics of the oxygen effect in irradiated mammalian cells. *Int. J. Radiat. Biol.* **33**, 195–9. [66]

Watts, M. E., Whillans, D. W. & Adams, G. E. (1975). Studies of the mechanisms of radiosensitization of bacterial and mammalian cells by diamide. *Int. J. Radiat. Biol.* **27**, 259–70. [76, 81, 196]

Webb, R. B. & Powers, E. L. (1963). Protection against actions of X-rays by glycerol in the bacterial spore. *Int. J. Radiat. Biol.* **7**, 481–90. [89]

Weichselbaum, R. R., Epstein, J., Little, J. B. & Kornblith, P. L. (1976). *In vitro* cellular radiosensitivity of human malignant tumors. *Europ. J. Cancer* **12**, 47–51. [36, 250]

Weichselbaum, R. R., Nove, J. & Little, J. B. (1978). Deficient recovery from potentially lethal radiation damage in ataxia telangiectasia and xeroderma pigmentosum. *Nature, Lond.* **271**, 261–2. [218, 238]

Weiss, H. & Santomasso, A. (1977). A post-irradiation effect of oxygen in anoxic wet spores exposed to intense 3 nsec duration pulses of electrons. *Radiat. Res.* **70**, 656–7. [66]

Whillans, D. W. & Hunt, J. W. (1978). Rapid-mixing studies of the mechanisms of chemical radiosensitization and protection in mammalian cells. *Br. J. Cancer* **37**, suppl. 3, 38–41. [77, 90, 91, 97]

Whillans, D. W. & Neta, P. (1975). Radiation chemical studies of the sensitizer diamide. *Radiat. Res.* **64**, 416–30. [77]

Whitmore, G. F., Gulyas, S. & Botond, J. (1965). Radiation sensitivity throughout the cell cycle and its relationship to recovery. In *Cellular Radiation Biology*, pp. 423–61. Baltimore: Williams & Wilkins Co. [153, 158]

Whitmore, G. F., Gulyas, S. & Kotalik, J. (1970). Recovery from radiation damage in mammalian cells. In *Time and Dose Relationships in Radiation Biology as Applied to Radiotherapy*, Carmel Conference, pp. 41–53. Brookhaven Nat. Lab. 50203 (c–57). [232]

Whitmore, G. F., Gulyas, S., & Varghese, A. J. (1978). Sensitizing and toxicity properties of misonidazole and its derivatives. *Br. J. Cancer* **37**, suppl. 3, 115–19. [83]

Whitmore, G. F., Till, J. E., Gwatkin, R. B. L., Siminovitch, L. & Graham, A. F. (1958). Increase of cellular constituents in X-irradiated mammalian cells. *Biochim. biophys. Acta* **30**, 583–90. [247]

Wienhard, I. & Kiefer, J. (1977). Split-dose and liquid-holding recovery after X-irradiation in diploid yeast *Saccharomyces cerevisiae*. I. Dependence on growth-phase. *Int. J. Radiat. Biol.* **31**, 477–84. [234]

Williams, J. F. & Till, J. E. (1966). Formation of lung colonies by polyoma-transformed rat embryo cells. *J. natn. Cancer Inst.* **37**, 177–83. [15]

Winans, L. F., Dewey, W. C. & Dettor, C. M. (1972). Repair of sublethal and potentially lethal X-ray damage in synchronous Chinese hamster cells. *Radiat. Res.* **52**, 333–51. [232]

Withers, H. R. (1967a). The dose-survival relationship for irradiation of epithelial cells of mouse skin. *Br. J. Radiol.* **40**, 187–94. [15, 254, 256]

Withers, H. R. (1967b). The effect of oxygen and anaesthesia on radiosensitivity of epithelial cells of mouse skin. *Br. J. Radiol.* **40**, 335–43. [261]

Withers, H. R. (1967c). Recovery and repopulation *in vivo* by mouse skin epithelial cells during fractionated irradiation. *Radiat. Res.* **32**, 227–39. [192, 251, 252, 256]

Withers, H. R. (1970). Capacity for repair in cells of normal and malignant tissues. In *Time and Dose Relationships in Radiation Biology as Applied to Radiotherapy*, Carmel Conference, pp. 54–69. Brookhaven Nat. Lab. 50203 (C–57). [199]

Withers, H. R. (1975a). Responses of some normal tissues to low doses of γ-radiation. In *Cell Survival after Low Doses of Radiation*, ed. T. Alper, pp. 369–75. London: Institute of Physics and John Wiley & Sons. [255]

Withers, H. R. (1975b). Cell cycle redistribution as a factor in multifraction irradiation. *Radiology*, **114**, 199–202. [257]

Withers, H. R., Brennan, J. T. & Elkind, M. M. (1970). The response of stem cells of intestinal mucosa to irradiation with 14 MeV neutrons. *Br. J. Radiol.* **43**, 796–801. [119, 120, 182, 252, 254]

Withers, H. R., Chu, A. M., Reid, B. U. & Hussey, D. H. (1975). Response of mouse jejunum to multifraction radiation. *J. Radiat. Oncol. Biol. Physics* **1**, 41–52. [255, 257]

Withers, H. R. & Elkind, M. M. (1969). Radiosensitivity and fractionation response of crypt cells of mouse jejunum. *Radiat. Res.* **38**, 598–613. [15]

Withers, H. R. & Elkind, M. M. (1970). Microcolony survival assay for cells of mouse intestinal mucosa exposed to radiations. *Int. J. Radiat. Biol.* **17**, 261–7. [15]

Withers, H. R., Hunter, N., Barkley, H. T. & Reid, B. O. (1974a). Radiation survival and regeneration characteristics of spermatogenic stem cells of mouse testis. *Radiat. Res.* **57**, 88–103. [252, 257]

Withers, H. R., Mason, K., Reid, B. O., Dubravsky, N., Barkley, H. T., Brown, B. W. & Smathers, J. B. (1974b). Response of mouse intestine to neutrons and gamma rays in relation to dose fractionation and division cycle. *Cancer* **34**, 39–47. [182]

Wold, E. & Brustad, T. (1974). Pulse radiolytic investigation of reactions of diamide with hydrated electrons and OH radicals. *Int. J. Radiat. Biol.* **25**, 407–11. [77]

Wolff, S. (1968). Chromosome aberrations and the cell cycle. *Radiat. Res.* **33**, 609–19. [220]

Wolstenholme, G. E. W. & O'Connor, M. (eds.) (1969). *Mutation as a Cellular Process.* London: J. A. Churchill Ltd. [21]

Wright, E. A. (1962). The influence of combining hypoxia and cysteamine treatments on whole body irradiation of mice. *Br. J. Radiol.* **35**, 361. [88]

Wright, E. A. (1963). Chemical protection at the cellular level. In *Radiation Effects in Physics, Chemistry and Biology*, ed. M. Ebert and A. Howard, pp. 276–89. Amsterdam: North-Holland Publishing Co. [88]

Wright, E. A. & Batchelor, A. L. (1959). The change in the radiosensitivity of the intact mouse thymus produced by breathing nitrogen. *Br. J. Radiol.* **32**, 168–73. [259]

Yatagai, F. & Matsuyama, A. (1977). LET-dependent radiosensitivity of *Escherichia coli* K12 rec and uvr mutants. *Radiat. Res.* **71**, 259–63. [215]

Yatvin, M. B. (1976). Evidence that survival of γ-irradiated *Escherichia coli* is influenced by membrane fluidity. *Int, J, Radiat. Biol.* **30**, 571–5. [223]

Yatvin, M. B., Wood, P. G. & Brown, S. M. (1972). 'Repair' of plasma membrane injury and DNA single strand breaks in γ-irradiated *Escherichia coli* B/r and B_{s-1}. *Biochim. biophys. Acta* **287**, 390–403. [239, 240]

Youngs, D. A. & Smith, K. C. (1973). X-ray sensitivity and repair capacity of a *polA1 exrA* strain of *Escherichia coli* K12. *J. Bacteriol.* **114**, 121–7. [239]

Zimmer, K. G. (1961). *Studies on Quantitative Radiation Biology.* Edinburgh and London: Oliver & Boyd. [18]

Zirkle, R. E. (1940). The radiological importance of the energy distribution along ionization tracks. *J. cell. comp. Physiol.* **16**, 221–35. [106]

Zirkle, R. E., Marchbank, D. F. & Kuck, D. D. (1952). Exponential and sigmoid survival curves resulting from alpha and X-irradiation of *Aspergillus* spores. *J. cell. comp. Physiol.* Supplement 1, **39**, 75–85. [106]

INDEX

Pages on which terms are defined are marked with an asterisk. References to tables in bold type, to figures in italics. Abbreviations used in index:

AT   ataxia telangiectasia
eqn   equation
expt   experiment
IR   ionizing radiation
irrad   irradiated
irradn irradiation
LET   linear energy transfer
no.(s.) number(s)
o.e.r. oxygen enhancement ratio

P_{O_2} partial pressure of oxygen
phage   bacterial virus, bacteriophage
pld   potentially lethal damage
radn   radiation
RBE   relative biological effectiveness
UV   ultraviolet (radiation, irradiation)
–SH   sulphydryl group

307